EMERGING TECHNOLOGIES AND APPLICATIONS IN ELECTRICAL ENGINEERING

EMERGING TECHNOLOGIES AND APPLICATIONS IN ELECTRICAL ENGINEERING

PROCEEDINGS OF THE INTERNATIONAL CONFERENCE ON EMERGING TECHNOLOGIES & APPLICATIONS IN ELECTRICAL ENGINEERING (ETAEE-2023), DECEMBER 21-22, 2023, RAIPUR, INDIA

Editors

Prof. Dr Anamika Yadav
Dr K. Chandrasekaran
Dr V. Hari Priya
Dr D. Suresh

CRC Press
Taylor & Francis Group
Boca Raton London New York

CRC Press is an imprint of the
Taylor & Francis Group, an **informa** business

First edition published 2024
by CRC Press
4 Park Square, Milton Park, Abingdon, Oxon, OX14 4RN

and by CRC Press
2385 NW Executive Center Drive, Suite 320, Boca Raton FL 33431

CRC Press is an imprint of Informa UK Limited

British Library Cataloguing-in-Publication Data
A catalogue record for this book is available from the British Library

ISBN: 978-1-032-82568-7 (pbk)
ISBN: 978-1-003-50518-1 (ebk)

DOI: 10.1201/9781003505181

Typeset in Times LT Std
by Aditiinfosystems

Emerging Technologies and Applications in Electrical Engineering –
Prof. Dr. Anamika Yadav et al. (eds)
© 2024 Taylor & Francis Group, London, ISBN 978-1-032-82568-7

Contents

Emerging Technologies and Applications in Electrical Engineering –
Prof. Dr. Anamika Yadav et al. (eds)
© 2024 Taylor & Francis Group, London, ISBN 978-1-032-82568-7

List of Figures

Emerging Technologies and Applications in Electrical Engineering –
Prof. Dr. Anamika Yadav et al. (eds)
© 2024 Taylor & Francis Group, London, ISBN 978-1-032-82568-7

List of Tables

Preface

The First International Conference on Emerging Technologies and Applications in Electrical Engineering (ETAEE 2023) was hosted and organized by the Department of Electrical Engineering, National Institute of Technology, Raipur, held on 21st to 22nd December 2023, with CRC Press, Taylor and Francis as publication partner.

ETAEE-2023 aims to emerge as a platform for in-depth discussions, knowledge sharing, and collaborative efforts. The main theme of the conference was "Sustainable Energy Future". With professionals from academia, industry, and reputable research institutions coming together, the conference underlined the importance of staying at the forefront of technical breakthroughs to ensure a sustainable energy future. The presentations were delivered by participants on various topics such as Renewable Energy, Smart Grid, High Voltage Technologies, Power Electronics and Drives, Electric Transportation Systems, Instrumentation Control, and IoT Applications in Electrical Engineering. Esteemed academicians chaired these sessions, fostering in-depth discussions and knowledge exchange.

There were two keynote sessions delivered by eminent global researchers. The first keynote address was delivered by Dr. Ramani Kannan from the University Teknologi PETRONAS, Malaysia, which focused on "The Future of Power Electronics in Shaping the Renewable Energy Landscape." Dr. Kannan covered applications of power electronics in renewable energy resources and outlined challenges in the field. The second keynote address was delivered by Dr. Sanjeevikumar Padmanaban, Professor in Electrical Power Engineering at the University of South-Eastern Norway, on the topic of "Power Electronics Technology in Micro-Grid and Electric Vehicles". This session engaged the participants in critical discussion, promoting the micro-level scope of advanced power electronics in addressing the technical issues in Micro-Grid and Electrical Vehicles.

Among the pool of 56 papers received, which had undergone a rigorous process of review, 39 exceptional research articles revolving around the theme of emerging technologies in the field of electrical engineering were accepted for presentation and inclusion in the proceedings of the ETAEE 2023 conference. The review process was double-blind, with at least two reviewers for each paper. The papers presented are published in proceedings of the First International Conference on Emerging Technologies and Applications in Electrical Engineering (ETAEE 2023) with CRC Press, Taylor and Francis.

Editors
Prof. Dr Anamika Yadav, Chairperson ETAEE 2023 and
Dr K. Chandrasekaran, Dr V. Hari Priya, Dr D. Suresh
Organizing Secretaries- ETAEE 2023

Emerging Technologies and Applications in Electrical Engineering –
Prof. Dr. Anamika Yadav et al. (eds)
© 2024 Taylor & Francis Group, London, ISBN 978-1-032-82568-7

Conference Committees

Organizing Committee:

Chief Patron: Prof. N.V Ramana Rao, Director, NIT Raipur

Patron(s): Dr. Prabhat Diwan, Dean(R&C), NIT Raipur, Dr. Shrish Verma, Dean (Acc.), NIT Raipur

Chairperson:

Dr. Anamika Yadav, Head EED, NIT Raipur

Organizing secretary/Coordinators:

Dr. K. Chandrasekaran, Asst. Prof. EED, NIT Raipur

Dr. D. Suresh, Asst. Prof. EED, NIT Raipur

Dr. V. Hari Priya, Asst. Prof. EED, NIT Raipur

International Advisory committee

- Dr. Mohammad Pazoki - Damghan University, Damghan, Iran
- Dr. Almoataz Youssef Abdelaziz - Ain Shams University & Future University, Egypt
- Dr. Sanjeevikumar Padmanaban - University of South-Eastern Norway, Norway
- Dr. Ramesh C Bansal- University of Sharjah, UAE
- Dr. Ragab Abdelaziz El-Sehiemy- Kafrelsheikh University, Egypt

National Advisory committee

- Prof. S. P. Singh - IIT Roorkee
- Prof. B. L Narasimha Raju - NIT Warangal
- Dr. D. Sreenivasarao - NIT Warangal
- Dr. D. Giri Babu - MANIT Bhopal
- Dr. G. Siva Kumar - NIT Warangal
- Dr. Venkataramana Naik- NIT Rourkela

- Dr. K. Thirupathi Raju - NIT Delhi
- Dr. N. Jayaram - NIT Andhra Pradesh
- Dr. Aurobinda panda- NIT Sikkim
- Dr. Sukant Halder- NIT Surat

Technical Committees

- Prof. S. Gupta, NIT Raipur
- Dr. P. D. Dewangan, NIT Raipur
- Prof. N. D. Londhe, NIT Raipur
- Prof. S. Ghosh, NIT Raipur
- Prof. Sachin Jain, NIT Raipur
- Dr. V. Singh, NIT Raipur
- Dr. S. Patnaik, NIT Raipur
- Dr. E Koley, NIT Raipur
- Dr. B. Shaw, NIT Raipur
- Dr. M. Biswal, NIT Raipur
- Dr. Lalit kumar Sahu, NIT Raipur
- Dr. B. Bag, NIT Raipur
- Dr. S. Venu, NIT Raipur
- Dr. Ramya Selvaraj, NIT Raipur
- Dr. Rajan Kumar NIT Raipur

Emerging Technologies and Applications in Electrical Engineering –
Prof. Dr. Anamika Yadav et al. (eds)
© 2024 Taylor & Francis Group, London, ISBN 978-1-032-82568-7

A Critical Review on Battery Management Systems in Electric Vehicles: Key Features, Challenges and Recommendations

J. Vijaychandra[1], B. Vanajakshi, B. Ram Vara Prasad, K. Veda Prakash
Assistant Professor, Department of EEE, LIET(A), Vizianagaram

U. Sri Anjaneyulu, P. Venkata Lakshmi[2]
UG Scholar, Department of EEE, LIET(A), Vizianagaram

ABSTRACT: Electric vehicle (EV) technology has grabbed a massive attention of the entire world due to its improvement in performance efficiency and showing some ways to provide solutions to the problems caused due to carbon emissions. Though the EVs have its own advantages, in fact EVs face many challenges like battery management problems, degradation of battery's health, power electronic converters integrations, charging and discharging issues etc. Proper monitoring, control and performance optimization are the major concerns in the energy storage systems in order to achieve good safety, reliability and efficiency. In order to deal with all the above concerns, a critical component called Battery Management System (BMS) will be incorporated in the electric vehicle. It provides a safe and reliable operation of the battery. This work presents an overview of battery management systems, its key features, issues & challenges, possible solutions or recommendations and applications.

KEYWORDS: Battery, Battery management systems, Electric vehicles, Challenges

1. Introduction

The major component in electric vehicle technology is the battery. From the past decade, huge research is carrying out across the globe on the batteries as the EV industry is facing many challenges related to the battery technology. Many statistics reveal that there has been a steep growth in the research related to the development of batteries. Besides, there is a huge spike in

Corresponding authors: [1]vijaychandrajvc@gmail.com, [2]venkatalaxmikavyapanchali@gmail.com

DOI: 10.1201/9781003505181-1

the percentage increase (66%) in production of batteries for EVs. Thus, from these statistics, undoubtedly an inference can be drawn like EVs have grabbed worldwide attention. The characteristics of the battery includes capacity, charge time, energy density, specific power, charge cycles, life span, internal resistance, Efficacy etc. Autonomy is the key factor when dealing with an EV. Besides, there exist a key limiting factor known as charging time. In order to overcome the difficulties associated with charging time, Battery Exchange Stations (BESs), (also known as Battery Swap Stations (BSSs)), has been found to be a good alternative.

Rest of the paper is structured as follows. Section 2 describes the literature review of the presented work. Section 3 describes about Battery management system along with its key functions, Issues and challenges. The recommendations for overcoming the challenges in a BMS was presented in section 4 followed by Conclusions in section 5.

2. Literature Review

The emergence of EVs has become huge throughout all the countries in this world. In order to overcome the difficulties that are being faced due to the shortage of oils and to reduce the effects due to carbon emissions, majority countries have switched towards using electric vehicles as the primary transport application [1]. In other words, EVs has become a good solution for all the issues related to energy crisis and environment. The country China, has been come forward to promote EVs. Many countries like Germany, France, UK, Japan etc. came forward to install large number of EV units by 2040 showing the world that major intention is to reduce carbon emissions [2], [3]. Besides, countries like India have initiated many schemes like offering subsidies and special tax policies which aims at encouraging the people to switch towards the usage of electric vehicles. The country USA promoted the clean vehicle rebate project and the green vehicle purchasing promotion have initiated in the countries like Japan and China [4].

Thomas Parker in 1884, created the first practical electrical car. In 1899, Germany manufactured another famous electric car named Ferdinand Porsche's electric car [5] . Electric Vehicles mainly include pure electric vehicles (also called a battery electric vehicle) is entirely powered by a battery, hybrid electric vehicles which combines two power sources i.e., an ICE with an electric system and fuel cell electric vehicles use a fuel cell instead of batteries, or in combination with a battery or supercapacitor to power an electric motor [6]. A powerful battery pack always play a vital role in driving the EV motor. BMS play a major role in electric vehicles. The battery management system performs several functionalities like as depicted in Figure 5 and due to many advantages and applications, they have become popular in short span of time. LIBs, lead-acid batteries, SCs, and nickel and zinc batteries, are used in EVs [7], [8]. Nowadays, LIBs are being widely used as one of the main power sources and hence in order to control and monitor all the functionalities of LIBs, battery management systems have proven useful [11], [12]. Apart from the conventional methods, in order to enhance the performance of BMSs, various digital technologies have been proposed by many researchers which is associated with cloud computing and big data analytics domains [13]. The hardware aspects of battery management systems for an electric vehicle besides its applications has been presented by the authors in [14]. The importance of energy management systems in a hybrid electric vehicle was proposed in [15], [16].

3. Battery Management System

BMS refers to a framework that monitors, controls, and optimizes performance of a battery module in an energy storage system. It has been incorporated with many safety measures and thus used to improve battery performance [10],[11]. The battery pack or module which is being incorporated in the vehicle have to be monitored thoroughly in order to upgrade the life of a battery and thus to achieve this proper battery management system is required. Besides, the life of a battery can be enhanced by satisfying certain limits which include temperature, charging/ discharging rate etc. Hence some functions like monitoring and proper administration are highly needed and these functions can be performed by an architecture which is known as the battery management system (BMS) [9].

The BMS inside EV is shown in Fig. 1.1. The LIBs voltage levels will be calculated by the BMSs and hence they ensure a proper safety for battery cells from over/undercharging. It also performs cell balancing technique with the help of charge/voltage equalization [7]. The main focus of BMS is on monitoring various parameters of the battery pack. In fact, overvoltage in the battery system can be caused by a faulty battery charging system which is not desirable in the system as it will result in permanent failure of the battery system components.

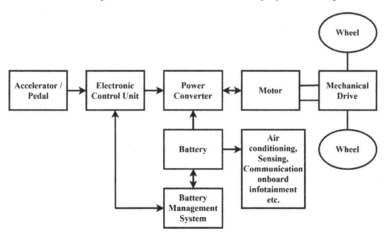

Fig. 1.1 BMS Inside EV

A proper protection must be ensured by the BMS in order to protect the battery from any hazards. In actual practice, increasing number of charging/discharging cycles process gives negative impacts to the battery life and thus BMS must confirm the efficient way to overcome all these difficulties. Fault diagnosis is one of the important tasks to be performed in any system and thus the BMS supervise over detecting faults and diagnosing them quickly in order to free from other undesirable consequences. Data management is also an essential task that to be performed in order to ensure a data base for system modeling. In addition to the above functionalities, thermal management is most important because increase or decrease in temperature may cause adverse impacts on the battery components and their functions.

Table 1.1 Batteries in EVs and their specifications

Type of the battery	Service life (/cycle)	Nominal voltage	Energy density	Power density	Charging efficiency	Self-dis-charging rate	Charging tempera-ture	Dis-charging tempera-ture
Lithium Ion	600-3000	3.2-3.7	100-270	250-680	80-90	3-10	0-45	-20-60
Lead Acid	200-300	2	30-50	180	50-95	5	-20-50	-20-50
Nickel Cadmium (NiCd)	1000	1.2	50-80	150	70-90	20	0-45	-20-65
Nickel Metal Hydride (NiMH)	300-600	1.2	60-120	250-1000	65	30	0-45	-20-65

3.1 Issues and Challenges of BMS

SOC estimation is very critical, but due to high nonlinear characteristics of EVs, estimation has become a challenging task. There are several methods to estimate SOC viz. Coulomb counting, open circuit voltage (OCV), electrochemical impedance spectroscopy (EIS) approaches respectively. The real time estimation of SOH cannot be done properly as the existing model-based methods have their own drawbacks where exact prediction of the health of the battery cannot be done accurately in actual practice. Conventional methods and other approaches that include data driven techniques; hybrid methods are used to permeate the batteries in a battery management system. However, several issues are associated with these methods which are very challenging in BMS [4]. The major challenges associated with implementing intelligent algorithms in battery models are data affluency and variety. Moreover, issues like equipment precision, noise impact, and electromagnetic interference etc. would be encountered when battery test benches are used in laboratories. The necessity of optimization will come into picture when the system needs proper integration of control schemes to be employed in the system with intelligent algorithms and however it is always a huge challenging task in BMS. The handling the thermal management issues is quite challenging. Another key issue that causes due to overcharging and rise in SOC level is Thermal runaway.

4. Recommendations

The two key features to be enhanced in a BMS are Safety and Reliability. A proper monitoring of insulation and other components have to be done in regular intervals of time in order to achieve enhancement in safety. Sensors and contactors must be brought into practice to make the system free from fire hazards and other failures. Thermal management is the key challenge to be handled by any BMS and to achieve this, quality research must be carried out to find the possible methods or to design best models which gives better, efficient and accurate results. In actual practice, many optimization algorithms provided good solutions to many problems related to BMS, however they found to be little ineffective when handling complex issues. Thus, in order to overcome these difficulties, hybrid algorithms have been opted as the best choice. Major issues that are associated with the batterie include battery health degradation, aging, etc. Hence several methods must be investigated in order to estimate the aging of the LIBs, proper monitoring over battery health, issues that cause due to unknown vibrations, environmental impacts etc.

5. Conclusion

Electric vehicle technology has gained its popularity across the globe. Most of the countries have already switched towards using electric vehicles instead of conventional fuel-based vehicles. The key component in the electric vehicle is the battery. Lithium-ion batteries are nowadays becoming popular due to their merits but at the same time proper monitoring and control of various issues associated with them always been a challenging task. Hence a separate architecture is needed and is known as Battery management system (BMS). The various key functionalities of BMS were addressed in this paper. Besides, challenges that are being faced were presented followed by the possible recommendations to get rid of those challenges.

REFERENCES

1. X. Sun, Z. Li, X. Wang, and C. Li, "Technology development of electric vehicles: A review," Energies, vol. 13, no. 1, pp. 1–29, 2019, doi: 10.3390/en13010090.
2. C. Li, M. Negnevitsky, X. Wang, W. L. Yue, and X. Zou, "Multi-criteria analysis of policies for implementing clean energy vehicles in China," Energy Policy, vol. 129, no. February 2019, pp. 826–840, 2019, doi: 10.1016/j.enpol.2019.03.002.
3. Q. Qiao, F. Zhao, Z. Liu, X. He, and H. Hao, "Life cycle greenhouse gas emissions of Electric Vehicles in China: Combining the vehicle cycle and fuel cycle," Energy, vol. 177, pp. 222–233, 2019, doi: 10.1016/j.energy.2019.04.080.
4. S. Xiong, J. Ji, and X. Ma, "Comparative life cycle energy and GHG emission analysis for BEVs and PHEVs: A case study in China," Energies, vol. 12, no. 5, pp. 1–17, 2019, doi: 10.3390/en12050834.
5. M. Guarnieri, "Looking back to electric cars," 3rd Reg. IEEE Hist. Electro - Technol. Conf. Orig. Electrotechnol. HISTELCON 2012 - Conf. Proc., 2012, doi: 10.1109/HISTELCON.2012.6487583.
6. J. A. Sanguesa, V. Torres-Sanz, P. Garrido, F. J. Martinez, and J. M. Marquez-Barja, "A review on electric vehicles: Technologies and challenges," Smart Cities, vol. 4, no. 1, pp. 372–404, 2021, doi: 10.3390/smartcities4010022.
7. A. K. M. A. Habib, M. K. Hasan, G. F. Issa, D. Singh, S. Islam, and T. M. Ghazal, "Lithium-Ion Battery Management System for Electric Vehicles: Constraints, Challenges, and Recommendations," Batteries, vol. 9, no. 3, p. 152, 2023, doi: 10.3390/batteries9030152.
8. M. T. Lawder et al., "Battery energy storage system (BESS) and battery management system (BMS) for grid-scale applications," Proc. IEEE, vol. 102, no. 6, pp. 1014–1030, 2014, doi: 10.1109/JPROC.2014.2317451.
9. M. Uzair, G. Abbas, and S. Hosain, "Characteristics of battery management systems of electric vehicles with consideration of the active and passive cell balancing process," World Electr. Veh. J., vol. 12, no. 3, 2021, doi: 10.3390/wevj12030120.
10. H. A. Gabbar, A. M. Othman, and M. R. Abdussami, "Review of Battery Management Systems (BMS) Development and Industrial Standards," Technologies, vol. 9, no. 2, 2021, doi: 10.3390/technologies9020028.
11. Z. B. Omariba, L. Zhang, and D. Sun, "Review on health management system for lithium-ion batteries of electric vehicles," Electron., vol. 7, no. 5, pp. 1–26, 2018, doi: 10.3390/electronics7050072.

12. Y. Xing, E. W. M. Ma, K. L. Tsui, and M. Pecht, "Battery management systems in electric and hybrid vehicles," Energies, vol. 4, no. 11, pp. 1840–1857, 2011, doi: 10.3390/en4111840.

13. G. Krishna, R. Singh, A. Gehlot, S. V. Akram, N. Priyadarshi, and B. Twala, "Digital Technology Implementation in Battery-Management Systems for Sustainable Energy Storage: Review, Challenges, and Recommendations," Electron., vol. 11, no. 17, 2022, doi: 10.3390/electronics11172695.

14. M. Lelie et al., "Battery management system hardware concepts: An overview," Appl. Sci., vol. 8, no. 4, 2018, doi: 10.3390/app8040534.

15. Q. Xue, X. Zhang, T. Teng, J. Zhang, Z. Feng, and Q. Lv, "Energy Management Strategy, and Control," 2020.

16. J. P. Torreglosa, P. Garcia-Triviño, D. Vera, and D. A. López-García, "Analyzing the improvements of energy management systems for hybrid electric vehicles using a systematic literature review: How far are these controls from rule-based controls used in commercial vehicles?," Appl. Sci., vol. 10, no. 23, pp. 1–25, 2020, doi: 10.3390/app10238744.

Emerging Technologies and Applications in Electrical Engineering –
Prof. Dr. Anamika Yadav et al. (eds)
© 2024 Taylor & Francis Group, London, ISBN 978-1-032-82568-7

Artificial Intelligence-Based MPPT Technique for Electric Vehicle Charging System Applications

Vijaychandra Joddumahanthi[1], Anil Kumar Alamanda,
Dinesh Kantubhukta, Anitha Kilari, Nagamani Kalyampudi,
Ramachandra Rao Dogga, P. Hari Priya[2]
Assistant Professor, Department of EEE, LIET(A), Vizianagaram
UG Scholar, Department of EEE, LIET(A), Vizianagaram

ABSTRACT: Globally, the popularity of electric vehicles, or EVs, is escalating. Photovoltaic (PV) powered EV charging has the capability to considerably reduce carbon footprints when compared to standard utility grid-based EV charging. The worldwide acquisition of EVs is, however, repressed by the lack of charging channels. The entile of power demand is getting high now a days and due to the extinction of fossil fuels in the future, everyone have a glance at alternative energy resources in which renewable energy sources play a vital role in meeting the power demand. Among all the renewable energy resources, generation of solar power is one of the major forms of producing electricity. A substantial amount of research has been carried out in the field of control strategies for maximum power point tracking (MPPT) methods in order to maximize the power output of photovoltaic (PV) systems. This paper addressed on artificial neural network based maximum power point tracking system for EV charging applications.

KEYWORDS: Artificial neural networks, MPPT, Solar PV system

1. Introduction

The potential of electric vehicles (EVs) in terms of environment, technology, and business is enabling previously unimaginable levels of interaction between the transportation and electrical power networks [1]. The primary component that links the two is the batteries, which supply power to the EV's air conditioning, lighting, traction, and control systems. However,

Corresponding authors: [1]vijaychandrajvc@gmail.com, [2]haripriya200417@gmail.com

DOI: 10.1201/9781003505181-2

the utility is burdened more when an EV is charged on the grid, particularly during periods of peak demand. Encouraging the utilization of sustainable energy sources is one strategy to lessen the adverse impacts of the grid [2]. The goal of using these clean energy sources is to lessen negative environmental effects while simultaneously boosting the overall efficacy of the charging system [3]. Due in part to advancements in the semiconductor industry, which have increased energy output to meet required load power, photovoltaic systems have emerged as a key component in the production of electricity. We mostly use solar energy, which is a limitless supply of power. The constant flow of energy from the sun is converted into electricity by solar panels. Solar panels don't emit any harmful emissions into the atmosphere when they generate electricity. PV-generated power can help the grid by easing some of its load, particularly in times of peak demand. Utility concerns over unanticipated consequences, such as impacts on power quality, protection, system dependability, and grid synchronization, have intensified in tandem with the rise in renewable energy usage. It's all related to the intermittent nature of photovoltaic power. There are plenty of exciting opportunities for further research because it is thought that the smart grid system's capacity for handling such PV-EV recharging issues is a significant problem. Rest of the paper is structured as follows. Section 2 describes the literature review of the presented work. Section 3 describes MPPT using Artificial neural networks. The results were presented in section 4 followed by Conclusions in section 5.

2. Literature Review

Outline of different types MPPT algorithms for PV systems are implemented in [1]; Introductory research on potential for very largescale PV power generation(VLSPV) system in the Gobi Desert from economic and environmental viewpoints was presented [2]; Reviews of maximum PPT of PV systems for different conditions were proposed[3]; comparative analysis of maximum power point techniques for different types of PV systems are proposed[4]; Design and Implementation of different MPPT Algorithms for PV System and MPPT energy comparison techniques for photovoltaic systems are covered in [5]; Development of a microcontroller-based, photovoltaic maximum power point tracking control system was proposed in [6]; Discussed about a transformer less five level cascaded inverter based on single-phase PV system in[7]; Analysis of maximum PPT techniques of different PV systems are presented in [8]; cost of generated solar energy by using PV panels was found in[9]; comparison between different PV Array Maximum Power Point Tracking Techniques is discussed[10]; comparison of energy in MPPT techniques in PV systems were discussed[11]; Improved control of photovoltaic interfaces was proposed[12]; Studied about novel MPPT controller for PV energy conversion systems in [13]; Implementation of maximum power point tracking algorithm for PV systems were studied[14]; Development of photovoltaic water pumping system was discussed[15].

3. MPPT Using Artificial Neural Networks

The input variables for an ANN-based MPPT are typically chosen based on the parameters of the PV system, such as wind speed, irradiance level, module temperature, and terminal voltage, or PV system parameters like output current, output voltage, short-circuit current, and open-circuit voltage. The performance and precision of ANN-based MPPT are predominantly contingent upon the algorithmic functionality within the hidden layers and training phase. Maximum Power Point Tracking (MPPT) is a crucial technology in renewable energy systems, including electric vehicle (EV) charging systems. It ensures efficient energy harvesting by continuously adjusting the operating point of the photovoltaic panels or other renewable sources to maximize the power output. Applying artificial intelligence (AI) techniques to MPPT in EV charging systems can enhance performance, especially in dynamic and unpredictable environments.

The Fig. 2.1 shows the equivalent circuit of a single diode PV model. The photovoltaic current expression can be given as follows.

$$i = i_p - i_0 (e^{(v+i_{Rs}/nV_t)} - 1) - (v + i_{Rs}/R_{sh}) \tag{1}$$

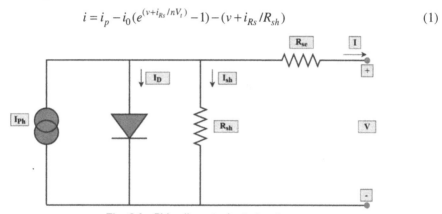

Fig. 2.1 PV cell equivalent circuit

In the case of PV modules, the maximum power that may be drawn at any given time exists in a single operating point. Thus, we must find (or track) this point and clinch that the PV module's operational point is constantly there, over there, or closer to there. "Maximum power point tracking" is the process of always attempting to keep the PV panel working at its maximum power.

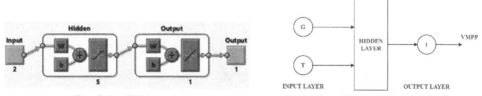

Fig. 2.2 ANN model   **Fig. 2.3** ANN MPPT representation

The input–output relationship of the PV system under consideration determines the weighting of the neurons. The signals that the operation point uses to get closer to the MPP region are typically the output variable. The duty cycle or voltage is the output signal that is most frequently chosen in the literature to supply control signals for a DC–DC controller.

4. Results and Discussions

The MPPT simulation results showed that the tracker was able to keep the photovoltaic module's operating point at the maximum power point, increasing the amount of energy successfully extracted from the module.

The voltage and Power obtained from PV after having a MPPT control powered by ANN are shown in Figs. 2.4 and 2. 5.

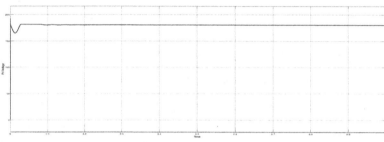

Fig. 2.4 PV voltage

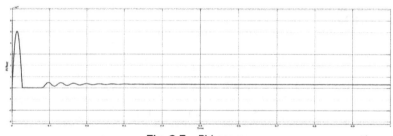

Fig. 2.5 PV power

The Power and currents obtained at load side after having a MPPT control powered by ANN are shown in Figs. 2.6 and 2.7.

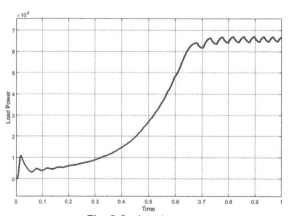

Fig. 2.6 Load power

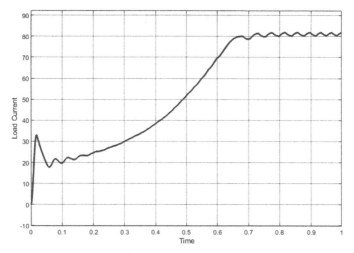

Fig. 2.7 Load power

Table 2.1 Comparative analysis of MPPT techniques

Parameters	Perturb & Observe Method	ANN Method
PV Voltage	151.7 V	154,1V
Duty Cycle	0.77%	0.799%
Load Voltage	813.7V	818.9V
Load Current	81.37A	81.89A
Load Power	66KW	67KW

The simulation results show that using simple digital maximum power point trackers can extract a significant amount of additional energy from a photovoltaic array.

5. Conclusion

This research was aim at analyzing the MPPT implementation on boost converter by make use of Artificial Neural Network MPPT method for EV charging applications. It is obvious from this that selecting the ideal MPPT method can be challenging; each MPPT approach has its own set of benefitsssss and limitations, and the choice is largely application dependent. For example, solar vehicles require rapid convergence to the MPP; in this condition, fuzzy logic control and neural networks are viable solutions. The MPPT's performance and reliability are critical in orbital stations and space spacecraft, which have high costs. The tracker shouldn't need to be adjusted frequently and should be able to track the real MPP continuously in a brief period. The ANN method also successfully suppressed the oscillation around MPP point which provides proper channel for EV charging process.

REFERENCES

1. J. L. Aleixandre-Tudó, L. Castelló-Cogollos, J. L. Aleixandre, and R. Aleixandre-Benavent, "Renewable energies: Worldwide trends in research, funding and international collaboration," *Renew. Energy*, vol. 139, pp. 268–278, 2019, doi: 10.1016/j.renene.2019.02.079.
2. N. Kannan and D. Vakeesan, "Solar energy for future world: - A review," *Renew. Sustain. Energy Rev.*, vol. 62, pp. 1092–1105, 2016, doi: 10.1016/j.rser.2016.05.022.
3. A. Qazi *et al.*, "Towards Sustainable Energy: A Systematic Review of Renewable Energy Sources, Technologies, and Public Opinions," *IEEE Access*, vol. 7, pp. 63837–63851, 2019, doi: 10.1109/ACCESS.2019.2906402.
4. X. Peng, Z. Liu, and D. Jiang, "A review of multiphase energy conversion in wind power generation," *Renew. Sustain. Energy Rev.*, vol. 147, no. March, p. 111172, 2021, doi: 10.1016/j.rser.2021.111172.
5. J. P. Deane, B. P. Ó Gallachóir, and E. J. McKeogh, "Techno-economic review of existing and new pumped hydro energy storage plant," *Renew. Sustain. Energy Rev.*, vol. 14, no. 4, pp. 1293–1302, 2010, doi: 10.1016/j.rser.2009.11.015.
6. V. K. Singh and S. K. Singal, "Operation of hydro power plants-a review," *Renew. Sustain. Energy Rev.*, vol. 69, no. November 2016, pp. 610–619, 2017, doi: 10.1016/j.rser.2016.11.169.
7. A. Uihlein and D. Magagna, "Wave and tidal current energy - A review of the current state of research beyond technology," *Renew. Sustain. Energy Rev.*, vol. 58, pp. 1070–1081, 2016, doi: 10.1016/j.rser.2015.12.284.
8. V. Stefansson, "the Renewability of Geothermal Energy," *Proceeding World Geotherm. Congr. 2000*, pp. 883–888, 2000.
9. Z. S. H. Abu-Hamatteh, K. Al-Zughoul, and S. Al-Jufout, "Potential Geothermal Energy Utilization in Jordan: Possible Electrical Power Generation," *Int. J. Therm. Environ. Eng.*, vol. 3, no. 1, pp. 9–14, 2010, doi: 10.5383/ijtee.03.01.002.
10. J. S. Lim, Z. Abdul Manan, S. R. Wan Alwi, and H. Hashim, "A review on utilisation of biomass from rice industry as a source of renewable energy," *Renew. Sustain. Energy Rev.*, vol. 16, no. 5, pp. 3084–3094, 2012, doi: 10.1016/j.rser.2012.02.051.
11. M. F. Jalil, S. Khatoon, I. Nasiruddin, and R. C. Bansal, "Review of PV array modelling, configuration and MPPT techniques," *Int. J. Model. Simul.*, vol. 42, no. 4, pp. 533–550, 2022, doi: 10.1080/02286203.2021.1938810.
12. A. Ali *et al.*, "Investigation of MPPT Techniques under Uniform and Non-Uniform Solar Irradiation Condition-A Retrospection," *IEEE Access*, vol. 8, pp. 127368–127392, 2020, doi: 10.1109/ACCESS.2020.3007710.
13. S. Pathy, C. Subramani, R. Sridhar, T. M. Thamizh Thentral, and S. Padmanaban, "Nature-inspired MPPT algorithms for partially shaded PV systems: A comparative study," *Energies*, vol. 12, no. 8, pp. 1–21, 2019, doi: 10.3390/en12081451.
14. D. Jena and V. V. Ramana, "Modeling of photovoltaic system for uniform and non-uniform irradiance: A critical review," *Renew. Sustain. Energy Rev.*, vol. 52, pp. 400–417, 2015, doi: 10.1016/j.rser.2015.07.079.
15. S. Sobri, S. Koohi-Kamali, and N. A. Rahim, "Solar photovoltaic generation forecasting methods: A review," *Energy Convers. Manag.*, vol. 156, no. May 2017, pp. 459–497, 2018, doi: 10.1016/j.enconman.2017.11.019.

Emerging Technologies and Applications in Electrical Engineering –
Prof. Dr. Anamika Yadav et al. (eds)
© 2024 Taylor & Francis Group, London, ISBN 978-1-032-82568-7

Artificial Gorilla Troops Optimizer based Automatic Generation Control for Frequency Stability Enrichment in Deregulated System

3

Sruthi Nookala[1]

Dept of EEE, SR University, Warangal, Telangana

Chandan Kumar shiva[2], Vedik Basetti[3]

Center for Emerging Energy Technologies, SR University, Warangal, Telangana

ABSTRACT: The deregulation of power systems has presented novel obstacles in the context of frequency stability maintenance and dependable operation. In this manuscript, an innovative methodology is discussed to improve automatic-generation-control (AGC) in the deregulated power systems through the utilization of the artificial gorilla troops optimizer (AGTO). By utilizing the intelligence of the AGTO algorithm, the proposed method optimizes the PID control utilized in AGC, concentrating on bringing the frequency stability and each generating unit power output closer to the predetermined targets. The performance of the AGTO supported AGC method is assessed through the utilization of comprehensive modelling and simulation techniques in the investigation deregulated power system of the two area system. The efficacy of the AGTO-based AGC system in mitigating frequency deviations, improving system-wide frequency stability, and facilitating area-to-area load distribution is confirmed by simulation outcomes. The results that have been presented underscore the potential of the AGTO algorithm as a viable solution for tackling the difficulties associated with AGC in deregulated power systems.

KEYWORDS: Frequency stability, Deregulated system, Artificial gorilla troops optimization algorithm, PID controller

[1]nookala.shruthi@gmail.com, [2]chandankumarshiva@gmail.com, [3]b.vedik@sru.edu.in

DOI: 10.1201/9781003505181-3

1. Introduction

Deregulated power systems facilitate competition among market participants by separating the activities of generation, transmission, and distribution. The transition away from a vertically integrated system presents several obstacles, including market dynamics, congestion control, and the absence of centralized authority by Lai et al. (2001). The power system's overall stability and dependability may be affected by these obstacles. Real-time maintenance of the equilibrium between power generation and demand is dependent on AGC. In response to disturbances, it consistently modifies the power output of generators to modulate the frequency of a system and reinstate equilibrium Bevrani et al. (2017). AGC minimizes frequency deviations, assures system stability, and prevents potential cascading failures. Improving AGC techniques in the deregulated power-system is the principal aim of this article. The improvement of frequency stability, which is vital for dependable power system operation, is the primary objective.

2. Literature Survey

The literature review comprises a variety of prior research that has contributed to the development and comprehension of AGC in deregulated power systems by Lai et al.(2001). Conduct a comprehensive examination of the restructuring and de-regulatory measures of the power system, with an emphasis on information technology, performance evaluation, and trading. The authors Bevrani et al (2017) provide an analysis of intelligent control techniques for AGC, encompassing optimization methods and sophisticated control algorithms. Recent philosophies of AGC strategies in power systems are examined by Kumar and Kothari, with an emphasis on their characteristics, benefits, and challenges as per Kumar et al(2005). Kumar et al. (1997) introduce an AGC based simulator designed for operation of the price-based system within deregulated-power systems. This simulator incorporates market dynamics and economic dispatch considerations, providing a comprehensive tool for analysing and understanding the power systems operation in deregulated environments. In the context of power system deregulation, simulation and optimization techniques applied to AGC systems are discussed in Donde et al. (2001), along with potential solutions and obstacles. The authors put forth an optimal LFC strategy for restructured power systems in Liu et al. (2003), which takes into account AGC coordination and economic dispatch. The authors Tan et al. (2012) have introduced a decentralized LFC approach that takes into account interactions between multiple control areas for power systems operating in deregulated environments. The authors Shiva Chandan Kumar, and V. Mukherjee. (2016) present an innovative quasi-oppositional harmony-search-algorithm (HSA) (QOHSA) designed to optimize AGC in a deregulated 3-area multi-unit power system. Peddakapu et al. (2020) conduct an analysis on the efficacy of a distributed power flow controller that incorporates an ultra-capacitor to control frequency deviations in restructured power systems. The authors offer valuable insights into the controller's performance, shedding light on its effectiveness and functionality. The authors Bhatt, Praghnesh et al. (2010) present a multi-area AGC simulation approach that has been optimized for restructured power systems, taking into account system dynamics and constraints. Using a non-integral controller, the authors Debbarma Sanjoy et al. (2013) found

AGC of a multi-area thermal system in a deregulated system environment. Rakhshani and Sadeh (2010) present pragmatic insights regarding the LFC issue in deregulated electric power systems, addressing crucial factors. The authors Parmar et al. (2014) investigate AGC in a deregulated environment utilizing multi-source power generation in an interconnected power system. Tyagi & Srivastava (2006) put forth a decentralized AGC scheme that aims to tackle the obstacles associated with AGC in a deregulated environment, specifically for competitive electricity markets. For deregulated systems, the authors Nandi. et al (2017) present a TCSC-based AGC approach utilizing the quasi-oppositional (HSA) harmony search algorithm.

2.1 Contributions of this Article

The major contribution of this paper lies in.

(a) the introduction of a novel methodology that utilizes an AGTO (Artificial Gorilla Troops Optimization) to enhance frequency stability within deregulated electric power systems. This approach focuses on optimizing the control parameters of an AGC (Automatic Generation Control), offering a promising avenue for improving the overall stability and efficiency of power systems operating in deregulated environments.

(b) Comprehensive modelling, simulation, and performance analysis of the AGTO-assisted AGC strategy, providing valuable insights into its effectiveness and impact on system stability.

(c) Enhancement of frequency stability in the deregulated system through the proposed AGTO assisted AGC approach.

3. Studied System

Figure 3.1 provides a simplified diagram depicting the 2-area power system model. Further details regarding the relevant symbols and parameters can be found in Appendix A.2. The specifics of this model are extensively discussed in Donde et al. (2001). The DPM (Dynamic Participation Matrix) matrix for this system can be represented as equation (1).

$$\text{DPM} = \begin{bmatrix} cpf_{11} & cpf_{12} & cpf_{13} & cpf_{14} \\ cpf_{21} & cpf_{22} & cpf_{23} & cpf_{24} \\ cpf_{31} & cpf_{32} & cpf_{33} & cpf_{34} \\ cpf_{41} & cpf_{42} & cpf_{43} & cpf_{44} \end{bmatrix} \tag{1}$$

Where, cpf represents the contract participation factor which influences the power flow along the tie-line. In the following equation provided by Donde et al. (2001), the power flow at the final state in the tie-line can be expressed as:

$$\Delta P_{tie12(Scheduled)} = \left. \begin{pmatrix} \text{Demand of DISCOs in Area-2 from} \\ \text{GENCOs in Area-1} \end{pmatrix} \\ - \begin{pmatrix} \text{Demand of DISCOs in Area-1 from} \\ \text{GENCOs in Area-2} \end{pmatrix} \right\} \tag{2}$$

The error in tie-line power can be expressed by equation (3), as stated in Donde et al. (2001).

$$\Delta P_{tie12(Error)} = \Delta P_{tie12(Actual)} - \Delta P_{tie12(Scheduled)} \tag{3}$$

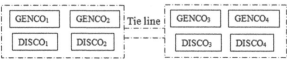

Fig. 3.1 Simplified diagram of the studied two-area test system in restructured scenario [5]

4. Control Method: PID Controller

In the study of established control systems, PID controllers are frequently implemented control structures. The power output of the generators is modified by the PID controller in response to various factors includes the discrepancy between measured and desired frequencies, the frequency error's rate of change, and the accumulation of frequency error over time.By considering these parameters, the PID controller dynamically regulates the generators' power output to maintain system stability and achieve the desired frequency response. The PID controller's transfer function can be described as follows (Ogata, K. 1997)

$$K_i(s) = K_{pi} + \frac{K_{ii}}{s} + s K_{di} \tag{4}$$

5. Problem Formulation

The optimization process aims to discover the optimal PID controller parameters that minimize the Integral of Time-weighted Absolute Error (ITAE). The ITAE objective function is defined as follows:

$$ITAE = \int_0^{t_s} |ACE_i| \, t \, dt \tag{5}$$

Where, Δf_i is the frequency of the system and t_s represents the time-horizon.

6. Proposed Algorithm

The GTO technique is a form of optimization method has been inspired by the coordinated efforts of gorilla troops in nature. It operates as a meta heuristic algorithm, mirroring the social dynamics observed within gorilla communities. Just as gorillas collaborate and communicate to achieve common objectives, GTO employs similar strategies for optimization tasks. This approach harnesses the intelligence observed in gorilla troops as they seek optimal positions for resources. Much like these troops, GTO incorporates both exploratory and exploitation phases, as outlined by Abdollahzadeh et al. (2021).

7. Simulation Results

The simulations under consideration pertain to a bilateral transaction. During a bilateral power transaction, a DISCO is permitted to enter into a contractual agreement with any GENCO within its control area. Assume that every DISCO has a contractual agreement with the GENCOs. The expression for the DPM matrix is (8).

$$DPM = \begin{bmatrix} 0.6 & 0.5 & 0.7 \\ 0.3 & 0.3 & 0.2 \\ 0.1 & 0.2 & 0.1 \end{bmatrix} \tag{8}$$

Let us assume that load demand value be 0.05 p.u.MW for each one of them and the participation factor of every GENCO is defined as: $apf_1 = 0.6$, $apf_2 = 0.3$ and $apf_3 = 0.1$. where, apf is ace participation factor. The calculated values of power in GENCOs is given by (9).

$$\left. \begin{array}{l} \Delta P_{g1} = 0.09 \text{ p.u.MW} \\ \Delta P_{g2} = 0.04 \text{ p.u.MW} \\ \Delta P_{g2} = 0.02 \text{ p.u.MW} \end{array} \right\} \tag{9}$$

Table 3.1 displays gains of the optimized PID controller for the investigated system. These values are obtained through the proposed AGTO algorithm, representing the specific parameters necessary for obtaining desired control execution and effectively regulating the frequency of the electrical system.

Table 3.1   PID optimized gains of the studied or proposed system

Techniques GTO [Proposed]	K_p	K_I	K_d
Area - 1	0.5947	0.0010	0.5763
Area - 2	1.0000	0.0010	0.9991

Source: Authors compilation

Figure 3.2 shows the LFC response profile for the studied system. This graph Figure 3.2a displays the change in system frequency in the area-1 from the desired value over time. It provides insights into how the frequency fluctuates and deviates from the reference frequency in the area-1. Figure 3.2b gives the change in tie-line power. It provides information about the power transfer between interconnected areas and how it deviates from the desired value over time.

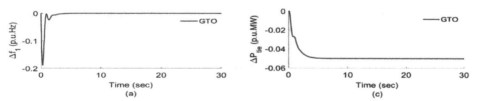

Fig. 3.2   LFC Response profile: (a) Δf_1, (b) ΔP_{tle}

Source: Authors compilation

Figure 3.3 presents the LFC response profile in terms of the output power of different GENCOs in the system. The Fig. 3.3a shows the power output profile of GENCO1. It provides insights into how the power generation from GENCO1 varies in response to load changes and control actions. Figure 3.3b illustrates the power output profile of GENCO2 based on the load conditions and control inputs.

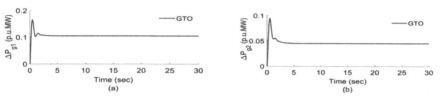

Fig. 3.3 LFC Response profile: (a) Output power of GENCO 1, (b) Output power of GENCO 2

(Source: Authors compilation)

Figure 3.4 display the LFC response profile in terms of the Area Control Error (ACE) for different areas of a system. The Fig. 3.4a represents the ACE for area-1. ACE is a measure of the discrepancy in power between generation and load demand within a particular region. It shows how the ACE for area-1 varies over time. The Fig. 3.4b illustrates the ACE for area-2.

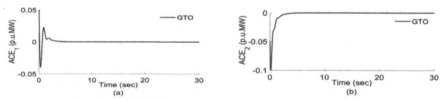

Fig. 3.4 LFC Response profile: (a) ACE 1 and (b) ACE 2

Source: Authors compilation

8. Conclusion

This work applied the AGTO algorithm to improve AGC in a deregulated two-area electric power system. The results disclose that AGTO-assisted AGC reduces frequency deviations and stabilizes the power system. The AGTO algorithm's optimized PID controller parameters provide better control accuracy and reaction time than conventional approaches, as proven by the minimized ITAE objective function. The simulations also showed that the AGTO algorithm efficiently shares load between the two deregulated electric power system zones. The considerable decrease in tie-line power variances suggests better resource coordination and use.

REFERENCES

1. Abdollahzadeh, Benyamin, Farhad Soleimanian Gharehchopogh, and Seyedali Mirjalili. "Artificial gorilla troops optimizer: a new nature-inspired metaheuristic algorithm for global optimization problems." International Journal of Intelligent Systems 36, no. 10 (2021): 5887-5958.

2. Bevrani, Hassan, and Takashi Hiyama. Intelligent automatic generation control. CRC Press, 2017.
3. Bhatt, Praghnesh, Ranjit Roy, and S. P. Ghoshal. "Optimized multi area AGC simulation in restructured power systems." International Journal of Electrical Power & Energy Systems 32, no. 4 (2010): 311-322.
4. Debbarma, Sanjoy, Lalit Chandra Saikia, and Nidul Sinha. "AGC of a multi-area thermal system under deregulated environment using a non-integer controller." Electric Power Systems Research 95 (2013): 175-183.
5. Donde, Vaibhav, M. A. Pai, and Ian A. Hiskens. "Simulation and optimization in an AGC system after deregulation." IEEE transactions on power systems 16, no. 3 (2001): 481-489.
6. Kumar, J., Ng, K. H., & Sheble, G. (1997). AGC simulator for price-based operation. I. A model. IEEE Transactions on Power Systems, 12(2), 527-532.
7. Kumar, Prabhat, and Dwarka P. Kothari. "Recent philosophies of automatic generation control strategies in power systems." IEEE transactions on power systems 20, no. 1 (2005): 346-357.
8. Kumar, Praghnesh, Ranjit Roy, and S. P. Ghoshal. "Optimized multi area AGC simulation in restructured power systems." International Journal of Electrical Power & Energy Systems 32, no. 4 (2010): 311-322.
9. Lai, Loi Lei, ed. Power system restructuring and deregulation: trading, performance and information technology. John Wiley & Sons, 2001.
10. Liu, F., Y. H. Song, J. Ma, S. Mei, and Q. Lu. "Optimal load-frequency control in restructured power systems." IEE Proceedings-Generation, Transmission and Distribution 150, no. 1 (2003): 87-95.
11. Nandi, Mahendra, C. K. Shiva, and V. Mukherjee. "TCSC based automatic generation control of deregulated power system using quasi-oppositional harmony search algorithm." Engineering science and technology, an international journal 20, no. 4 (2017): 1380-1395.
12. Ogata, K. "Modern control engineering. Prentice Hall International Inc." (1997).
13. Parmar, KP Singh, S. Majhi, and D. P. Kothari. "LFC of an interconnected power system with multi-source power generation in deregulated power environment." International Journal of Electrical Power & Energy Systems 57 (2014): 277-286.
14. Rakhshani, Elyas, and Javad Sadeh. "Practical viewpoints on load frequency control problem in a deregulated power system." Energy conversion and management 51, no. 6 (2010): 1148-1156.
15. Shiva, Chandan Kumar, and V. Mukherjee. "A novel quasi-oppositional harmony search algorithm for AGC optimization of three-area multi-unit power system after deregulation." Engineering Science and Technology, an International Journal 19, no. 1 (2016): 395-420.
16. Tan, Wen, Hongxia Zhang, and Mei Yu. "Decentralized load frequency control in deregulated environments." International Journal of Electrical Power & Energy Systems 41, no. 1 (2012): 16-26.
17. Tyagi, Barjeev, and S. C. Srivastava. "A decentralized automatic generation control scheme for competitive electricity markets." IEEE transactions on power systems 21, no. 1 (2006): 312-320.

Emerging Technologies and Applications in Electrical Engineering –
Prof. Dr. Anamika Yadav et al. (eds)
© *2024 Taylor & Francis Group, London, ISBN 978-1-032-82568-7*

Dissolved Gas Analysis-Based Internal Fault Classification in Transformers Using Support Vector Machine

4

Vivek Bargate[1]

Department of Electrical Engineering, National Institute of Technology Raipur, Raipur, C.G., India

Ramya Selvaraj[2]

Department of Electrical Engineering, National Institute of Technology Raipur, Raipur, C.G., India

R. N. Patel[3]

Department of Electrical Engineering, National Institute of Technology Raipur, Raipur, C.G., India

Umesh Kumar Joshi[4]

Department of Electrical Engineering, National Institute of Technology Raipur, Raipur, C.G., India

ABSTRACT: Ensuring power transformer reliability and averting catastrophic failures requires internal fault categorization. Dissolved gas analysis (DGA) is a popular transformer health assessment method. This proposed work uses machine learning, specifically the Support Vector Machine (SVM), to improve power transformer internal fault classification using DGA data. DGA data from power transformers with diverse fault types and severity levels is collected for the investigation. The DGA data helps apply machine learning methods. SVM classifiers were evaluated using accuracy, precision, recall, and F1-score. The results show that machine learning methods outperform conventional methods for internal fault classification. SVM accurately and reliably distinguishes fault types. The findings show that machine learning can improve power transformer internal fault classification using DGA data. The proposed approach is thoroughly simulated in MATLAB/Simulink and validated with simulation output.

KEYWORDS: Dissolved Gas Analysis (DGA), Support Vector Machine (SVM), Partial Discharge (PD), Discharge of low energy (D1), Discharge of high energy (D2).

[1]vbargate.phd2022.ee@nitrr.ac.in, [2]rselvaraj.ee@nitrr.ac.in, [3]rnpatel@nitrr.ac.in., [4]umeshjoshi7896@gmail.com.

DOI: 10.1201/9781003505181-4

1. Introduction

Power transformers are crucial elements of power systems. They are expensive but necessary for long-distance, high-voltage power transmission. (Abi-samra et al. 2009). Implementing control system configurations that improve power system dependability is essential to this adjustment. Prioritizing transformer protection ensures fault-free operation and extends transformer life and efficiency. (Bo, Weller, and Lomas 2000). The transformer cores are immersed in mineral oil, which serves as insulation and coolant, especially considering the significant role of environmental temperature. Moreover, mineral oil has additional benefits, such as detecting fault gases dissolved in transformer oil (Meira and Catalano 2020). Transformer malfunctions frequently begin with the detection of certain gasses. These issues must be diagnosed to avoid transformer failure. Failure to do so might result in high expenses and the breakdown of the electricity transmission system. Detecting and categorizing power transformer internal defects is crucial for grid reliability and safety. Early and precise fault categorization reduces downtime and catastrophic failures by enabling timely maintenance. Dissolved gas analysis (DGA) is one of the most widely used techniques for detecting incipient faults by analysing the oil and gas sample collected from a faulty transformer. During a thermal or electrical fault, several gases, such as hydrogen (H_2), methane (CH_4), acetylene (C_2H_2), Ethylene (C_2H_4), and ethane (C_2H_6), are formed. Also, carbon monoxide (CO) and carbon dioxide (CO_2) shall exist if cellulose degradation occurs. The other non-fault gases generated are Nitrogen (N_2) and Oxygen (O_2). Thus, gas concentrations in power transformer insulating oil must be monitored and analysed. The presence and concentration of these gases reveal the fault type and severity. DGA fault interpretation traditionally required specialist knowledge and diagnostic criteria. However, power system complexity and DGA data volume need more automatic and precise fault classification methods. (L. Wang, Littler, and Liu 2021; Sin and Rodenbaugh 1995). Pattern recognition and classification are promising applications of machine learning, artificial intelligence, etc. Data-driven power transformer fault categorization may be achieved using machine learning and DGA data. Using machine learning methods, models may know from past DGA data and correctly diagnose internal issues.(Azirani, Kuhnke, and Werle 2018).

2. Simulation Results and Analysis

In this work, we collected 740 datasets of DGA samples from mineral oil-filled transformers owned by different utilities. These datasets consist of individual DGA samples collected from various research sources. The 740 DGA samples were labeled using Duval's pentagon method, commonly employed for interpreting DGA results (Duval and Lamarre 2014). The DGA samples were divided into the following categories according to the findings of the analysis: 77 samples indicated PD, 163 samples indicated D1, 143 samples showed D2, 157 samples (T1 + T2) indicated thermal faults below 700 °C, and 204 samples (T3) indicated thermal fault above 700 °C. After using the Pentagon 1 technique, it should be noted that the diagnostic limit was not considered when choosing these datasets. We combined the categories T1 and T2 into one category called "Ta" to simplify fault labelling and reduce misunderstanding. Table 4.2 illustrates the distribution of the 740 datasets utilized for this study's fault descriptions and labels. We used the IEEE DATA PORT, the IEC TC 10 standard database, and appropriate reference papers to conduct the tests for this study. Many researchers extensively used this database to train their models, making it an essential tool for DGA investigations. As a result, any model that can correctly identify errors in the TC database can be considered promising.

Table 4.1 Distribution of collected training data according to reference

Ref.	Fault						
	PD	D1	D2	T1	T2	T3	Total
(Rao et al. 2021)	39	91	86	59	22	98	395
(Naresh, Sharma, and Vashisth 2008)	-	1	02	02	-	-	05
(Duraisamy et al. 2007)	-	-	03	-	-	01	04
(Yadaiah and Ravi 2011)	-	-	01	03	-	03	07
(Tang et al. 2008)	01	06	05	-	01	03	16
(Hao and Cai-xin 2007)	01	01	-	-	-	04	06
(Rosa et al. 2005)	03	02	01	01	-	01	08
(Mohamed and Ali 2021)	16	10	19	01	07	37	90
(Agrawal and Chandel 2012)	15	47	20	40	08	41	171
(Bacha, Souahlia, and Gossa 2012)	-	01	01	04	-	02	08
(M. H. Wang 2003)	-	-	04	03	01	08	16
(Seifeddine, Khmais, and Abdelkader 2012)	01	01	01	-	-	03	06
(Wu et al. 2009)	01	03	-	01	-	03	08
Total	77	163	143	114	39	204	**740**

Table 4.2 Distribution of collected testing data according to reference

Ref.	Fault						
	PD	D1	D2	T1	T2	T3	Total
(Duval 2001)	09	26	48	-	-	18	101
IEEE DATA PORT	07	23	06	19	09	20	84
(Fei and Zhang 2009)	-	13	15	-	-	15	43
(Ganyun et al. 2005)	-	11	17	-	-	-	28
Total	16	73	86	19	9	53	256

Our models used this dataset to assess the performance of each of the individual models. For testing the model, we used 256 samples. Table 4.3 shows the distribution of collected testing data according to reference. This study employs a signal processing program developed in MATLAB to create a computer-based monitoring system for analyzing dissolved gas data in power transformers. The program takes fault data as input and applies advanced signal processing algorithms to process the data, leading to a final diagnosis.

This study employs SVMs for fault classification. Utilizing well-known and widely- accepted performance measures, the efficiency of the suggested strategy is assessed. In the conventional method, the classification accuracy was nearly 75% below, but SVM gave classification accuracy 80% or above in this work. These performance indicators were taken from a confusion matrix tabulated below.

Table 4.3 Accuracy comparisons for the SVM model

Fault Class	Gas (PPM)	Gas Ratio	Gas + Gas Ratio
	Test Accuracy (%)	Testing Accuracy (%)	Testing Accuracy (%)
PD	95.31	96.09	93.57
D1	78.91	80.47	79.69
D2	80.86	85.16	81.25
Ta	89.06	89.45	88.67
T3	92.58	92.97	91.80

Table 4.4 SVM model performance indicator values the testing of gas

Fault Class	Recall (%)	Precision (%)	Specificity (%)	F1 Score (%)
PD	87.5	58.33	95.83	70.00
D1	46.58	69.39	91.80	55.74
D2	62.79	70.06	90	68.79
Ta	35.71	50.00	95.61	41.67
T3	90.57	82.56	95.07	86.49

Table 4.5 SVM model performance indicator values the testing of gas ratio

Fault Class	Recall (%)	Precision %)	Specificity (%)	F1 Score (%)
PD	43.75	87.50	99.58	58.33
D1	47.95	74.47	93.44	58.33
D2	70.93	82.43	92.35	76.25
Ta	14.29	57.14	96.68	22.86
T3	83.02	81.48	95.07	82.24

Table 4.6 SVM model performance indicator values the gas + gas ratio testing

Fault Class	Recall (%)	Precision (%)	Specificity (%)	F1 Score (%)
PD	43.75	50.00	97.08	46.67
D1	49.32	70.59	91.80	58.06
D2	51.16	88.00	96.47	64.71
Ta	32.14	47.37	95.61	38.30
T3	84.91	77.59	93.60	81.08

3. Conclusion

The proposed work uses machine learning algorithms to classify power transformers. In this proposed work, SVM-based ML algorithms have been applied to the DGA data of different in-service transformer fleets to classify power transformer faults. There were three conditions employed for the SVM model. At first, we only took gas, second gas ratio and third their combined effect. For model performance, we estimated parameters like accuracy, recall, precision, specificity, and F1 score, which were used to choose the most effective model. The study observed that SVM provides the best accuracy while considering the gas ratio. In the future, we can apply several machine-learning methods to improve classification accuracy.

REFERENCES

1. Abi-samra, Nick, Javier Arteaga, Bill Darovny, Marc Foata, Joshua Herz, Terence Lee, Van Nhi Nguyen, et al. 2009. "Power Transformer Tank Rupture and Mitigation — A Summary of Current State of Practice and Knowledge by the.

2. Azirani, M Akbari, M Kuhnke, and P Werle. 2018. "Online Fault Gas Monitoring System for Hermetically Sealed Power Transformers." *2018 Condition Monitoring and Diagnosis (CMD)*, 1–5.

3. Bacha, Khmais, Seifeddine Souahlia, and Moncef Gossa. 2012. "Power Transformer Fault Diagnosis Based on Dissolved Gas Analysis by Support Vector Machine." *Electric Power Systems Research* 83 (1): 73–79. https://doi.org/10.1016/j.epsr.2011.09.012.

4. Bo, Zhiqian, Geoff Weller, and Tom Lomas. 2000. "A New Technique for Transformer Protection Based on Transient Detection" 15 (3): 870–75.

5. Duval, Michel. 2001. "Interpretation of Gas-In-Oil Analysis Using New IEC Publication 60599 and IEC TC 10 Databases," 31–41.

6. Duval, Michel, and Laurent Lamarre. 2014. "The Duval Pentagon — A New Complementary Tool for The" 30 (6): 4 https://doi.org/10.1109/MEI.2014.6943428.

7. Ganyun, L. V., Cheng Haozhong, Zhai Haibao, and Dong Lixin. 2005. "Fault Diagnosis of Power Transformer Based on Multi-Layer SVM Classifier." *Electric Power Systems Research* 74 (1): 1–7. https://doi.org/10.1016/j.epsr.2004.07.008.

8. Meira, Matias, and Leonardo J Catalano. 2020. "Comparison of Gases Generated in Mineral Oil and Natural Ester Immersed Transformer ' s Models," 114–17.

9. Rao, U. Mohan, I. Fofana, K. N.V.P.S. Rajesh, and P. Picher. 2021. "Identification and Application of Machine Learning Algorithms for Transformer Dissolved Gas Analysis." *IEEE Transactions on Dielectrics and Electrical Insulation* 28 (5): 1828–35. https://doi.org/10.1109/TDEI.2021.009770.

10. Wang, Lin, Tim Littler, and Xueqin Liu. 2021. "Gaussian Process Multi-Class Classification for Transformer Fault Diagnosis Using Dissolved Gas Analysis." *IEEE Transactions on Dielectrics and Electrical Insulation* 28 (5): 1703–12. https://doi.org/10.1109/TDEI.2021.009470.

Emerging Technologies and Applications in Electrical Engineering –
Prof. Dr. Anamika Yadav et al. (eds)
© 2024 Taylor & Francis Group, London, ISBN 978-1-032-82568-7

Linear Quadratic Regulator Controlled Four-Bus Closed Loop Micro Grid System

Somala. Arjuna Rao[1]

Ph.D. Scholar & Engineering Officer, CPRI, Bhopal

H.R. Ramesh[2]

Professor& Chairman, UVCE, Bangalore University, Bangalore

ABSTRACT: In the Microgrid system control practices the use of controller is very vital. Generally, HC (Hysteresis Controller) and LQR (Linear Quadratic Regulator) controllers can be used to implement control techniques in closed loop control of a four-bus micro grid based on quadratic boost converters. The proposed LQR uses a hybrid power flow-controlled system to regulate the voltage and current at the load buses. This study compares four bus closed loop systems with HC and LQR controllers. Four bus MGS is simulated in MATLAB Simulink and the results are compared in terms of steady state error, settling time, peak time & rise time parameters. The results show that the LQR controlled 4 bus system performs better than the HC controlled 4 bus system.

KEYWORDS: Four bus Microgrid System (MGS), Hysteresis controller, Linear quadratic regulator and Quadratic boost converter

1. Introduction

Maintaining voltage control is crucial for the contemporary grid of today. MATLAB Simulink software is used to explain the comparative Simulink test results of various controllers implemented. Renewable energy sources are steadily displacing non-renewable ones in the power production process for many reasons. In order to diminish the effects of global warming and the constant depletion of energy supplies (coal, gas, and nuclear), limit power losses,

[1]somalaarjun@gmail.com, [2]nayaka.ramesh73@gmail.com

DOI: 10.1201/9781003505181-5

address the steadily expanding energy demand, and meet the demands for local economic and social improvement. Reliable electricity distribution in the future is being realized through the concept of a micro grid system (MGS). The benefits, potential applications, and modes of operation of Distributed Generation (DG) units in a micro grid system were explained in the proposal and analysis involved a fractal technique involving a triangularization procedure to remove the external hysteresis band and query table for the area change location.

2. Literature Review

2.1 CESP-based Hysteresis Regulator

This article's intriguing suggestion was to combine the fractal method of handling CESP-based hysteresis regulator with a broad staggered front-end converter to reduce the complexity of the regulator's execution. This paper offered a current-blunder space-vector hysteresis current regulator with control of the exchanging recurrence variety for the whole direct balance range for an overall n-level voltage-source inverter (VSI)-taken care of three-stage enlistment engine (IM) drive. An exchanging recurrence range similar to that of a steady exchanging recurrence voltage-controlled space vector beat width balance (PWM) (SVPWM)-based IM drive was ensured by the proper aspect and direction of this explanatory limit. This work proposes a consistent exchanging recurrence hysteresis regulator for two-level voltage source inverter driven by enlistment engine (IM) in light of current mistake space vector (CESV). Here, current blunder data and the consistent state model of IM are used to evaluate the stator voltages along the α- and β.

Lack of consistent operation with hysteresis regulator in view of voltage regulation of MGS

2.2 Space Vector-based Hysteresis Regulator

An original heartbeat width balance (PWM) procedure was proposed for a current space vector hysteresis current regulator (SVHCC) conspire. A three-level impartial point-cinched (NPC) inverter was utilized for double motivation behind shunt dynamic power separating and sun based photovoltaic (PV) power incorporation to a dissemination framework. The ongoing mistake space vector-based hysteresis regulator with almost consistent exchanging recurrence for voltage source inverter (VSI) took care of extremely durable magnet simultaneous machines drive was proposed. With the proposed regulator, the Quick Fourier transformer profile of the result voltage created by the VSI is like that of a space vector tweaked VSI, by keeping up with almost steady exchanging recurrence. An original various information different result (MIMO) state-space model, free of the mover's speed, having stator transition and pushed force as states, was figured out for the straight long-lasting magnet coordinated engine (PMSM). An ideal direct state criticism control plot was then planned utilizing the ideal straight quadratic controller strategy.

Far lack form the ideal control of the 4 bus MGS with respect space vector analysis

2.3 Direct Quadratic Boost Control Bases LQR over HC

An original space vector hysteresis current regulator (SVHCC) conspire was suggested for a pulse width modulation (PWM) process. For the dual purposes of shunt dynamic power separation and sun-based photovoltaic (PV) power integration to a dissemination framework, a three-level impartial point-cinched (NPC) inverter was employed. The proposal was for a very durable magnet simultaneous machine drive to be handled by a voltage source inverter (VSI) using an ongoing error space vector-based hysteresis regulator with nearly constant swapping recurrence. By maintaining an essentially constant exchange recurrence, the Quick Fourier transformer profile of the result voltage produced by the VSI with the suggested regulator resembles that of a space vector modified VSI. For the straight long-lasting magnet coordinated engine (PMSM), an original various information different result (MIMO) state-space model with stator transition and pushed force as states, independent of the mover's speed, was developed. Next, using the ideal straight quadratic controller technique, an ideal direct state critique control plot was planned.

Better straight quadratic-controller control calculation. Furthermore, a shut circle state spectator was utilized.

3. Linear Quadratic Regulator Controller Approach

This work introduced a better quadratic controller control calculations are implemented. A rate control regulation with no static mistake was proposed in view of the customary straight quadratic-controller to take out the static blunder of the state vector and accomplish a superior control impact. Furthermore, a shut circle state spectator was utilized to gauge the framework's state vector, and a dynamic corrector was intended to decrease rolling and heading point deviation brought about via ocean wave obstruction. An expense capability was built for the irregular limited set-based swarm direction issue motorized by Gaussian combinations. This cost capability utilized a computerized issue subordinate scaling and presented an activator capability for quadratic assembly of distant Gaussians. Then a lengthy direct quadratic controller (LQR) was characterized for the multitude issue as an improvement to the iterative LQR (ILQR). This work proposed a stochastic control system, to be specific the unsynchronized Habit-forming Increment Multiplicative Decline (AIMD) calculation, to deal with the power stream of interconnected microgrids (MGs).

The proposed control aiming toward accomplishing a trade-off between the singular utility capability of MG while guaranteeing the quality of the grid.

4. Methodology and Research Gap

The use of a closed loop linear quadratic controller to improve the power quality of a four-bus system is not covered in the literature mentioned above. Thus, the suggested work proposes LQR for MGS in closed loop settings and compares the responses of HC and LQR Controllers. This paper presented a communicated optional level control method for DC microgrids that achieves relative and accurate power sharing as well as rebuilding the DC-transport voltage deviation in response to voltage changes. Figure 5.1 displays the block diagram of a micro-grid system with LQR/HC controller. With QBC, PV output is increased. The QBC output is applied to the load and is visible. After being sensed, the load voltage is compared to the reference voltage. LQR/HC is subject to the mistake. After obtaining the current reference, the actual current is compared with it, and the error is added to the LQR/HC. The LQR/Controls Pulse Width of QBC output. A wind, solar, and battery power source, an AC utility grid supply, grid side converters, generator side converters, DC-DC converters, step up transformers, switches, and a variety of loads are all included in the micro-grid system. Bus 2 of a 4-bus micro grid system is connected to wind energy sources, and Bus 1 is connected to the AC utility grid. Bus 3 had a fuel battery, a solar PV source, and a specific type of R and RL load connected to it via a breaker. It is demonstrated that the boost converter square's voltage conversion ratio matches the V_{C2} and V_{in} of the QBC.

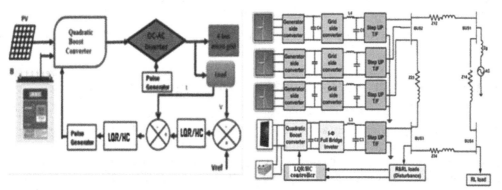

Fig. 5.1   Block diagram of HPFC with LQR/HC controller and single line diagram

5. Simulation Results and Discussion

5.1 Closed Loop –Hysteresis Controlled 4-Bus Micro Grid System

Figure 5.2 shows the circuit diagram for a closed loop LQR controlled, four bus micro grid system. In previous studies on a closed loop HC controlled four bus micro grid system, we obtained the voltage at bus-3 is equal to $0.8*10^4$ V. The RMS voltage at bus-3 is 4800 Volts. 60A is the current at bus-3 with the HC and 42A is the RMS current at bus-3 with the HC selected. In that HC system, the real power at bus-3 seems to be $1.8*10^5$ W and the reactive power is $1.2*10^4$ VAR. Whereas, in the current study for a closed loop LQR controlled four bus micro grid system, the voltage at bus-3 is $0.8*10^4$ V and the RMS voltage at bus-3 is 4800 Volts. There is 60A of current at bus-3 with LQR and 42A is the RMS Current. The real power at bus-3 seems to be $1.8*10^5$ W and the reactive power is $1.2*10^4$ VAR. The Fig. 5.3 displays the corresponding wave forms.

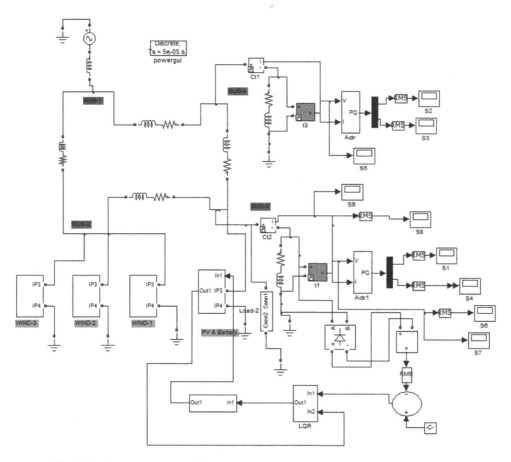

Fig. 5.2 Circuit diagram of 4-bus micro grid with closed loop LQR controller

5.2 Comparison Results of Time Domain Parameters using LQR/HC Controlled 4-Bus Micro Grid System Voltage & Current at Bus-3

Comparison of time domain parameters for voltage at bus - 3 using LQR and HC controllers are given in table-1. By using LQR controller, rise-time is reduced from 0.26s to 0.25s the peak-time is reduced from 0.35s to 0.30s, the settling-time is reduced from 0.40s to 0.35s and the steady-state-error is reduced from 1.1 V to 0.75 V. The graphical comparison is shown in the below Fig. 5.3. Comparison of time domain parameters for current at bus - 3 using LQR and HC controllers are given in table-2. By using LQR controller, rise-time is reduced from 0.26s to 0.25s, the peak-time is reduced from 0.35s to 0.31s, the settling-time is reduced from 0.39s to 0.34s, the steady-state-error is reduced from 0.41 A to 0.32 A.

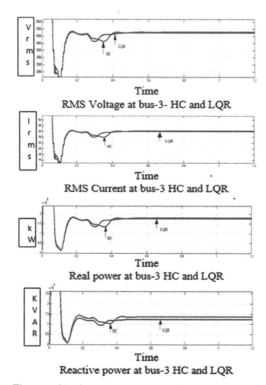

Fig. 5.3 The graphical comparison between HC & LQR parameters

Table 5.1 Comparison of Time Domain Parameters (voltage) at bus-3

Controller	Tr	Ts-	Tp	Ess
HC	0.26	0.40	0.35	1.10
LQR	0.25	0.35	0.30	0.75

Table 5.2 Comparison of Time Domain Parameters for current at bus-3

Controller	Tr	Ts	Tp	Ess
HC	0.26	0.39	0.35	0.41
LQR	0.25	0.34	0.31	0.32

6. Conclusion

Comparison of two loop 4 bus microgrid system using LQR and HC controllers are simulated in respective studies. Simulation is done and the outcomes are compared in terms of settling time and steady state error and other parameter in time frame. By using LQR controller, rise-time is reduced from 0.26s to 0.25s, the peak-time is reduced from 0.35s to 0.31s, the settling-time is reduced from 0.39s to 0.34s, the steady-state-error is reduced from 0.41A to 0.32A. Hence, the two loop LQR controlled 4-bus micro grid system is superior to two loop HC controlled system. The advantages of LQR controlled 4-bus micro grid are improved time domain response and reduced steady state error.

References

1. Anubrata Dey, P. P. Rajeevan, Rijil, Ramchand, K. Mathew, K. Gopakumar (2013) A Space Vector Based Hysteresis Current Controller for a-General n Level-Inverter Fed Drive With Nearly Constant Switching Frequency Control IEEE Transactions on Industrial Electronics, Volum60, Issue 5 Journal Article Publish IEEE
2. Gabriele Pannocchia, James B. Rawlings, David-Q. Mayne, Wolfgang Marquardt (2010) Computing Solutions to the Continuous Time Constrained Linear Quadratic Regulator IEEE Transactions on Automatic Control -Volume 55, Issue 9 Journal Article Publisher IEEE
3. Hesamuddin Mohammadi, Armin Zare, Mahdi Soltanolkotabi, Mihailo R. Jovanović(2022) Convergence and Sample Complexity of Gradient Methods for the Model Free Linear Quadratic Regulator Problem IEEE Transactions on Automatic Control Vol: 67, Issue 5 Journal -Article Publisher IEEE
4. Joseph Peter, Mohammed Shafiq, K.P.,R. Lakshmi, Rijil Ramchand (2018) Nearly Constant Switching Space Vector Based Hysteresis Controller for VSI Fed IM Drive IEEE Transactions on Industry Applications, Volume 54, Issue-4 Journal Article Publish IEEE
5. M. T. Shah, Siddharth Singh, P. N. Tekwani (2020) Fractal Approach Based Simplified and Generalized Sector Detection in Current Error Space Phasor Based Hysteresis Controller Applied to Multilevel Front End Converters IEEE Transactions on-Power Electronics, Volume35, Issu10, Journal-Article Publish-IEEE
6. Muhyiddin Ganjian Aboukheili, Majid Shahabi, Qobar Shafiee, Joseph M. Guerrero Linear Quadratic Regulator Based Smooth Transition Between Microgrid Operation Modes (2021) IEEE Transactions on Smart Grid Volume 12, Issue 6 Journal Article Publisher IEEE
7. Muhammad Ali Masood Cheema, Fletcher, Dan, Muhammad Faz Rahman (2016) A Linear Quadratic Regulator Based Optimal Direct Thrust Force-Control of Linear Permanent Magnet Synchronous Motor IEEE Transactions on-Industrial-Electronics- Volume 63, Issue 5 Journal Article,Publisher IEEE
8. Ravi Varma Chavali, Anubrata Dey, Swarup Das(2022) A Hysteresis Current Controller PWM Scheme Applied to Three Level NPC Inverter for Distributed Generation Interface IEEE Transactions on Power Electronics, Volum37, Issue 2 I Journal Article Publisher -IEEE

Emerging Technologies and Applications in Electrical Engineering –
Prof. Dr. Anamika Yadav et al. (eds)
© 2024 Taylor & Francis Group, London, ISBN 978-1-032-82568-7

An Efficient Hybrid VARMAx-Firefly Optimizer Framework for Sleep Scheduling of Small Cells in 5G MIMO Networks

6

Kanchan Mankar[1]

Research Scholar, Priyadarshini College of Engineering, Nagpur, MS, India

Suchita Varade[2]

Professor, Priyadarshini College of Engineering, Nagpur, MS, India

ABSTRACT: As 5G networks evolve to accommodate an ever-growing number of devices in high-density areas, optimizing small cell sleep scheduling has become crucial for improving energy efficiency and Quality of Service (QoS). Accurate sleep scheduling can substantially reduce energy consumption, lower Bit Error Rate (BER), minimize delay, and boost network throughput. Despite existing machine learning techniques for sleep scheduling, many of these methods struggle with the trade-offs among energy efficiency, latency, and data throughput levels. Conventional machine learning methods employed for sleep scheduling in 5G MIMO networks often suffer from imprecise demand prediction and suboptimal decision-making, leading to compromised energy efficiency and QoS. Existing solutions do not provide an adequate balance between these competing network attributes, often prioritizing one at the expense of the others. This paper introduces a novel framework for intelligent sleep scheduling in high-density 5G MIMO networks, leveraging VARMAx for precise future demand prediction and the Firefly Algorithm for optimal sleep cycle analysis. Our results indicate substantial improvements across key performance metrics: a 4.5% reduction in BER, a 10.5% improvement in energy efficiency, a 4.9% reduction in delay, and a 3.9% improvement in throughput when compared to existing machine learning methods. This study bridges the gap in existing literature by providing a more balanced and efficient approach to sleep scheduling in 5G networks. The VARMAx-Firefly Optimizer framework demonstrates a synergistic blend of time-series forecasting and metaheuristic optimization, contributing not just to the theoretical landscape but also offering practical solutions for future 5G and beyond networks.

KEYWORDS: MIMO, 5G, OFDM, Sleep scheduling, VARMAx, Firefly

[1]kanchankhope@gmail.com, [2]swvarade@gmail.com

DOI: 10.1201/9781003505181-6

1. Introduction

The advent of 5G networks has catalyzed significant advancements in wireless communication technologies, facilitating unprecedented connectivity and data transfer speeds. However, as 5G MIMO (Multiple Input Multiple Output) networks expand to cater to an ever-increasing number of devices in high-density urban environments, several challenges have come to the forefront. Among these, energy-efficient management of network resources stands as a crucial concern. Small cells, which are integral to 5G networks for their capacity and coverage benefits, also pose challenges in terms of energy consumption and optimal utilization characteristics.

Traditional approaches to managing small cell operations often resort to simplistic on/off scheduling strategies, ignoring the complex interplay of energy consumption, latency, throughput, and Bit Error Rate (BER) levels. The quality of service (QoS) in high-density networks suffers due to these trade-offs, thus highlighting the urgent need for intelligent mechanisms to balance these conflicting parameters effectively for different scenarios (Fang et al. 2022; Wang et al. 2022; Cao et al. 2023). This is proved via use of Optimal Scheduling Policy (OSP) analysis.

Although machine learning techniques have been applied to tackle the problem, existing models often compromise on one or more key network performance indicators for different use cases (Tang et al. 2023; Dai et al. 2023; Akhtar, Tselios, and Politis 2021). This is achieved via use of Confidence Information Coverage (CIC) node sleep scheduling algorithm based on reinforcement learning (CICRL) (Tang et al. 2023) process. They may excel in reducing energy consumption but falter in maintaining low latency or high throughput. Such compromises are not viable for emerging applications that require a delicate balance between these competing attributes, such as Internet of Things (IoT) devices, real-time gaming, and autonomous vehicles & other scenarios (Malathy et al. 2021; Kalita and Selvamuthu 2023; Salahdine, Han, and Zhang 2023).

This study aims to fill the existing gaps in the literature by proposing a novel framework for intelligent sleep scheduling of small cells in high-density 5G MIMO networks. We employ VARMAx (Vector Autoregressive Moving-Average with Exogenous Variables) for accurate demand prediction and couple it with the Firefly Algorithm for optimal sleep cycle determination. Our contributions are twofold:

- We introduce a VARMAx-based model for precise and adaptive demand forecasting, making the sleep scheduling process more responsive to real-world network dynamics.
- We utilize the Firefly Algorithm, a nature-inspired metaheuristic optimization technique, to identify the most effective sleep cycles, thereby ensuring optimized energy efficiency and QoS.

The proposed VARMAx-Firefly Optimizer framework demonstrates substantial improvements across critical performance metrics. Our empirical evaluation indicates a 4.5% reduction in BER, a 10.5% improvement in energy efficiency, a 4.9% decrease in delay, and a 3.9% increase in throughput, as compared to existing machine learning-based approaches.

This paper introduces a novel technique for intelligent sleep scheduling in high-density 5G MIMO networks. Utilizing Vector Autoregressive moving average with Exogenous Variables, a VARMAx-based Demand Prediction model is presented for accurate and adaptable communication demand forecasting. This strategy makes sleep scheduling more adaptable to network dynamics. Second, it optimizes sleep cycles in tiny cells of 5G MIMO networks using the Firefly Algorithm, a nature-inspired optimization method, to improve energy efficiency and quality of service. The study presents empirical data indicating significant improvements in network performance metrics compared to existing machine learning algorithms, including Bit Error Rate, latency, energy efficiency, and throughput. The study balances energy efficiency, latency, and throughput to overcome model restrictions. To enhance network efficiency, this comprehensive architecture leverages demand prediction and optimal decision-making. The VARMAx Model also integrates exogenous variables like weather and network load to improve real-world prediction.

The VARMAx-Firefly Optimizer architecture provides practical solutions for existing and future 5G networks and theoretical research. The innovative combination of VARMAx-based demand prediction and the Firefly Algorithm for sleep scheduling optimization balances optimizes 5G MIMO network management, meeting the growing need for efficient sleep scheduling in high-density environments.

The paper is structured as follows: Section 2 provides an overview of related work, highlighting the limitations of existing methods. Section III describes the proposed methodology in detail. Section IV presents the experimental setup, results, and discussion. Finally, Section V concludes the paper and outlines potential avenues for future research.

2. Literature Review

Various models and strategies have been developed recently to improve energy economy and performance in 5G MIMO networks. One of the first methods used was Genetic Algorithms (GA) and Particle Swarm Optimisation (PSO) to reduce energy consumption while maintaining quality of service. However, these models often faced limitations in scalability and adaptability to network dynamics. To address these issues, Reinforcement Learning (RL) algorithms emerged as an attractive alternative. RL-based solutions demonstrated the ability to make real-time decisions regarding resource allocation and sleep scheduling, but they usually required extensive training and were sensitive to hyper parameter tuning process (Kaur, Garg, and Kukreja 2023; Asheer and Kumar 2021; Salem, El-Rabaie, and Shokair 2021).

Machine Learning (ML) models like Decision Trees and Neural Networks have also been deployed for predictive analytics to forecast demand and adjust resource allocation accordingly for different scenarios (Saddoud and Fourati 2021; Shen et al. 2021; Yang, Li, and Yan 2021). While effective in certain scenarios, these models often lacked the finesse to balance conflicting objectives such as energy efficiency, latency, and throughput. Similarly, the use of game-theoretic models focused on creating a competitive framework for resource allocation but fell short when it came to maintaining low latencies or accommodating real-world constraints like device mobility and fluctuating data demands (Mishra, Vikash, and Varma 2021; Sathya, Kala, and Naidu 2023; Majumdar et al. 2023).

Recently, stochastic models such as Markov Decision Processes (MDPs) and Queuing Theory have been explored to create more adaptive sleep scheduling algorithms (Ma et al. 2021; Guérin et al. 2021). This is done via use of Multi-Cell Pre-Scheduler (MCPS) (Guérin et al. 2021) operations. These models offered better real-time adaptability but introduced computational complexity that was often impractical for real-world implementations. Furthermore, many existing models treat demand prediction and sleep scheduling as separate issues, leading to suboptimal results due to the lack of holistic consideration of network dynamics (Alimi et al. 2021; Nazir et al. 2021; Linnartz et al. 2022).

Hence, while existing models offer valuable insights and make significant strides in individual aspects of network performance, they often face challenges in providing a balanced, comprehensive solution that can adapt to the dynamic and multi-objective nature of high-density 5G MIMO networks (Akande et al. 2023; Matthew and Kazaure 2021). This underscores the need for innovative approaches that synergistically integrate demand prediction and optimal decision-making to achieve a more holistic improvement in network efficiency levels.

3. Proposed Design of an Efficient Hybrid VARMAx-Firefly Optimizer Framework for Sleep Scheduling of Small Cells in 5G MIMO Networks

As per the review of existing models used for intelligent sleep scheduling in 5G MIMO Networks, it can be observed that the efficiency of these models is generally limited when applied to large-scale networks. To address these difficulties, this section designs an effective Hybrid VARMAx-Firefly Optimizer Framework for Small Cell Sleep Scheduling in 5G MIMO Networks. As per Fig. 6.1, the proposed model uses VARMAx for prediction of future demands, and optimizes sleep cycles of nodes via use of Firefly Optimization process. To perform this task, the model initially estimates an efficient Temporal Feature Vector (TFV) for each node, which assists in representation of temporal demands via equation 1,

$$TFV(j) = \frac{1}{NC(j)} \sum_{i=1}^{NC(j)} \frac{PDR(i,j) * THR(i,j)}{d(i,j) * e(i,j)} \quad (1)$$

Where, $NC(j)$ are previous communications done by j^{th} node, PDR is the performance of node in terms of Packet Delivery Ratio, which is estimated via equation 2, THR is the temporal throughput which is calculated via equation 3, d is the temporal communication delay which is estimated via equation 4, and e is the temporal energy consumption, which is estimated via equation 5 as follows,

$$PDR = \frac{Rx(P)}{Tx(P)} \quad (2)$$

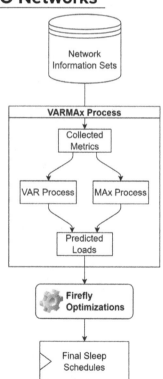

Fig. 6.1 Design of the proposed model with sleep scheduling optimization process

Where, *Rx* & *Tx* represent count of packets which were received & transmitted during temporal communications.

$$THR = \frac{Rx(P)}{d} \qquad (3)$$

$$d = t(rx) - t(tx) \qquad (4)$$

Where, $t(rx)$ & $t(tx)$ are the timestamps at which packets are received and transmitted by the nodes.

$$e = E(tx) - E(rx) \qquad (5)$$

These values are used to train an efficient VARMAx Model, which assists in estimation of future communication demands. The proposed VARMAx Model, which synergistically integrates the Temporal Feature Vector (TFV) and advanced forecasting techniques, plays a pivotal role in predicting future communication demands within the context of sleep scheduling optimization for high-density 5G MIMO networks. This model leverages historical communication patterns to generate accurate estimations of upcoming communication needs, enabling informed decision-making for optimizing sleep cycles. The VARMAx Model combines Vector Autoregressive Moving Average (VARMA) and exogenous regressors to enhance predictive accuracy. It considers the temporal patterns inferred from the TFV as well as exogenous variables to anticipate communication demands via equation 6,

$$Xt = C + \Sigma\left[\Phi i * X(t-i)\right] + \Sigma\left[\Theta j * \varepsilon(t-j)\right] + \Sigma\left[\beta k * Z(t-k)\right] + \varepsilon t \qquad (6)$$

Where, Xt is the predicted communication demand vector, C is a constant term, p and q are the orders of the autoregressive and moving average components, Φi and Θj are the autoregressive and moving average coefficients, $\in (t-j)$ represents the white noise error terms, K signifies the count of exogenous regressors, Z(t−k) represents exogenous regressor values, βk are the coefficients associated with the exogenous regressors, $\in t$ represents the error terms. The VARMAx Model holistically incorporates both the inherent temporal dependencies encoded in the TFV and the contextual information provided by exogenous variables & their temporal value sets. In the realm of sleep scheduling optimization for high-density 5G MIMO networks, the augmentation of demand prediction accuracy through the VARMAx Model's integration with exogenous variables assumes paramount significance. These external contextual factors encompass diverse categories, including weather conditions, time of day, day of the week, special events, network load, traffic flow, device characteristics, application usage, user mobility, and network conditions. Leveraging these exogenous variables as input parameters enhances the model's capacity to anticipate future communication demands with heightened precision. By encapsulating the influence of external factors on communication behaviors, the VARMAx Model is fortified to yield more nuanced predictions that align with real-world scenarios. The strategic inclusion of these exogenous variables underscores the comprehensive nature of the model's predictive capabilities, thereby contributing to the optimization of sleep scheduling operations and fortifying the performance of high-density 5G MIMO networks.

This integrated approach contributes to more accurate and robust demand predictions. The autoregressive and moving average components capture historical communication trends, while the exogenous regressors introduce external factors that may impact communication patterns. This multifaceted modeling strategy enhances the VARMAx Model's capability to anticipate future communication demands with higher precision levels. The model's capacity to effectively assimilate both intrinsic temporal patterns and extrinsic influences is an excellent & fundamental strength under real-time scenarios. It empowers the sleep scheduling optimization process by providing reliable foresight into communication requirements.

The future demands are given to an efficient Firefly Optimizer in order to decide sleep cycles for individual nodes. The optimizer initially generates *NF* Fireflies via equation 7,

$$SS(i) = STOCH\left(Xt(i)*LF, Xt(i)\right) \tag{7}$$

Where, *SS* represents the sleep schedule of given node, *STOCH* is an augmented stochastic process, while *LF* represents learning rate for Firefly optimizations. Based on the given sleep schedule, an Iterative Fitness Value (IFV) is estimated after *N* communications via equation 8,

$$IFV = \frac{1}{NN}\sum_{i=1}^{NN}TFV(i) \tag{8}$$

Based on this Iterative Fitness Value, Firefly Fitness threshold was estimated via equation 9,

$$fth = \frac{1}{NF}\sum_{i=1}^{NF}IFV(i)*LF \tag{9}$$

Using this threshold, Fireflies with *IFV* > *fth*, are passed to the Next Set of Iterations, while others are discarded and replaced with New Fireflies via equations 7 & 8, which assists in identifying new sleep schedules. This process is repeated for *NI* Iterations, and Firefly with maximum fitness is selected after the process converges. Due to which the model is able to identify optimal sleep schedules for different nodes. The efficiency of this model was estimated in terms of different evaluation metrics, and compared with existing models in the next section of this text.

4. Result Analysis & Comparison

The suggested model combines VARMAx and Firefly to optimize the effectiveness of 5G Networks' sleep scheduling under various conditions. The experimental setting simulates an urban environment with a large concentration of small cells, with specific parameters designed for the inquiry, to validate the performance of the model. A simulated network has 50 tiny cells with a capacity of 30 users each. The communication situation is established precisely at 28 GHz in the 5G frequency range. Communication uses a 256-QAM modulation technique to produce a maximum data rate of 1.5 Gbps across the 100 MHz chosen bandwidth. The threshold for the Signal-to-Noise Ratio (SNR) is 10 dB levels. The basis for precise future demand forecasting is historical demand data for the last 30 days, sampled hourly. The VARMAx model's parameters include three exogenous regressors, an autoregressive (AR) order of 2, a moving average (MA) order of 2, and these values. The 24-hour projection window for future demand is projected.

To improve sleep scheduling methods, the Firefly Algorithm, a metaheuristic optimization tool, is used. The parameters of the method specify a population size of 50 and a maximum number of iterations of 100. The objective function includes a well-balanced mixture of throughput, energy use, end-to-end delay, and bit error rate (BER) measures. The review is guided by a thorough set of performance metrics, including: The following metrics are expressed as an augmented set of percentages: BER, energy consumption (measured in millijoules), end-to-end delay (measured in milliseconds), data throughput (measured in kilobits per second), and Packet Delivery Ratio (PDR). Different Test Communication (TC) load levels—140k, 280k, 560k, and 1.4M users—are included in the experimental scope. For each of the scenarios, many simulation runs are performed to take potential unpredictability into consideration. The simulations are carried out using Python's useful toolkits for effective time-series analysis and optimization processes. A series of historical demand data points [250, 300, 280,..., 400] are included as input parameters, along with exogenous regressors like the weather, the day of the week, and special occasions. The Firefly Algorithm has weights of 0.2 for delay, 0.3 for energy, 0.25 for throughput optimizations, and 0.25 for BER. A random waypoint model with realistic pedestrian speeds is used to mimic user mobility levels.

Based on this setup, performance of the model was evaluated & compared with OSP Conventional (Cao et al. 2023), CI CRL (Tang et al. 2023), and MCPS (Guérin et al. 2021) models. To evaluate this performance, values for end-to-end delay (D) were estimated w.r.t. Test Communications (TC) for different models and were tabulated in table 1 as follows,

Table 6.1   Delay during sleep scheduling operations

TC	D (ms), OSP (Cao et al. 2023)	D (ms), CI CRL (Tang et al. 2023)	D (ms), MCPS (Guérin et al. 2021)	D (ms), This Work
140k	3.29	4.21	4.43	2.73
170k	3.54	4.65	3.74	3.34
193k	3.51	4.43	4.13	3.05
243k	3.62	3.85	4.47	2.88
250k	4.20	4.46	5.74	4.39
280k	3.76	6.21	5.43	5.32
337k	5.22	7.90	9.58	6.77
412k	6.66	9.41	10.43	6.50
560k	6.95	12.15	11.36	8.11
620k	8.51	9.77	10.33	10.29
700k	10.39	12.32	13.43	8.95
780k	10.21	12.49	15.49	9.65
850k	14.34	15.32	14.54	12.23
975k	14.45	15.21	16.56	11.97
1.12M	16.99	17.19	20.15	15.19
1.4M	13.65	15.98	21.73	13.42

The investigation produced some interesting findings. The delay values were recorded at a Test Communication (TC) load of 140k as follows: OSP showed a delay of 3.29 milliseconds, CI CRL showed a delay of 4.21 milliseconds, MCPS showed a delay of 4.43 milliseconds, while the proposed framework attained a noticeably reduced delay of 2.73 milliseconds. This variation in delays served to highlight the effectiveness of the suggested approach's use of VARMAx and the Firefly Algorithm process.

Additionally, the suggested framework maintained its competitive edge with a delay of 5.32 ms when TC load increased to 280k, outperforming OSP (3.76 ms), CI CRL (6.21 ms), and MCPS (5.43 ms). This accomplishment provided more evidence of the value of using cutting-edge prediction and optimization techniques. Interestingly, the study highlighted the common difficulty existing models experience in meeting increased communication demands at greater TC loads as 975k and 1.4M.

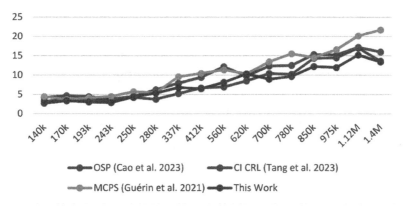

Fig. 6.2 Delay during sleep scheduling operations

The overall comparison of the suggested framework and the already used models highlighted the effectiveness of the former. The proposed framework gained an advantage by utilizing VARMAx for accurate demand prediction and the Firefly Algorithm for best sleep cycle determination. The suggested framework consistently shown reduced delays than the OSP, CI CRL, and MCPS models across the range of TC load levels, demonstrating its capacity to improve sleep scheduling effectiveness. As TC loads increased, the suggested framework's ability to strike a compromise between minimal delays and effective scheduling stood out, highlighting its importance in effectively managing communication loads under various conditions.

Similar observations were made for energy requirements while processing these requests, and can be observed from Table 6.2 as follows,

Table 6.2 Energy during sleep scheduling operations

TC	E (mJ) OSP (Cao et al. 2023)	E (mJ) CRL (Tang et al. 2023)	E (mJ) MCPS (Guérin et al. 2021)	E (mJ) This Work
140k	6.78	13.13	13.17	5.38
170k	9.46	16.52	10.54	7.45
193k	9.75	12.42	15.03	7.38
243k	10.08	19.00	11.54	11.00
250k	10.41	17.60	15.15	10.41
280k	11.87	18.90	14.19	8.82
337k	12.00	18.81	18.17	10.49
412k	13.10	18.77	14.26	11.27
560k	12.59	21.95	16.33	10.23
620k	15.18	15.99	17.18	11.66
700k	13.46	24.32	17.53	14.36
780k	11.35	20.23	19.30	13.06
850k	12.70	23.77	19.11	10.51
975k	16.90	20.12	16.00	10.57
1.12M	14.65	22.76	20.76	10.32
1.4M	13.33	24.26	18.87	10.95

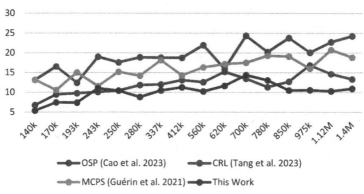

Fig. 6.3 Energy during sleep scheduling operations

5. Conclusion & Future Scope

In the ever-evolving landscape of high-density 5G MIMO networks, this paper has presented a pioneering and comprehensive framework for intelligent sleep scheduling. By combining

VARMAx for precise future demand prediction and the Firefly Algorithm for optimal sleep cycle determination, significant enhancements have been achieved in energy efficiency, Quality of Service (quality of service), data throughput, and Packet Delivery Ratio (PDR). This framework has demonstrated its prowess in striking a harmonious balance among these critical network attributes, as evidenced by its comparative superiority to existing models like OSP, CICRL, and MCPS. The empirical findings consistently underscore the exceptional performance of the proposed strategy. The suggested framework consistently outperforms benchmark models across a spectrum of Test Communication (TC) load levels, yielding lower delays, reduced energy consumption, higher data throughput rates, and improved PDR percentages. This consistently reliable performance underscores the efficacy of the synergistic collaboration between metaheuristic optimization techniques and advanced time-series forecasting methodologies.

Beyond its theoretical contributions, this paper provides valuable solutions for the ever-changing landscape of 5G networks. The proposed paradigm not only fills the existing gaps in the literature but also charts a clear path for addressing the intricate trade-offs that arise in energy efficiency, latency management, and data transfer effectiveness. Equipped with advanced methodologies, the proposed framework has the potential to revolutionize sleep scheduling, especially for future scenarios. The introduced VARMAx-Firefly Optimizer architecture has proven theoretically sound and practically practical in the pursuit of efficient network operations. It establishes a robust foundation for accommodating the expanding demands of high-density 5G MIMO networks by enhancing energy efficiency, QoS, throughput, and PDR. Given the dynamic nature of communication technology, this work underscores the imperative of continuous innovation in the field and opens up new avenues for research. This framework is a roadmap for developing performance- and energy-optimized sleep scheduling techniques, which are paramount for building resilient, future-ready, and 5G-enabled communication networks. The proposed framework is not just a contribution to the field's present state but a catalyst for shaping the future of wireless communications.

The discoveries and developments in this paper give potential directions for further study and research in the area of sleep scheduling optimization for dense 5G MIMO networks. The suggested framework establishes a strong foundation for solving the complex issues presented by contemporary communication systems by integrating VARMAx for demand prediction with the Firefly Algorithm for optimization. The improvement of the integrated techniques is one prospective research subject. The VARMAx model and the Firefly Algorithm's settings could be adjusted extensively, resulting in even higher gains in energy efficiency, latency reduction, data throughput, and PDR. Improving optimization accuracy and effectiveness could entail investigating adaptive parameter tuning techniques and investigating more sophisticated metaheuristic algorithm modifications.

The proposed framework's applicability can also be expanded to include a variety of real-world situations. The resilience and generalization capabilities of the framework may be revealed by examining how well it can adapt to various network designs, traffic patterns, and user behaviors. Additionally, investigating how resource allocation and load balancing interact with sleep scheduling holds the promise of developing synergistic solutions that further improve overall network performance. Given the quick development of communication technologies, the suggested framework may be expanded to include the difficulties and possibilities of new developments like 6G. Future communication systems may become more innovative and relevant by having heterogeneous networks, dynamic spectrum sharing, and unique communication paradigms.

Additionally, the suggested framework could be modified to meet the particular characteristics and needs of Internet of Things (IoT) networks as the deployment of IoT devices increases. Under real-world conditions, optimizing sleep scheduling for various IoT devices with diverse communication patterns and strict energy limits is an intriguing area for investigation. To sum up, this paper opens the door to a wide range of fascinating research paths. The proposed framework is a strong contender for future developments that enhance the evolution of communication networks due to its adaptability, robustness, and variety. Researchers and practitioners may advance the development of more effective, sustainable, and high-performing communication networks that meet the needs of current and upcoming communication ecosystems by forging ahead with these new research scopes.

REFERENCES

1. Akande, Akinyinka Olukunle, Cosmas Kemdirim Agubor, Folasade Abiola Semire, Olusola Kunle Akinde, and Zachaeus Kayode Adeyemo. 2023. "Intelligent Empirical Model for Interference Mitigation in 5G Mobile Network at Sub-6 GHz Transmission Frequency." *International Journal of Wireless Information Networks* 30 (4): 287–305. doi:10.1007/s10776-023-00603-z.
2. Akhtar, Tafseer, Christos Tselios, and Ilias Politis. 2021. "Radio Resource Management: Approaches and Implementations from 4G to 5G and Beyond." *Wireless Networks* 27 (1): 693–734. doi:10.1007/s11276-020-02479-w.
3. Alimi, Isiaka. A., Romilkumar K. Patel, Akeem O. Mufutau, Nelson J. Muga, Armando N. Pinto, and Paulo P. Monteiro. 2021. "Towards a Sustainable Green Design for Next-Generation Networks." *Wireless Personal Communications* 121 (2): 1123–1138. doi:10.1007/s11277-021-09062-2.
4. Asheer, Sarah, and Sanjeet Kumar. 2021. "A Comprehensive Review of Cooperative MIMO WSN: Its Challenges and the Emerging Technologies." *Wireless Networks* 27 (2): 1129–1152. doi:10.1007/s11276-020-02506-w.
5. Cao, Xianghui, Jia Wang, Yu Cheng, and Jiong Jin. 2023. "Optimal Sleep Scheduling for Energy-Efficient AoI Optimization in Industrial Internet of Things." *IEEE Internet of Things Journal* 10 (11): 9662–9674. doi:10.1109/JIOT.2023.3234582.
6. Dai, Guangli, Weiwei Wu, Kai Liu, Feng Shan, Jianping Wang, Xueyong Xu, and Junzhou Luo. 2023. "Joint Sleep and Rate Scheduling With Booting Costs for Energy Harvesting Communication Systems." *IEEE Transactions on Mobile Computing* 22 (6): 3391–3406. doi:10.1109/TMC.2021.3135865.
7. Fang, Zhengru, Jingjing Wang, Yong Ren, Zhu Han, H. Vincent Poor, and Lajos Hanzo. 2022. "Age of Information in Energy Harvesting Aided Massive Multiple Access Networks." *IEEE Journal on Selected Areas in Communications* 40 (5): 1441–1456. doi:10.1109/JSAC.2022.3143252.
8. Guérin, Nicolas, Malo Manini, Rodolphe Legouable, and Cédric Guéguen. 2021. "High System Capacity Pre-Scheduler for Multi-Cell Wireless Networks." *Wireless Networks* 27 (1): 13–25. doi:10.1007/s11276-020-02441-w.

9. Kalita, Priyanka, and Dharmaraja Selvamuthu. 2023. "Stochastic Modelling of Sleeping Strategy in 5G Base Station for Energy Efficiency." *Telecommunication Systems* 83 (2): 115–133. doi:10.1007/s11235-023-01001-9.

10. Kaur, Preetjot, Roopali Garg, and Vinay Kukreja. 2023. "Energy-Efficiency Schemes for Base Stations in 5G Heterogeneous Networks: A Systematic Literature Review." *Telecommunication Systems* 84 (1): 115–151. doi:10.1007/s11235-023-01037-x.

11. Linnartz, J. P. M. G., C. R. B. Corrêa, T. E. B. Cunha, E. Tangdiongga, T. Koonen, X. Deng, M. Wendt, et al. 2022. "ELIoT: Enhancing LiFi for next-Generation Internet of Things." *EURASIP Journal on Wireless Communications and Networking* 2022 (1): 89. doi:10.1186/s13638-022-02168-6.

12. Ma, Shengcheng, Xin Chen, Zhuo Li, and Ying Chen. 2021. "Performance Evaluation of URLLC in 5G Based on Stochastic Network Calculus." *Mobile Networks and Applications* 26 (3): 1182–1194. doi:10.1007/s11036-019-01344-1.

13. Majumdar, Parijata, Diptendu Bhattacharya, Sanjoy Mitra, and Bharat Bhushan. 2023. "Application of Green IoT in Agriculture 4.0 and Beyond: Requirements, Challenges and Research Trends in the Era of 5G, LPWANs and Internet of UAV Things." *Wireless Personal Communications* 131 (3): 1767–1816. doi:10.1007/s11277-023-10521-1.

14. Malathy, S., P. Jayarajan, Henry Ojukwu, Faizan Qamar, MHD Nour Hindia, Kaharudin Dimyati, Kamarul Ariffin Noordin, and Iraj Sadegh Amiri. 2021. "A Review on Energy Management Issues for Future 5G and beyond Network." *Wireless Networks* 27 (4): 2691–2718. doi:10.1007/s11276-021-02616-z.

15. Matthew, Ugochukwu O., and Jazuli S. Kazaure. 2021. "Chemical Polarization Effects of Electromagnetic Field Radiation from the Novel 5G Network Deployment at Ultra High Frequency." *Health and Technology* 11 (2): 305–317. doi:10.1007/s12553-020-00501-x.

16. Mishra, Lalita, Vikash, and Shirshu Varma. 2021. "Seamless Health Monitoring Using 5G NR for Internet of Medical Things." *Wireless Personal Communications* 120 (3): 2259–2289. doi:10.1007/s11277-021-08730-7.

17. Nazir, Mohsin, Aneeqa Sabah, Sana Sarwar, Azeema Yaseen, and Anca Jurcut. 2021. "Power and Resource Allocation in Wireless Communication Network." *Wireless Personal Communications* 119 (4): 3529–3552. doi:10.1007/s11277-021-08419-x.

18. Saddoud, Ahlem, and Lamia Chaari Fourati. 2021. "Joint Microcell-Based Relay Selection and RRM for LTE-A toward 5G Networks." *Telecommunication Systems* 78 (3): 363–375. doi:10.1007/s11235-021-00817-7.

19. Salahdine, Fatima, Tao Han, and Ning Zhang. 2023. "5G, 6G, and Beyond: Recent Advances and Future Challenges." *Annals of Telecommunications* 78 (9–10): 525–549. doi:10.1007/s12243-022-00938-3.

20. Salem, A. Abdelaziz, S. El-Rabaie, and Mona Shokair. 2021. "Survey on Ultra-Dense Networks (UDNs) and Applied Stochastic Geometry." *Wireless Personal Communications* 119 (3): 2345–2404. doi:10.1007/s11277-021-08334-1.

21. Sathya, Vanlin, Srikant Manas Kala, and Kalpana Naidu. 2023. "Heterogenous Networks: From Small Cells to 5G NR-U." *Wireless Personal Communications* 128 (4): 2779–2810. doi:10.1007/s11277-022-10070-z.

22. Shen, Aiguo, Qiubo Ye, Guangsong Yang, and Xinyu Hao. 2021. "M2M Energy Saving Strategy in 5G Millimeter Wave System." *Telecommunication Systems* 78 (4): 629–643. doi:10.1007/s11235-021-00836-4.

23. Tang, Yong, Xianjun Deng, Lingzhi Yi, Yunzhi Xia, Laurence T. Yang, and Xiao Tang. 2023. "Collaborative Intelligent Confident Information Coverage Node Sleep Scheduling for

6G-Empowered Green IoT." *IEEE Transactions on Green Communications and Networking* 7 (2): 1066–1077. doi:10.1109/TGCN.2022.3193996.

24. Wang, Jia, Xianghui Cao, Bo Yin, and Yu Cheng. 2022. "Sleep–Wake Sensor Scheduling for Minimizing AoI-Penalty in Industrial Internet of Things." *IEEE Internet of Things Journal* 9 (9): 6404–6417. doi:10.1109/JIOT.2021.3112211.

25. Yang, Mao, Bo Li, and Zhongjiang Yan. 2021. "MAC Technology of IEEE 802.11ax: Progress and Tutorial." *Mobile Networks and Applications* 26 (3): 1122–1136. doi:10.1007/s11036-020-01622-3.

Emerging Technologies and Applications in Electrical Engineering –
Prof. Dr. Anamika Yadav et al. (eds)
© 2024 Taylor & Francis Group, London, ISBN 978-1-032-82568-7

Analysis of Rogowski Coil Shielding Effectiveness with External Vertical Magnetic Field for Horizontal Airgap Using FEM

7

Priti Bawankule[1] and Kandasamy Chandrasekaran[2]
Department of Electrical Engineering
National Institute of Technology, Raipur-(C. G)-492010

ABSTRACT: Rogowski coil (RC) play a significant role in detecting the condition of electrical power equipment and measuring current in high end application. In order to achieve high level of accuracy, magnetic shielding is crucial, because it reduces leakage fields and influence of external magnetic fluxes on RC. To ensure the effective design of magnetic shielding for the RC, it is essential to conduct calculations for shielding effectiveness. The analysis of shield effectiveness employs the magnetostatics approach. This paper explores a suggested magnetic circuit model, specifically addressing a hollow cylindrical structure with shielding structures of rectangular, circular, and hexagonal cross-sections. The magnetic shielding effectiveness of the Rogowski coil is examined in the presence of an external vertical magnetic field with a horizontal air gap in the shield. The Rogowski coil is designed in ANSYS Maxwell software, and the Finite Element Method (FEM) is employed for validation. The various design parameters are considered, and simulation results are reported. The magnetic field distribution for rectangular, circular and hexagonal structure was analyzed using FEM simulation. From simulation result, it is clear that shield effectiveness for circular or hexagonal cross sections are nearly double that of rectangular cross-sectional RC. As the shield thickness altered by 5mm with a fixed distance between the shield and coil the shield effectiveness is changed rapidly. The effect of horizontal airgap in shield and relative permeability of shield material on shielding effectiveness is analyzed. The simulation result indicates that a notable reduction in shield effectiveness with an increase in the horizontal airgap. Beyond an airgap of 1.2mm, there is only a minor fluctuation in shield effectiveness. The simulation result indicate that the shield effectiveness remains unaffected when designed with materials possessing more relative permeability.

KEYWORDS: Finite Element Method, Horizontal Airgap, Magnetic shield effectiveness, Rogowski coil, External magnetic field

[1]pritinbawankule@gmail.com, [2]vrkchandran@gmail.com

DOI: 10.1201/9781003505181-7

1. Introduction

Rogowski coil (RC) are mostly employed to measure alternating current (AC), high amplitude currents and transient currents. One of the most significant applications of them is current measurement of industrial electronic devices. Since this is a non-magnetic, they are not saturable (B. M. H. Samimi, 2015). The external magnetic interference has significantly influenced the performance of the RC at the industrial site. Thus, the RC needs to have a suitable magnetic shielding structure. Magnetic shielding is an important factor in reducing magnetic disturbances in coil. It is necessary to analyze the shielding effectiveness of the magnetic shield when designing shield for current comparators (H. Shao, 2013), (IEEE Std 299, 1998), (X. Wang, 2013) and (S. Ren,2016). Reference (G. A. Kyriazis, 2002) evaluates the effectiveness of high permeability shells as magnetic shields for current comparator cores. Using the magnetostatic approach, magnetic shielding reduced leakage fields (radial, dipole, and axial) generated by ratio windings in current comparators. For specific field configurations that emulate ratio windings. Research in the literature has investigated shielding effectiveness for external horizontal and vertical magnetic field without air gaps based on magnetic circuit approaches (S. Ren,1995), numerical methods, and experimentation (K. Draxler, 2018). The paper (B. Ayhan, 2020) presented magnetic circuit solutions for magnetic shielding are explored, assuming both rectangular and circular shell structure under vertical external magnetic fields. An improved magnetic circuit model is also proposed, incorporating hollow cylindrical shell of shielding to calculating magnetic shielding effectiveness in Rogowski coils. In (M. Rezaee, 2008), (S., Deng, 2021) and (P. Bawankule, 2022) investigated the influence of coil cross-sectional geometry, specifically comparing circular and rectangular cross-sections, on the mutual inductance of Rogowski coils. Mostly used RCs now are with circular or rectangular core's structure. For the purpose of increasing the sensitivity of Rogowski coil measurements as well as smooth wire winding, hexagonal cross section of Rogowski coil has examines in (S. Al-Sowayan, 2014).

This paper deliberates proposed model of magnetic circuit, involve of hollow structure, with different cross section of RCs. Magnetic shielding for RCs can also be examined and designed in ANSYS Maxwell. Magnetic shield effectiveness of proposed model of RCs for the rectangular, circular and hexagonal cross while considering vertical external magnetic field was analyzed using FEM. The horizontal air gap in shield is consider for calculation of magnetic shield.

2. Magnetic Shield with Horizontal Airgap

The various cross sections of the RC are enveloped by tape winding on magnetic cores considering horizontal airgaps, illustrated in Fig. 7.1. Tape-wound magnetic cores are widely used in power system applications due to their straightforward design and construction.

Figure 7.1 depicts a cross-section view of a magnetic shield, consisting of a nonmagnetic Rogowski coil core enclosed by a rectangular shell. Rogowski coils (RCs) can effectively measure the primary current of a high-voltage power system operating at a fundamental

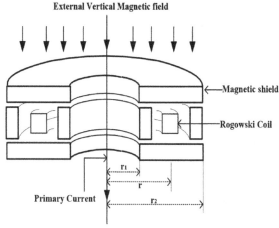

Fig. 7.1   3D view of rectangular cross section of RC enclosed by the horizontal airgap of shield

frequency of 50 Hz. The proposed method is applicable to cylindrical magnetic shielding structures with horizontal air gaps, making it versatile for various device configurations.

A magnetic shield's effectiveness is generally determined by comparing a magnetic field with a shield and without shield (H. shao, 2013). The magnitude and direction of magnetic field intensity vectors may change depending on the location of magnetic shielding. In this investigation, the average magnetic field intensities are taken into account when calculating shield effectiveness both before and after magnetic shielding.

The shielding effectiveness S is given by,

$$S = \frac{\overline{H_0}}{\overline{H_T}} \tag{1}$$

The direction of external magnetic field to the Rogowski coil aligns parallel to the shield axis, influencing the magnetic shielding effectiveness as elucidated in (2) when there is no airgap [9].

Equitation (2) for Shield effectiveness for rectangular structure of shield with vertical external magnetic field without airgap S_{Va}.

$$S_{Va} = \frac{8}{3} \frac{\mu_s}{\mu_0} \frac{ab}{c(c-a)} - \frac{2a}{c} + 1 \tag{2}$$

The assume an equal flux distribution inner and outer shields for a vertical external magnetic field. The magnetic shield effectiveness with horizontal has explained in [2]. Then (2) has to be improved for horizontal airgap. Equation (3) provides the shield effectiveness for a vertical magnetic field with a horizontal airgap.

$$S_V = \frac{8ab}{c} \frac{\mu_s}{\mu_0 \left(3c - 3a - 4g + 4g\right) + 4\mu_s g} - \frac{2a}{c} + 1 \tag{3}$$

2.1 Rectangular and Circular Cross- section of RC

The magnetic shielding involves of four tape-wound magnetic cores in a hollow cylindrical structure surrounding a rectangular non-magnetic core with horizontal air gaps as shown in Fig. 7.2. The dimensions in this scenario are as follows: the thickness of shield is denoted by 'a,' the shield and coil distance denoted by 'b', the total length of the rectangular shield is 'c,' the height of the Rogowski coil is 'h,' and the vertical or horizontal airgap is 'g.' The permeability of the shield and air is represented by μ_s and μ_0, respectively. Assuming the Rogowski coil's axis aligning with the external magnetic flux, the airgaps on both the right and left sides are represented by 'g'. The Rogowski coil's non-magnetic core possesses a relative permeability of unity μ_0

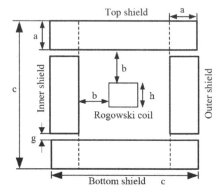

Fig. 7.2 Rectangular cross section of RC with horizontal airgap

The magnetic shield is composed of a toroidal hollow structure enclosed the circular cross-section of the magnetic shield for the Rogowski coil, as illustrated in Fig. 7.3. Rogowski coils employ a non-magnetic core for the measurement of high currents. The shield thickness, distance between RC and shield, diameter of RC and diameter of circular structure including magnetic shield are a, b, h and c respectively. Horizontal air gap for left and right is g. The relative permeability of shield material is μ_s and for air is μ_0.

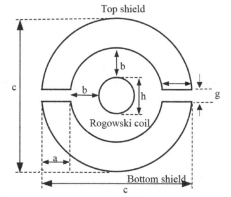

Fig. 7.3 Circular cross section (toroidal) of RC with horizontal airgap

2.2 Hexagonal Cross-section of RC

In general, we examine circular or rectangular cross section of RC. The mutual inductance for shapes hexagonal and rectangular is different that compared in (S. Al-Sowayan, 2014), the advantage of employing hexagonal core is smooth and easy winding of copper wires around the core compared to rectangular which will change coil longevity. The area of cross section of rectangular, circular and hexagonal are different, that will be effect on magnetic field intensity, as a result shield effectiveness are also different. The hexagonal have a larger cross section than circular as shown in Fig. 7.4 which will definitely more magnetically flux linkages, that effect on sensitivity of coil.

In this paper, for analysis purposes, a hexagonal cross section of a coil is considered. Hexagonal cross section of RC is surrounded by toroidal hollow shell hexagonal structure of magnetic

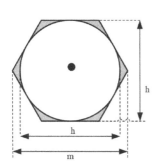

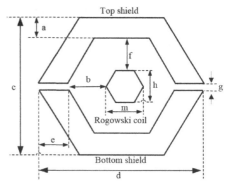

Fig. 7.4 Comparison of cross-sectional area of circular and hexagonal structure of coil

Fig. 7.5 Hexagonal cross section of RC with horizontal gap

shield. Rogowski coil as illustrated in Fig. 7.5. The thickness of hexagonal of shield is 'a', Distance between corner of RC and corner of shield is 'b'. Distance between edge of RC to edge of shield is 'f'. The thickness of corner of shield is 'e'. The relation between 'a' and 'e' as described in (4). The length of horizontal and vertical of magnetic shield is not same in hexagonal structure, which is 'd' and 'c' respectively as given in (5) and (6) respectively. similarly horizontal and vertical length of RC is 'm' and 'h' respectively.

$$e = (2/\sqrt{3})a = ka \tag{4}$$

Multiplication factor, $k = 1.155$

$$d = (2b + 2e + m = (2b + 2ka + m) \tag{5}$$

$$c = (2a + 2f + h) = \left(2a + 2\left(\frac{b}{k}\right) + \left(\frac{m}{k}\right)\right) \tag{6}$$

3. FEM Simulation Result

The simulation was conducted using ANSYS Maxwell (Ansys maxwell software V.18). Rather than conducting 3-D simulations, 2-D simulations are conducted to maximize computer memory and speed up simulations. Using a magnetostatic solver, the two-dimensional Rogowski coil structure is designed and simulated to calculate shield effectiveness. An external vertical magnetic field of 1000 A/m is created in presence of magnetic shield and a nonmagnetic core in the Rogowski coil.

3.1 Impact of Shield Thickness (b) and Distance between the Shield and RC (a) on Shield Effectiveness

To calculate the effectiveness of the shield, specifically the thickness of shield 'a' and separation distance between shield and RC 'b' are varied in RC model using FEM simulations. For simulation, assume g=0.2mm and relative permeability of shield material $\mu_s = 4000$. The magnetic field distribution for rectangular, circular and hexagonal cross section of RC for vertical external field considering horizontal air gap in shield by keeping, g=0.2mm, a=10mm and b=10mm and is illustrated in Fig. 7.6, Fig. 7.7 and Fig. 7.8.

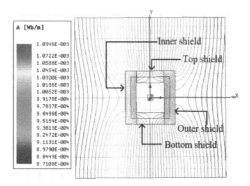

Fig. 7.6 Magnetic field distribution for rectangular cross section of RC with horizontal airgap

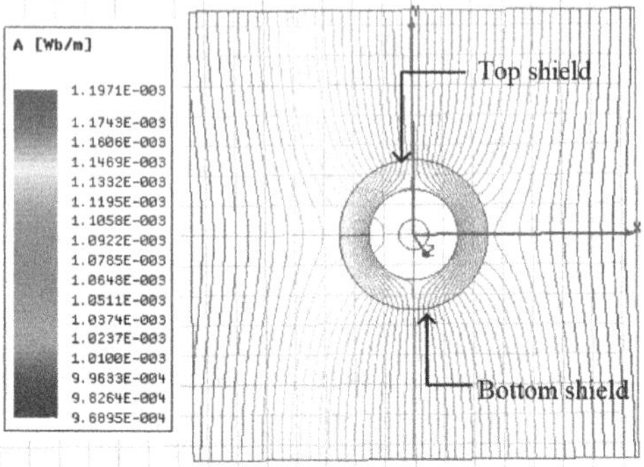

Fig. 7.7 Magnetic field distribution for circular cross section of RC with horizontal air gap

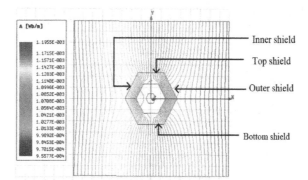

Fig. 7.8 Magnetic field for hexagonal cross section of RC with horizontal air gap

Figure 7.6 illustrated that magnetic field distribution for a rectangular cross-section with a horizontal airgap ranging from of 0.8712 to 1.0923 *mWb/m*. For circular cross section, the magnetic field distribution for horizontal airgap between 0.96895 and 1.1971 *mWb/m* is shown in Fig. 7.7. Similarly for Hexagonal cross section of coil, the magnetic field distribution varies from 0.9558 to 1.1955 *mWb/m* as presented in Fig. 7.8. It is observed in Fig. 7.6, Fig. 7.7 and Fig. 7.8 that the outer and inner shields have nonuniform magnetic fields. Field lines arrive the shield through the upper surface of the top shield, the sides of the top shield, the inner shield's interior surface, and the outer shield's exterior surface. The shield effectiveness is calculated as per Eq. (1), using FEM as illustrated in Table 7.1, Table 7.2 and Table 7.3. The shield thickness (a) and the distance between shield and the RC (b) are altered in 5mm steps.

Table 7.1 Magnetic shield effectiveness for rectangular cross section of RC model with Horizontal airgap g=0.2mm

a (mm)	b (mm)					
	5	10	15	20	25	30
5	8.096	8.991	9.456	9.861	10.006	10.287
10	13.297	14.419	15.358	15.915	16.398	16.708
15	18.089	19.477	20.508	21.362	22.172	22.675
20	22.691	24.236	25.529	26.553	27.412	28.114
25	27.166	28.951	30.364	31.416	32.530	33.334
30	31.884	33.658	35.038	36.245	37.313	38.323

Table 7.2 Magnetic shield effectiveness for circular cross section of RC model with Horizontal airgap g=0.2mm

a (mm)	b (mm)					
	5	10	15	20	25	30
5	14.884	15.54497	15.81658	15.995	16.074	16.210
10	25.527	26.72909	27.44255	27.908	28.323	28.514
15	35.173	36.87655	38.10004	38.953	39.523	39.858
20	44.261	46.5565	48.1098	49.221	50.004	50.574
25	53.303	55.81685	57.61947	58.923	59.943	60.728
30	62.034	64.756	66.765	68.319	69.503	70.452

Table 7.3 Magnetic shield effectiveness for hexagonal cross section of RC model with horizontal airgap g=0.2mm

a (mm)	b (mm)					
	5	10	15	20	25	30
5	15.281	15.919	16.261	16.446	16.645	16.699
10	25.961	27.294	28.108	28.758	29.124	29.346
15	35.727	37.634	38.942	39.816	40.462	40.910
20	45.059	47.348	48.988	50.183	51.039	51.690
25	54.116	56.643	58.547	59.982	61.082	61.896
30	62.901	65.621	67.757	69.354	70.677	71.724

From Table 7.2 and Table 7.3, it is clear that S_V for horizontal air gap of circular cross section of coil is near same hexagonal cross section of coil as per simulation result. There will be nearly double the effectiveness of shields with circular or hexagonal cross sections compared to rectangular cross sections. The distance between the shield and coil will be kept same, and the magnetic shield thickness will increase, so the effectiveness of the shield will differ greatly.

Magnetic shield effectiveness does not differ much if shield thickness increases while shield and coil distances remain the same as presented in Table 7.1, Table 7.2 and Table 7.3.

3.2 Impact of Change in Horizontal Airgap on Shield Effectiveness

In shield effectiveness analysis, the airgaps hold significant importance in determining shielding effectiveness. For simulation, consider the relative permeability of shield material μ_s is 4000. The thickness of shield is 10mm and a distance between the shield and the coil is also 10mm. The air gap within the shield is varied within the range of 0.2 to 1.6 mm. As per Finite Element Method (FEM) simulation results presented in Table 7.4, the shielding effectiveness with a horizontal air gap is observed to be lower for the rectangular structure compared to circular and hexagonal structure. Additionally, there is a notable decline in shielding effectiveness with an increase in the horizontal air gap; beyond g=1.2mm, there is minimal variation, as depicted in Fig. 7.9. Conversely, there is not much variation in shielding effectiveness when considering a horizontal gap for circular and hexagonal structures, as shown in Fig. 7.9.

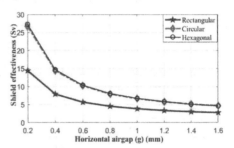

Fig. 7.9 The correlation between horizontal air gap and shield effectiveness

Table 7.4 The correlation between shield effectiveness and Horizontal airgap

S. No.	Air gap (g) (mm)	Shielding Effectiveness (S_V) for Horizontal airgap		
		Rectangular	Circular	Hexagonal
1	0.2	14.419	26.729	27.294
2	0.4	7.934	14.422	14.692
3	0.6	5.701	10.210	10.366
4	0.8	4.581	7.976	8.081
5	1	3.891	6.6775	6.755
6	1.2	3.394	5.812	5.853
7	1.4	3.062	5.181	5.135
8	1.6	2.822	4.710	4.649

3.3 Impact of Change in Relative Permeability on Shield Effectiveness

The relative permeability of shield influences the shielding effectiveness of a horizontal airgap.

Moreover, an exploration into the effect of relative permeability on shielding effectiveness (S_V) was conducted by altering it within the range of 1000 to 100000, by keeping 'b' and 'a' both are 10mm. The Shielding Effectiveness (S_V) changes as the relative permeability of shield (μ_s) increases, but up to the certain point after then it will remain constant is described in Table 7.5. To enhance clarity, a graph depicting between the relative permeability of the shield

Table 7.5   Correlation between relative permeability of shielding material and shield effectiveness

S. No.	Relative permeability (μ_s)	Shielding effectiveness (S_V) for horizontal airgap		
		Rectangular	Circular	Hexagonal
1	1000	13.6518	24.319	24.6769
2	4000	14.4184	26.729	27.2944
3	10,000	14.5823	27.270	27.8853
4	20,000	14.6377	27.457	28.0878
5	40000	14.6656	27.548	28.1902
6	60000	14.6750	27.578	28.2245
7	80000	14.6796	27.593	28.2417
8	100000	14.6824	27.090	28.2519

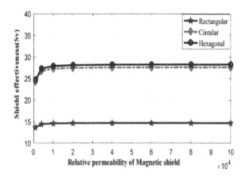

Fig. 7.10   The correlation between the relative permeability of the magnetic shield and the shield effectiveness

and magnetic shielding effectiveness is presented in Fig. 7.10. In both circular and hexagonal structures, the shield effectiveness is to be almost the same while considering the change in (μ_s).

4. Conclusion

An ANSYS Maxwell model for Rogowski coil is developed using hollow rectangular, circular, and hexagonal cross-sections of magnetic shield with horizontal airgap. For a wide range of design parameters, the FEM is employed for evaluation of magnetic shielding effectiveness. Simulations shows that, hexagonal RC model results are nearly similar to circular RC model for different airgap, relative permeability, and design parameters under study. The value of 'b' and 'a' was altered from 5 to 30 mm, the simulation results show that circular and hexagonal shields have greater shield effectiveness than rectangular shields. The shield effectiveness for rectangular structure is approximately half of that of circular and hexagonal with horizontal airgap. An analysis of the effectiveness of shielding in magnetic shields based on the horizontal airgap and relative permeability. The shield effectiveness is observed only for horizontal airgaps up to 1.2 mm; above that, there are no such changes. Moreover, simulations indicate that shields designed with higher relative permeability materials do not affect shielding effectiveness.

REFERENCES

1. ANSYS Maxwell 2D/3D Field Simulator, v.18. Available online: https://www.ansys.com
2. B. M. H. Samimi, A. Mahari, M. A. Farahnakian, and H. Mohseni, (2015). The rogowski coil principles and applications: A review. IEEE Sens. Journal., vol. 15, no. 2, pp. 651–658.
3. B. Ayhan and C. Uçak (2020). Improved Magnetic Circuit Model for Magnetic Shielding Effectiveness in Rogowski Coil. IEEE Trans. on Magn, vol 56, no. 3. Art. no. 8500109
4. G. A. Kyriazis. (2002). The effectiveness of current-comparator magnetic shields against leakage fields at power frequencies. in Proc. Dig. CPEM, pp. 34–35.
5. H. Shao et al. (2013). Magnetic shielding effectiveness of current comparator. IEEE Trans. Instrum. Meas., vol. 62, no. 6, pp. 1486–1490.
6. IEEE Standard Method for Measuring the Effectiveness of Electromagnetic Shielding Enclosures, IEEE Std 299-1997, 1998.
7. K. Draxler and R. Styblikova, (2018). Magnetic shielding of Rogowski coils. IEEE Trans. Instrum. Meas., vol. 67, no. 5, pp. 1207–1213.
8. M. Rezaee and H. Heydari. (2008). Mutual Inductances Comparison in Rogowski Coil with Circular and Rectangular Cross-Sections and Its Improvement. 3rd IEEE Conf. on Industrial Electronics and Application.
9. P. Bawankule and K. Chandrasekaran. (2022). Rogowski Coil with an Active Integrator for Impulse Current Measurement. IEE Global conference for Adnavncement in Technology, 9–13.
10. S. Al-Sowayan. (2014). Improved Mutual Inductance of Rogowski Coil Using Hexagonal Core. International Conference on Electrical, Computer and Communication Engineering, (ICECCE) Barcelona, Spain.
11. S., Deng, E., Peng, C., Zhang, G., Zhao, Z. and Cui, X. (2021). Method of Turns Arrangement of Noncircular Rogowski Coil with Rectangular Section. IEEE Transactions on Instrumentation and Measurement, 70.
12. S. Ren, S. Guo, X. Liu, and Q. Liu. (2016). Shielding effectiveness of double layer magnetic shield of current comparator under radial disturbing magnetic field. IEEE Trans. Magn., vol. 52, no. 10, Oct. 2016, Art. no. 9401907.
13. S. Ren, H. Ding, M. Li, and S. She. (1995). Magnetic shielding effectiveness for comparators. IEEE Trans. Instrum. Meas., vol. 44, no. 2, pp. 422–424.
14. X. Wang and S. Ren (2013). Calculating magnetic shielding effectiveness for high-power DC comparator by magnetic circuit method. J. Chongqing Univ., vol. 6, no. 2, pp. 113–118.

Emerging Technologies and Applications in Electrical Engineering –
Prof. Dr. Anamika Yadav et al. (eds)
© 2024 Taylor & Francis Group, London, ISBN 978-1-032-82568-7

Partial Discharge Localization for a Single Source Using Acoustic Sensors and XG-Boost Algorithm

8

Anna Baby[1]
Assistant Professor, Adi Shankara Institute of Engineering and Technology, Kalady

Nasirul Haque[2]
Assistant Professor, Electrical Engineering Department, National Institute of Technology, Calicut, Kerala, India

Preetha P.[3]
Professor, Electrical Engineering Department, National Institute of Technology, Calicut, Kerala, India

ABSTRACT: Proper monitoring of power transformer insulation is very essential for the reliability of power system. Partial discharges are a very common fault that occurs in power transformers. Continuous presence of them over a prolonged period may lead to complete failure and insulation breakdown of the power transformer. In this work, for detecting and localizing PD activities a new method is proposed in the power transformer. This method uses acoustic sensors which are kept on the outer body of the transformer tank. For this purpose, PD activities were intentionally produced inside a tank emulating real-life transformer and the location of PD is varied. For each location, a sensor is used to measure the generated acoustic signals at a time and it is recorded with an oscilloscope. Data is recorded corresponding to each location, and suitable features are extracted. The training and testing of XG based machine learning framework is done with the help of the collected features. The test results shows that, the proposed framework can find partial discharge sources which gives an accuracy of over 90% when they operate standalone in the insulation system.

KEYWORDS: Acoustic sensor, Feature extraction, Partial discharge

[1]anna_p210126ee@nitc.ac.in, [2]nasirul@nitc.ac.in, [3]preetha@nitc.ac.in

DOI: 10.1201/9781003505181-8

1. Introduction

Early diagnosis of a damage in a high voltage equipment plays an important role in the power system stability. Partial discharge is an insulation degradation process very much common in power transformers. The localized electrical discharge that take place normally within the insulation is called as Partial discharge. Properties like charge movement, acoustic emission, electromagnetic radiation, and chemical reactions are different for partial discharge. The conventional PD measurement systems have a major drawback since they cannot discriminate the internal PD pulses and external pulse shaped interferences, especially in online and on-site conditions (Sharifinia Sajjad et al, 2022).

Acoustic pressure waves are produced by partial discharge in electrical equipment and the propogation of these waves is the basis of acoustic detection technique. To record the acoustic signals, acoustic sensors are located on the outer surface of the transformer tank, which makes acoustic detection method a non-invasive method(S. Biswas et al, 2016).Cross wavelet transform is used to extract the features from acoustic PD signals. For classifying and detecting the extracted features, the author proposed ensemble support vector machine based classifier.

PD source recognition using conventional methods are done with the help of time domain analysis. Characteristics such as location of the phasor, repetition behavior of the pulse and the ratio of amplitude of the partial discharge signals are monitored in this case (Haresh Kumar et al,2021). In this paper, the measurement and characterization of an acoustic wave is done which is generated due to a partial discharge. For the simultaneous measurement of the acoustic wave, more number of acoustic sensors are needed, that makes the system more complicated and costly. In this work, localizing PD sources is done by applying XG boost Algorithm, a very popular machine learning technique.

2. Experiment

2.1 Experimental Arrangement

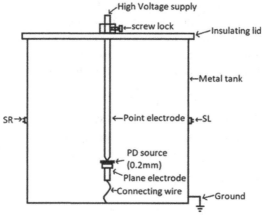

Fig. 8.1 Experimental arrangement

For doing the experiment, a model of actual transformer tank is emulated with partial discharge sources kept within the transformer tank. 3 acoustic sensors are kept at the 3 outer walls of the transformer tank. The location of the partial discharge source is changed and the acoustic signals are estimated by placing one sensor for an instant and it is connected to an oscilloscope, corresponding to each location. A machine learning based model is developed with the help of the data collected from the experimental set up and this data can be used to automatically anticipate the position of the partial discharge source from its learning experience. The algorithm selected for this objective is XG boost algorithm.

The experimental arrangement is shown in Fig. 8.1. It has dimensions of 30x30x30cm. A wooden insulating lid is kept on the top of the emulated transformer tank. With the help of a screw lock arrangement on the insulating lid, a point-plane electrode is kept inside the transformer tank. The point-plane electrode is kept at an adjustable height and multiple horizontal positions. An insulation is made of polypropylene material with a thickness of 0.1mm and it is placed between both the electrodes. Three sensors are named as SR, SL and SD and are kept on the 3 side faces of the emulated transformer tank. Another sensor SU is positioned on the top of the tank. The response of Acoustic sensors is first filtered using High-pass filter circuit and then recorded through an oscilloscope. The photograph of the experiment conducted in high voltage lab is shown in Fig. 8.2. A total of 20 datasets were recorded from each of the three sensors whereas the PD location was kept fixed. In this way a total of 60 datasets were recorded corresponding to one PD location. After this the PD location was varied and same procedure was repeated. All total, three positions were considered in present study and 180 datasets were recorded.

Fig. 8.2 Testing arrangement in High Voltage laboratory including transformer tank, high voltage transformer and oscilloscope

2.2 Position of PD Source

Figure 8.3 shows the positioning of PD sources. In all positions, the partial discharge source is kept at a height of about 7.5 cm (quarter height of the transformer tank). The partial discharge source is kept oil immersed. The acoustic sensors are kept at the middle of each face of the tank.

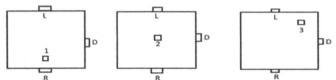

Fig. 8.3 Positioning of sensors and top view of PD location. The sensor positioned on top of the tank is not shown here.

3. Experimental Results

The acoustic waveforms data gathered from the three sensors corresponding to PD in position is shown in Fig. 8.4(a), (b), (c). Each signal is of $2\mu s$. From these recorded signals variables like, deviation of peak value, time needed for oscillation damping, steepness of damping etc are noted. With the help of statistical methods these quantities can be discriminated and certain features about these pulse can be obtained. The collected features are considered as the standard for facilitating the localization of partial discharge source inside the transformer tank.

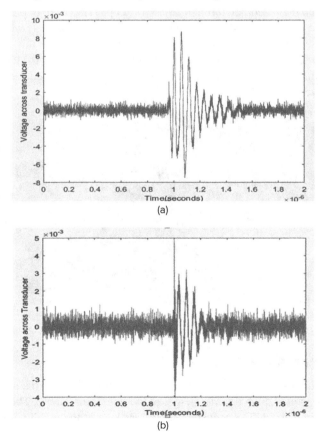

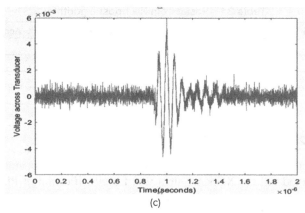

(c)

Fig. 8.4 (a) Position 1 sensor L (left of image), (b) Position 1 sensor R (right of image), (c) Position 1 sensor U (upper of image)

4. Discrimination of pd Position using XG Boost Algorithm

4.1 XG Boost Algorithm

XG Boost (eXtreme Gradient Boosting) is a well developed machine learning algorithm that belongs to the family of gradient boosting methods. The method behind XG Boost is to iteratively train a series of decision trees. Consecutive trees are built to correct the mistakes made by the previous trees. This iterative process is known as boosting. By incorporating the predictions of multiple trees, XG Boost improves the overall accuracy and generalization ability of the model. It can control both numerical and categorical features, and it incorporates regularization techniques to prevent overfitting. Some of the advantages of XG boost algorithm include scalability, efficiency, and the ability to handle large datasets. It also provides additional features like built-in cross-validation, early stopping, and handling missing values.

4.2 Application of XG Boost in Present Case

All the measured waveforms corresponding to different sensors and different PD locations were utilized for feature selection. A total of five features (Max, Mean, Standard deviation, Kurtosis, Skewness) were selected in presented work. After this, the implementation of XG boost was performed. Out of the 60 data sets for each PD position and sensor, 48 datasets are selected for training and remaining 12 for testing. The testing and training process was iterated five times, each time with different combinations of training and testing dataset so that every dataset is used once for training and once for testing. As a result, finally all 60 datasets corresponding to a particular PD location got tested. The final results are given in Table 8.1.

Table 8.1 Result from XG boost algorithm

Position	Total Test	Correct Prediction	Accuracy (%)
1	60	57	95
2	60	58	97
3	60	56	93
	Overall Accuracy		95

5. Conclusion

A method has been proposed for PD localization inside a power apparatus using acoustic detection. The method utilizes XGboost algorithm for classification of PD locations. Three different PD positions were considered in present case and an accuracy of 95% was achieved using the proposed methodology.

Acknowledgement

The authors are very much thankful to the Department of Electrical, NIT Calicut and Mechanical Department of NIT Calicut for their help in this project work.

REFERENCES

1. Haresh Kumar, Muhammad Shafiq, Ghulam Amjed Hussain, Kimmo Kauhaniemi. (2021). Comparison of Machine Learning Algorithms for classification of Partial Discharge Signals in Medium Voltage Components, 2021 IEEE PES Innovative Smart Grid Technologies, 978-1-6654-4875-8/21.
2. Ilkhechi, Hossein Dadashi, and Mohammad Hamed Samimi. (2021). Applications of the acoustic method in partial discharge measurement: A review. IEEE Transactions on Dielectrics and Electrical Insulation, vol. 28, no. 1, pp. 42-51.
3. Jiang, Jun, Judong Chen and Jiansheng Li. (2021). Partial discharge detection and diagnosis of transformer bushing based on UHF method. IEEE Sensors Journal, vol. 21, no. 15, pp. 16798-16806.
4. L. Pradeep, N. Haque, P.Preetha. (2022). Identification Of Partial Discharge Sources In Oil-Pressboard Insulation With Single Type Defect Through TF Mapping, IEEE International Conference on Condition Assesment Techniques in Electrical Systems (CATCON), pp. 250-253, December 17-19, Durgapur India.
5. S.Biswas, D.Dey, B.Chatterjee and Sivaji Chakravorti. (2016). Cross spectrum Analysis based Methodology for Discrimination and Localization of Partial Discharge Sources using Acoustic Sensors, IEEE Transactions on Dielectrics and Electrical Insulation Vol 23, No.6
6. Sharifinia, Sajjad, Mehdi Allahbakshi and Teymoor Ghanbari. (2022). Application of a Rogowski Coil Sensor for Separating Internal and External Partial Discharge Pulses in Power Transformers. IEEE Transactions on Industrial Electronics, no.1-8.

Emerging Technologies and Applications in Electrical Engineering –
Prof. Dr. Anamika Yadav et al. (eds)
© 2024 Taylor & Francis Group, London, ISBN 978-1-032-82568-7

Analytical Study on Inrush Charging Current in Switched Capacitor Multilevel Inverter

Pankaj Kumar Yadav[1], Hari Priya Vemuganti[2], Monalisa Biswal[3]

Department of Electrical Engineering, National Institute of Technology Raipur, India

ABSTRACT: Switched capacitor multilevel inverters (SCMLI) offer a promising solution for high voltage, high power applications, providing advantages like reduced component count, improved efficiency, a common DC link, and voltage boosting. The paper addresses severe issues of inrush charging current and voltage ripple in SC MLI. An analytical study conduct, discussing solutions proposed by various authors and researchers. The identified techniques include connecting an inductor in series with the capacitor to minimize inrush current. Another solution involves reducing voltage ripple through hybrid or LS-PWM modulation techniques, minimizing Long Discharge Period (LDP). The paper provides valuable insights into SC MLI issues, serving as inspiration for further research in this domain.

KEYWORDS: Switched capacitor-based multilevel inverter, Inrush current, Charging current, Hybrid pulse width modulation, Level shift pulse width modulation

1. Introduction

The SC MLI is appealing for its potential to enhance efficiency, reduce component count, and utilize a common DC source. These inverters employ switched capacitor networks to synthesize multi-level voltage waveforms, resulting in improved voltage output quality. Switched capacitor configurations achieve desired output voltage with a single DC source, using capacitors in series and parallel. Developed for enhanced voltage boosting and balancing, they attract attention for minimizing DC power needs and increasing voltage gain (Barzegarkhoo et al., 2022). However, a notable drawback common to many of these SC topologies is the issue

[1]pnkjkumrydv@gmail.com, [2]hpvemuganti.ee@nitrr.ac.in, [3]mbiswal.ele@nitrr.ac.in

DOI: 10.1201/9781003505181-9

of inrush currents. Inrush currents are characterized by high capacitor currents that exceed the steady-state operating current, and they occur due to the switching of large coupling capacitors in series with the DC source in the charging path. These currents are typically associated with the charging and discharging of capacitors within the inverter, and they have the potential to introduce adverse effects into the system (Barzegarkhoo et al., 2022), (Yadav et al., 2023).

Various solutions from existing literature are reviewed, including the use of an inductor in series with a capacitor to minimize inrush current in low-power inverters (Yadav et al., 2023) - (Khan et al., 2022). For high-power applications, researchers propose adding an antiparallel diode with an inductor to bypass it during capacitor discharge, mitigating voltage drop and power loss (Yadav et al., 2023), (Sabour, 2021). The literature also discusses the influence of capacitor voltage and switching frequency on inrush current, exploring techniques such as hybrid and modified PWM schemes to alleviate these issues (Ye et al., 2020) – (Chen et al., 2020).

The paper's objective is to comprehensively explore the origins and consequences of inrush currents in switched capacitor inverters, focusing on voltage imbalances and semiconductor device stress. Through systematic analysis, the study aims to deepen understanding and propose strategies to mitigate and manage inrush currents, optimizing these inverters for high-power applications and advancing power electronics. The paper is organized into sections covering the inrush current phenomenon, its impact on output voltage and inverter performance, reported solutions, and a conclusion that identifies research gaps and encourages further exploration in this field.

The paper is structured as follows: Section II focuses on the inrush current phenomenon, specifically its generation in the switched capacitor (SC) unit. In Section III, the paper elaborates on the impact of inrush charging current on output voltage and inverter performance. Section IV presents reported solutions for mitigating inrush charging current, including comparisons of their features and disadvantages. Section V serves as the conclusion, summarizing key points, identifying research gaps, and motivating further exploration in this field.

2. Inrush Current Phenomenon

The inrush charging current, occurring during the initial energization or voltage changes of capacitors in the switching process, results from rapid voltage fluctuations as the capacitors attempt to equalize. To understand the generation of pulsating charging current, an equivalent circuit of a SC unit is presented in Fig. 9.1(a) and (b). The circuit includes an internal resistance (R_{eq}) associated with the capacitor (C_{eq}). In Fig. 9.1(a), the charging path is illustrated, and in Fig. 9.1(b), the discharging of the capacitor to a load resistance (R) is shown. The charging and discharging currents are denoted as I_{ch} and I_{dis}, respectively (Barzegarkhoo et al., 2022). During charging, switches S_1 and S_3 are in the ON condition, causing the charging current to flow through both switches, thereby subjecting them to stress. Let the equivalent capacitor have an initial voltage of V_0 volts when connected to the DC source for charging up to voltage V_{dc} (Yadav et al., 2023). So, the charging current (Fig. 9.1(c)) of the capacitor can be expressed mathematically as given below:

$$i_{ch}(t) = \frac{V_{dc}(t) - V_0(t)}{R_{eq}} \tag{1}$$

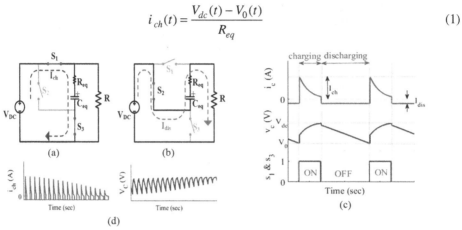

Fig. 9.1 Operation of basic SC unit (a) charging of capacitor (b) discharging path for capacitor, (c) Charging process of capacitor in SC unit, (d) Capacitor current and voltage during hard-charging

Figure 9.1(d) illustrates the inrush current during the hard-charging of the capacitor, revealing a spike in the charging current. This setup offers insights into the generation of inrush or pulsating charging current in the inverter. The simultaneous activation of switches S_1 and S_3 during the charging phase results in a surge in current, potentially impacting the stress, reliability of these switches, and overall inverter performance.

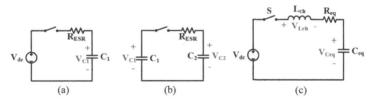

Fig. 9. 2 Hard charging of capacitor in SC MLI (a) charging of capacitor through DC source (b) Capacitor charging through another capacitor (Lei et al., 2015), (c) Charging of capacitor in SC MLI using inductor in charging path

3. Problem Associated with Inrush Charging Current

3.1 Power Loss

In the SC circuit, the charging process involves connecting a capacitor in parallel with the DC source or another charged capacitor, as shown in Fig. 9.3(a) and (b). Capacitors C_1 and C_2 are considered, with an assumption that, during switching operations, they may have different voltage levels (V_1 and V_2), where the initial voltage of C_1 is assumed greater than that of C_2. This configuration introduces potential power loss, known as ripple loss in multilevel inverters (equation 2), which is dependent on the switching frequency of the switch. Notably, the ripple loss is not influenced by the specific value of the inherent resistance (R_{ESR}) but is determined by the initial voltage difference between capacitors. The ripple loss is directly proportional to the charge consumed by the load and inversely related to capacitance values. Moreover, the change in charge within the capacitors is associated with the duration of the charge and discharge cycle, inversely related to the switching frequency as well (Barzegarkhoo et al., 2022), (Lei et al., 2015).

$$P_{loss} = \frac{1}{4}C_2(V_1 - V_2)^2 f_{sw} = \frac{1}{4}C_2\Delta V^2 f_{sw} \tag{2}$$

$$\Delta V \propto \frac{1}{f_{sw}} \frac{1}{C_{eq}} \tag{3}$$

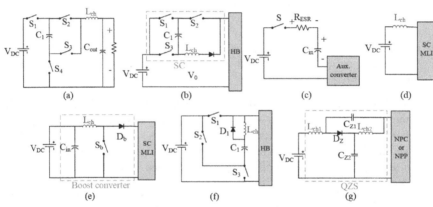

Fig. 9.3 Reported insertion techniques of an inductor in charging path of capacitor in SC MLI to reduce inrush charging current (a) Topology-1 (Lei et al., 2015), (b) topology-2 (Khan et al., 2022), (c) Topology-3 (Pilawa-Podgurski et al., 2008), (d) Topology-4 (Ye and Cheng, 2018), (Barzegarkhoo et al., 2021), (Khan et al., 2021), (Zeng et al., 2017), (Khoun-Jahan et al., 2021), (e) Topology-5 (Khan et al., 2020), (f) Topology-6 (Yadav et al., 2023), (Sabour et al., 2017) and (g) Topology-7 (Ye et al., 2021).

Stress on Components and Electromagnetic Interference (EMI)

In a basic SC unit, parasitic inductance is typically negligible, with the primary factor being the equivalent charging resistance (R_{eq}). This resistance is associated with the parasitic resistance when the charging path switch is in the ON state and the equivalent series resistance (ESR) of the capacitor. Depending on the values of R_{eq} and the switching frequency (f_{sw}), a significant discontinuous charging current (i_{ch}) occurs when the switch is ON. This can adversely impact the lifespan of switches and capacitors and raise concerns about electromagnetic interference (EMI), potentially causing false activation of switches not directly involved in the SC charging path. To address these issues, a complete charging process should operate within a defined fast switching limit (FSL) as long as the f_{sw} exceeds a critical frequency determined by the characteristics of the pure SC-based circuit (Barzegarkhoo et al., 2022), (Lei et al., 2015). It leads to incomplete charging, causing slow switching performance and undesirable higher charging current spikes or inrush current (Pilawa-Podgurski et al., 2008), (Cheng, 2018).

$$f_{sw} > \frac{1}{2\pi R_{eq} C_{eq}} \tag{4}$$

4. Mitigation Techniques for Inrush Charging Current

The prominent reported techniques to reduce the inrush charging current are discussed here with their advantages and disadvantages. The techniques are categorized into two categories for demonstration.

Soft Charging Techniques

A prominent solution to mitigate the charging inrush current in the SC unit is the transition from hard switching to soft switching of switches. This technique is known as soft switching (Yadav et al., 2023) - (Khan et al., 2022) or quasi-soft charging (QSC) (Barzegarkhoo et al., 2021). To introducing a very small charging inductor, denoted as L_{ch} into the charging circuit of the SC converter (shown in Fig. 9.3) is an effective approach for achieving a complete transition to soft charging operation (Ye and Cheng, 2018), (Barzegarkhoo et al., 2021), (Khan et al., 2021), (Zeng et al., 2017). As it is known to everyone that the inductor allows instantaneous change in its terminal voltage so, it can work as a controlled current load []. In (Khan et al., 2020), (Lee, 2021) used a buck or boost converter for the soft charging process, because it contains the inductor itself. By inserting a charging inductance in the SC unit, it works as a series RLC network shown in Fig. 9.2(c). So, for proper fast switching limit author in (Barzegarkhoo et al., 2022), (Lei et al., 2015) represents a condition that is given below:

$$f_{sw} > \frac{1}{2\pi \sqrt{L_{ch} C_{eq}}} \tag{5}$$

The condition mentioned above is applicable to a constant duty cycle network i.e., DC-DC converter (Assem et al., 2020). However, determining the value of the inductor is more complicated in a SC MLI with a variable duty cycle. The value for charging inductors can be calculated by different formulas given in the reported paper depending on their charging network. The fundamental calculation of the charging inductor is given below equation reported in (Yadav et al., 2023).

$$L_{ch} \geq \frac{\Delta V_{Ceq}}{F_{sw} \Delta I_{ch}} \tag{6}$$

Where ΔV_{Ceq} is the ripple voltage across the capacitor, ΔI_{ch} is an acceptable ripple on charging current and F_{sw} is the switching frequency.

However, the main problem associated with a large inductor to attenuate the inrush current is the voltage drop across this inductor. Which causes a voltage drop in the output end of the MLI. So, the authors of papers (Yadav et al., 2023) and (Sabour et al., 2017) suggested an anti-parallel diode across the inductor (shown in Fig. 9.3(f)) and each SC unit has a dedicated inductor. The purpose of this anti-parallel diode is to bypass the inductor during the discharging of a capacitor to load. But still, the anti-parallel diode has conduction power loss which decreases the efficiency of the inverter and it cannot address the longest discharging period issues of many SC MLI.

Modulation Strategies

When multiple SC-based units are cascaded together to achieve a larger voltage conversion gain, the issues become more severe. This is because the top positive and negative output voltage levels are generated by discharging all the capacitors in series, and any voltage ripple in these capacitors is magnified. Many existing SC-MLIs lack Reduced Switching States (RSSs), which could be useful for minimizing the LDP of the capacitors through hybrid Pulse-Width Modulation (PWM) techniques. SC-MLIs require the use of the well-established level-shifted-sinusoidal pulse-width modulation (LS-SPWM) technique to drive the switches. The longest discharging period (LDP) of the capacitors in the LS-SPWM technique can be managed by reducing the maximum value of the modulation index (Ye et al., 2020) - (Chen et al., 2020).

The primary objective of employing the multicarrier PWM technique in the seven-level inverter proposed in (Ye et al., 2020) is to diminish low-frequency harmonics and achieve a closer approximation to a sinusoidal output voltage. Notably, the PS-PWM method is employed to decrease voltage fluctuations and equalize the power distribution between capacitors. In this proposed technique combination of both LS PWM and PS PWM carrier waves is used (Tsunoda et al., 2014), (Karami et al., 2015). The PS PWM carrier wave is specially used for the generation of higher-level output voltage (shown in Fig. 9.4). To maintain a maximum voltage ripple of capacitors, the capacitance C_i must meet the following conditions (Ye et al., 2020), (Tsunoda et al., 2014):

$$C_i \geq \frac{6M_a - 3}{\delta f_c R} \qquad (7)$$

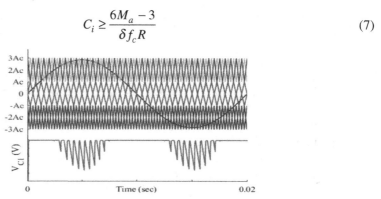

Fig. 9.4 Hybrid PWM scheme used in (Tsunoda et al., 2014), (Karami et al., 2015) for seven-level SC MLI to reduce the capacitor ripple voltage.

Where M_a is the modulation index, f_C is the frequency of the phase-shifted carriers, and R is the load, δ is the voltage ripple ratio, a dimensionless value between 0 and 1. According to the author of this proposed strategy (Karami et al., 2015), it is advisable to minimize the parasitic resistance in the charging loop to reduce the voltage ripple of capacitors. Consequently, it is recommended to use semiconductor devices with low ON resistance and capacitance with low ESR.

5. Conclusion

The proposed paper conducts an analytical study, consolidating insights from various authors and researchers. Notably, the identified techniques emphasize the use of an inductor in series with the capacitor to minimize inrush current, underscoring the importance of carefully determining the inductor value to manage charging voltage drop. Additionally, the paper promoted to reduce inrush current through the adoption of hybrid or LS-PWM modulation techniques, aiming to alleviate the LDP and enhance the significance of RSS in cascaded SC units. Ultimately, this paper serves as a valuable source of inspiration, shedding light on the challenges associated with SC MLI and encouraging further exploration by researchers in this field.

REFERENCES

1. Barzegarkhoo, R., Forouzesh, M., Lee, S. S., Blaabjerg, F. and Siwakoti, Y. P. (2022) Switched-capacitor multilevel inverters: a comprehensive review. IEEE Transactions on Power Electronics. 37(9):11209-11243.
2. Yadav, P. K., Vemuganti, H. P. and Biswal, M. (2023). A seven-level switched capacitor-based RSC-MLI topology with suppressed inrush currents for grid-connected applications. 2023 5th International Conference on Power, Control & Embedded Systems (ICPCES), Allahabad, India. (pp. 1-6).
3. Lei, Y. and Pilawa-Podgurski, R. C. N. (2015). A general method for analyzing resonant and soft-charging operation of switched-capacitor converters. IEEE Transactions on Power Electronics. 30(10):5650-5664.
4. Pilawa-Podgurski, R. C. N., Giuliano, D. M. and Perreault, D. J. (2008). Merged two-stage power converter architecture with soft charging switched-capacitor energy transfer. 2008 IEEE Power Electronics Specialists Conference, Rhodes, Greece. (pp. 4008-4015).
5. Ye, Y. and Cheng, K. W. E. (2018). Analysis and design of zero-current switching switched-capacitor cell balancing circuit for series-connected battery/supercapacitor. IEEE Transactions on Vehicular Technology. 67(2):948-955.
6. Khan, M. N. H., Forouzesh, M., Siwakoti, Y. P., Li, L. and Blaabjerg, F. (2020). Switched capacitor integrated (2n + 1)-level step-up single-phase inverter. IEEE Transactions on Power Electronics. 35(8):8248-8260.
7. Barzegarkhoo, R., Siwakoti, Y. P., Aguilera, R. P., Khan, M. N. H., Lee, S. S. and Blaabjerg, F. (2021). A novel dual-mode switched-capacitor five-level inverter with common-ground transformerless concept. IEEE Transactions on Power Electronics. 36(12):13740-13753.
8. Khan, S. A., et al. (2021). Topology, modeling and control scheme for a new seven-level inverter with reduced dc-link voltage. IEEE Transactions on Energy Conversion. 36(4):2734-2746.
9. Lee, S. S., Siwakoti, Y. P., Barzegarkhoo, R. and Lee, K. -B. (2021). Switched-capacitor-based five-level T-type inverter (SC-5TI) with soft-charging and enhanced dc-link voltage utilization. IEEE Transactions on Power Electronics. 36(12):13958-13967.

10. Zeng, J., Wu, J., Liu, J. and Guo, H. (2017). A quasi-resonant switched-capacitor multilevel inverter with self-voltage balancing for single-phase high-frequency ac microgrids. IEEE Transactions on Industrial Informatics. 13(5):2669-2679.
11. Sabour, S., Hassanifar, M., Choupan, R., Golshannavaz, S., Neyshabouri, Y. and Nazarpour, D. (2021). Voltage source boost multilevel inverter with high modularity: circuit configuration and modulation. IET Power Electronics. 13:4336-4347.
12. Khoun-Jahan, H. et al. (2021). Switched capacitor based cascaded half-bridge multilevel inverter with voltage boosting feature. CPSS Transactions on Power Electronics and Applications. 6(1):63-73.
13. Ye, Y., Zhang, Y., Wang, X. and Cheng, K. -W. E. (2021). Quasi-Z-source-fed switched-capacitor multilevel inverters without inrush charging current. IEEE Transactions on Industrial Electronics. 70(2):1115-1125.
14. Khan, M.N., Barzegarkhoo, R., Siwakoti, Y.P., Khan, S.A., Li, L. and Blaabjerg, F. (2022). A new switched-capacitor multilevel inverter with soft start and quasi resonant charging capabilities. International Journal of Electrical Power & Energy Systems. 135:107412.
15. Ye, Y., Peng, W., and Yi, Y. (2020). Analysis and optimal design of switched-capacitor seven-level inverter with hybrid PWM algorithm. IEEE Transactions on Industrial Informatics. 16(8):5276-5285.
16. Tsunoda, A., Hinago, Y. and Koizumi, H. (2014). Level- and phase-shifted PWM for seven-level switched-capacitor inverter using series/parallel conversion. IEEE Transactions on Industrial Electronics. 61(8):4011-4021.
17. Karami, B., Barzegarkhoo, R., Abrishamifar, A. and Samizadeh, M. (2015). A switched-capacitor multilevel inverter for high AC power systems with reduced ripple loss using SPWM technique. The 6th Power Electronics, Drive Systems & Technologies Conference (PEDSTC2015), Tehran, Iran. (pp. 627-632).
18. Chen S., Ye, Y. and Wang, X. (2020). Design of hybrid PWM algorithm for switched-capacitor multilevel inverter based on TMS320F28335. 2020 8th International Conference on Power Electronics Systems and Applications (PESA), Hong Kong, China. (pp. 1-6).
19. Assem, P., Liu, W. -C., Lei, Y., Hanumolu, P. K. and Pilawa-Podgurski, R. C. N. (2020). Hybrid Dickson switched-capacitor converter with wide conversion ratio in 65-nm CMOS. IEEE Journal of Solid-State Circuits. 55(9):2513-2528.

Emerging Technologies and Applications in Electrical Engineering –
Prof. Dr. Anamika Yadav et al. (eds)
© 2024 Taylor & Francis Group, London, ISBN 978-1-032-82568-7

Single Phase Shift Control of Dual Active Bridge DC-DC Converter

10

Adarsh Singh[1], Malothu Ganesh[2], Bhawesh Sahu[3], Hari Priya Vemuganti[4]
Department of Electrical Engineering, NIT Raipur, India

ABSTRACT: Isolated bidirectional DC-DC (IBDC) converters have attracted the attention of many researchers due to their various applications. Among these converters, Dual Active Bridge (DAB) topology stands out as the best choice. To improve the performance of the DAB converter, its operation needs to be thoroughly reviewed. This study aims to analyze DAB models using measurement and simulation-based methods. The aim is to analyze and compare these results to provide a better understanding of the how control operation of DAB converters occurs. Moreover, the analysis shows that operation and control of one of commonly used control method, Single Phase Shift (SPS) method. A single-phase, high-frequency Dual Active Bridge (DAB) DC-DC converter with a 10 kHz operating frequency is presented in this study. It also goes over how to use the DAB configuration for grid-to-vehicle (G2V) and vehicle-to-grid (V2G) applications. Using the simulation programmer MATLAB/Simulink, a careful investigation of the DAB's operating mode and the electrical analysis simulation findings have been carried out and verified.

KEYWORD: Bidirectional power flow, Dual active bridge, Single phase shift, Isolated DC-DC converter

1. Introduction

The increase in demand for renewable energy sources has also led to an increase in electricity use. DC-DC converters play an important role in controlling the output power of renewable energy sources and storing this energy in the power source. Among various DC-DC converters, isolated bidirectional DC-DC (IBDC) converters are known for their bidirectional power

[1]adarshubn@gmail.com, [2]malothuganesh21@gmail, [3]bhaveshsahu6626@gmail.com, [4]hpvemuganti.ee@nitrr.ac.in

DOI: 10.1201/9781003505181-10

efficiency. In particular, it stands out that the Dual Active Bridge (DAB) converter is mature and widely used. It has many applications in automobiles, especially electric, hybrid and fuel cell vehicles. As the demand for these vehicles continues, the need to improve the performance of DAB converters also increases (Sivakumar, Sathik, Manoj, & Sundararajan, 2016).

This paper aims to achieve this goal by examining operation of DAB converters in various applications. In addition, the zero voltage switching (ZVS) ability of the switch, which is directly related to switching, is also being investigated. The modulation scheme chosen for the DAB converter is phase-shift modulation because it is simple and helps measure how single-phase shifts affect the operation of the converter. This insight is important to study the performance of DAB converters under different operating conditions. The purpose of this analysis is to define the ZVS operating area of the DAB converter to help reduce loss (Everts, 2016). This work should increase the understanding of the performance of DAB converters under ZVS conditions.

Using MATLAB/Simulink, this paper conducted a comprehensive simulation of a DAB converter employing Single Phase Shift (SPS) control at a frequency of 10 kHz. The study focused on waveform analysis, emphasizing ZVS and efficiency evaluation. Insights gained from this research contribute to the optimization of bidirectional power flow systems, particularly in electric vehicles and renewable energy applications, by enhancing the understanding of SPS-controlled DAB converter performance in terms of stability and overall reliability.

2. Bidirectional DC-DC Converters

Renewable energy relies on bidirectional DC-DC converters to ensure energy flow in both directions, becoming an important component of integrated energy storage. These converters can affect different lines by increasing or decreasing the voltage to maintain a constant voltage (Mohan Bharathidasana, Sureshb, Jasińskib, & Leonowiczb, 2022). They have transformers that offer galvanic isolation and increase safety and efficiency. Bidirectional DC-DC converters are often used in power distribution systems (Chmielewski, Piórkowski, Bogdziński, & Mozaryn, 2023) and automotive applications such as electric vehicles (EV), hybrid electric vehicles (HEV) and fuel cell vehicles (Chiu & Lin, 2006). Increasing demand for low-cost vehicles is increasing the demand for these converters, which are also widely used in renewable energy, aerospace applications (Naayagi, Forsyth, & Shuttleworth, 2012) and personal electronics. This versatility is the key to its widespread use. The DAB converter is a notable topology frequently employed in high-power automotive applications (F. Krismer & Kolar, 2011). Its soft switching capability facilitates ZVS, resulting in minimal switching losses even at high operating voltages. This feature enhances efficiency and reliability, particularly during on-off phase transitions. Its lightweight design and high performance have stimulated further research, improved performance and expanded its application field. As renewable energy has a great future, DAB converters play an important role in driving these developments.

The DAB converter is a versatile and important topology in the field of power electronics. It is primarily used in applications that require bidirectional power flow with galvanic isolation, such as electric vehicles (Dini, Saponara, & Colicelli, 2023), renewable energy systems, and energy storage systems. Here's an overview of the theory and operation of the DAB converter using SPS method.

3. Dual Active Bridge

An isolation transformer and the single-phase, full bridge architecture are frequently used in isolated bidirectional DC–DC converters. According to (Chen, Nguyen, Yao, Wang, Gao, & Hu, 2021) these converters have a basic architecture depicted in Fig. 10.1, which features two H-bridges on the main and secondary sides of the high-frequency transformer. The high-frequency transformer is crucial for providing the required isolation and voltage alignment between the low- and high-voltage buses. Furthermore, the leakage inductance of the transformer serves as a part of the immediate energy storage system.

Fig. 10.1   Block diagram of DAB

The following are a few features of DAB converters: Power bidirectional transfer between two DC sources or loads is the main purpose of the DAB converter. This capability is particularly useful in applications like electric vehicles, where regenerative braking can transfer energy back to the battery. The transformer in the DAB converter provides galvanic isolation between the primary and secondary sides. This is crucial for applications where electrical isolation is required for safety or operational reasons. The heart of the DAB converter is its control strategy. The DAB converter may accomplish soft switching, including Zero Current Switching (ZCS) and ZVS, by carefully controlling the bidirectional switches' switching. This minimizes switching losses, reduces stress on the components, and improves overall efficiency.

4. Method for Controlling Power Flow

In a DAB Converter, power is directed from the leading H-bridge to the lagging H-bridge, and this power flow can be managed using one of four control methods shown in below Fig. 10.2. In which SPS method is the simplest phase-shifting control method and remains the predominant choice in current implementations.

From the circuit diagram Fig. 10.3, when a gate pulse is applied, the diagonal switches of both bridges such that switch on and off in unison. In other words, initially, for the first bridge, T1 and T4 receive a gate pulse with a 50% duty ratio. The signal from this gate pulse is then transmitted to the other switches, T2 and T3, ensuring that the diagonal switches operate concurrently, turning on and off simultaneously (Kumar, Kumar, Bhat, & Agarwal, 2017). As per circuit diagram Fig. 10.3, as explained above it consists of two H-bridge which are connected each other by High frequency transformer. Both bridges have two legs, each consists of two IGBT switch with their antiparallel diodes. H1-bridge has one leg consist of T1 & T2

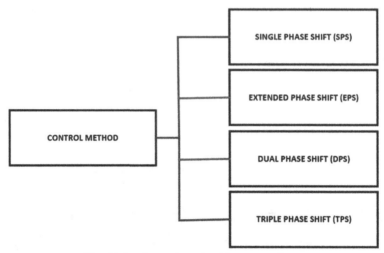

Fig. 10.2   Control method for power flow

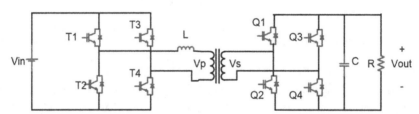

Fig. 10.3   Circuit diagram of DAB converter

switch with M1 & M2 antiparallel diode respectively. Other legs consist of T3 & T4 switch with M3 & M4 antiparallel diode respectively. Similarly, for H2- bridge one leg has Q1, Q2 and other leg has Q3, Q4 switches with N1, N2 & N3, N4 antiparallel diode respectively (Mi, Bai, Wang, & Gargies, 2008). Interval wise operation given below:

During t_0-t_1: In this interval, switches M1, M4 are conducted in H1-bridge and switches N2, N3 are conducted in H2-bridge. During this time period L starts to store energy. During t_1-t_2: In this interval, switches T1, T4 are conducted in H1-bridge and switches Q2, Q3 are conducted in H2-bridge. Voltage across L is same in previous period. During t_2-t_3: In this interval, switches T1, T4 are conducted in H1-bridge and switches Q1, Q4 are turned off in H2-bridge. Due to which primary voltage is same but secondary voltage become $+V_{out}$. During t_3-t_4: In this interval, switches T2, T3 are turned on and M2, M3 diode conducted in H1-bridge and diode N1, N4 are conducted in H2-bridge. During this time period L start to release stored energy During t4-t5: In this interval, switches T1, T4 are conducted in H1-bridge and switches Q2, Q3 are conducted in H2-bridge. Here voltage across L is same as in previous period but only difference is that the I_L (inductor current) become negative (Peng, Li, Su, & Lawler, 2004).

5. Simulation Result

The MATLAB/Simulink software is used in the construction and validation of the DAB converter model. This model is tailored for electric vehicle charging applications, featuring a power rating of 800W. The isolated DAB DC-DC converter, constructed using 8 no. of 1.2kV, 100A SiC IGBTs, can efficiently transfer 800 W of power at a switching frequency of 10 kHz. Various circuit parameters derived and optimized through an iterative process,

Table 10.1   Design parameter for DAB

Parameter	Values
Po	800 W
Vin	100 V
Vout	40 V
Fsw	10 KHz
L	10.2μH
Cout	600μH

as detailed in Table 10.1. Additionally, the isolation transformer has a ratio of 2:1, the auxiliary inductor has a value of 10.12μH, the high-frequency filter capacitor is 660μF, and the resistive load is 2 Ω.

Utilizing parameters from Table 10.1, the DAB converter produces V_{out}, depicted in Fig. 10.4, along with the input current waveform in Fig. 10.5 and output current in Fig. 10.6. Subsequent to DC to AC conversion, the primary voltage waveform, illustrated in Fig. 10.7, undergoes a step-up process with the aid of HF Transformer, as shown in Fig. 10.8. Figure 10.9 and Fig. 10.10 show the voltage and current waveforms over the inductor, respectively. A combined waveform of the input voltage, inductor voltage, and current waveform across the inductor is shown in Fig. 10.11.

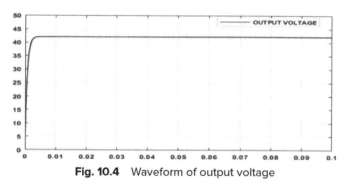

Fig. 10.4   Waveform of output voltage

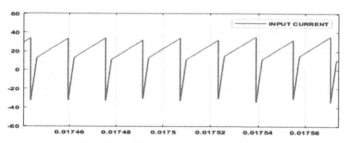

Fig. 10.5   Waveform of input current

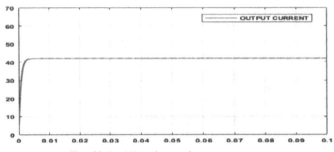

Fig. 10.6 Waveform of output current

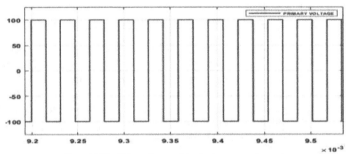

Fig. 10.7 Waveform - Vp (primary voltage)

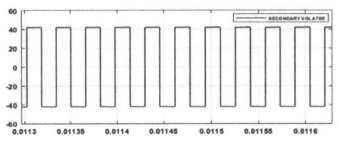

Fig. 10.8 Waveform of Vs (secondary voltage)

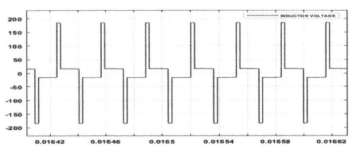

Fig. 10.9 Waveform of voltage across inductor

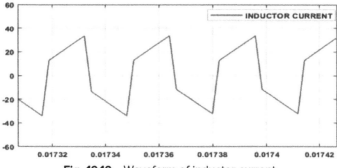

Fig. 10.10 Waveform of inductor current

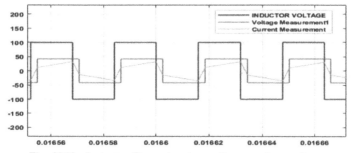

Fig. 10.11 Inductor Current, primary & secondary voltage

6. Conclusion

In-depth examination of the architecture, and features, of a high-switching, SPS control DAB converter is provided in this work. Specifically for battery charging, a rigorous design and validation procedure was conducted, with the goal of achieving a rated power of 800 W, a switching frequency of 10 kHz, and grid and battery voltages of 100 and 40 volts, respectively. The DAB converter proves particularly beneficial for high-frequency, rapid onboard battery charging applications in electric vehicles (EVs), contributing to decreased volume and heightened power densities.

REFERENCES

1. Chen, J., Nguyen, M., Yao, Z., Wang, C., Gao, L., & Hu, G. (2021). DC-DC converters for transportation electrification: topologies, control, and future challenges. *IEEE Electrification Magazine, 9(2)*, 10-22.
2. Chiu, H.-J., & Lin, L.-W. (2006). A Bidirectional DC–DC Converter for Fuel Cell. *IEEE Trans. on Power Electronics, 21*, 950-958.
3. Chmielewski, A., Piórkowski, P., Bogdziński, K., & Mozaryn, J. (2023). Application of a Bidirectional DC/DC Converter to Control the Power Distribution in the Battery–Ultracapacitor System. *Energies, 16*, 3687.

4. Dini, P., Saponara, S., & Colicelli, A. (2023). Overview on Battery Charging Systems for Electric Vehicles. *Electronics ,12(20)*, 4295.

5. Everts, J. (2016). Closed-Form Solution for Efficient ZVS Modulation. *IEEE transactions on Power Electronics, 32*, 7561-7576.

6. F.Krismer, & Kolar, J. (2011). Efficiency-optimized high-current dual active bridge converter for automotive applications. *IEEE Transactions on Industrial Electronics, 59(7)*, 2745-2760.

7. Kumar, B., Kumar, A., Bhat, A., & Agarwal, P. (2017). Comparative study of dual active bridge isolated DC to DC converter with single phase shift and dual phase shift control techniques. *Recent Developments in Control, Automation & Power Engineering (RDCAPE)*, 453-458.

8. Mi, C., Bai, H., Wang, C., & Gargies, S. (2008). Operation, design and control of dual H-bridge-based isolated bidirectional DC–DC converter. *IET Power Electronics, 1(4)*, 507-517.

9. Mohan Bharathidasana, V. I., Sureshb, V., Jasińskib, M., & Leonowiczb, Z. (2022). A review on electric vehicle: Technologies, energytrading, and cyber security. *Energy Reports, 8*, 9662-9685.

10. Naayagi, R., Forsyth, A., & Shuttleworth, R. (2012). High-power bidirectional DC–DC converter for aerospace applications. *IEEE Transactions on Power Electronics, 27(11)*, 4366-4379.

11. Peng, F., Li, H., Su, G., & Lawler, J. (2004). A new ZVS bidirectional DC-DC converter for fuel cell and battery application. *IEEE Transactions on power electronics, 19(1)*, 54-65.

12. Sivakumar, S., Sathik, M. J., Manoj, P., & Sundararajan, G. (2016). An assessment on performance of DC–DC converters for renewable energy applications. *Renewable and Sustainable Energy Reviews, 58*, 1475-1485.

Emerging Technologies and Applications in Electrical Engineering –
Prof. Dr. Anamika Yadav et al. (eds)
© 2024 Taylor & Francis Group, London, ISBN 978-1-032-82568-7

Short Term Load Forecasting using ANN for Chhattisgarh State

Suruchi Shrivastava[1], Anamika Yadav[2], Shubhrata Gupta[3]
Department of Electrical Engineering, National Institute of Technology Raipur, CG, India

ABSTRACT: The estimation of load in advance is generally referred to as Load Forecasting (LF). Load forecasting helps Electrical utilities and their load dispatch center (LDCs) to monitor and control the power flow and provide solutions for sustainable load management and smart grid operational planning and optimize operational efficiency and reliability. Short term load forecasting (STLF) is one of the most important tools of an energy management system. Several trending methods are available to predict the load demand such as Artificial Neural Network (ANN), Gaussian Process Regression (GPR), Time Series etc. This paper mainly focuses on the use of ANN to predict the future load demands. A day ahead prediction is shown for two different time frames in order to show the fitness of the LF model. The model is trained with an input dataset consisting of historical values of load demand of timeframe from 1st January 2021 to 30th September 2023. The whole dataset is splitted into 70-30 ratio for training and testing, and LF model is trained and tested and hence evaluated through statistic matrices such as MAPE (Mean Absolute Percentage Error) and MAE (Mean Absolute Error), in order to have the accurate prediction and ensures the fitness of a proposed model. At last, the result obtained showcases the forecasted value of load demand of Chhattisgarh state by ANN approach. The difference between the actual and predicted value is also shown to show the accuracy and efficiency of the proposed model.

KEYWORDS: Load forecasting, Short term load forecasting, Artificial neural network, Load demand

[1]suruchishrivastava55@gmail.com, [2]ayadav.ele@nitrr.ac.in, [3]sgupta.ele@nitrr.ac.in

DOI: 10.1201/9781003505181-11

1. Introduction

Planning for every power system company starts with an objective to forecast the future load and its related requirements, which helps to control and manage the upcoming need accordingly. Load forecasting mainly needs historical and available data to predict the future values [Akhtar et al 2023]. There are many parameters available such as historical load data previous day, last week data or similar day last week data or similar day last year data and weather data like temperature, wind speed etc. on which LF method depends for providing accurate result. Load forecasting is about estimating future load demand based on numerous data/information available and according to consumer behavior [Dhaval and Deshpande,2020]. The goal of short-term load forecasting is to provide load demand prediction with basic production planning functions with respect to the quality of power supply to the consumers at any time and timely information from the dispatcher. The load forecasting or demand estimation has a direct impact on the scientific community and society, as they are the real consumer of the electricity generated and thereby the load demand is from its LT and HT consumers only. Based on the load demand estimate and the estimated availability from different sources, SLDC shall plan demand management measures like load shedding, power cuts, exchanging power with neighboring regions, and deciding the electricity tariff. Load forecasting can help to estimate load flows and to make decisions that can prevent overloading. The timely implementation of such decisions leads to the improvement of power quality, network reliability and to the reduced occurrences of equipment failures and blackouts. Numerous research has been carried out on Load forecasting with different available methods using number of parameters majorly need to meet the accuracy of the prediction. In [Yadav et al,2023] Twelve GPR models along with Different kernel are functions and basic functions of GPR models is proposed. The experiment performed on hourly and monthly basis for 4 Indian cities in Maharashtra State and one Australian city. Further the comparison shows that the different kernel functions and basic functions of GPR models are evaluated to find the best combination of functions that give the least MAPE in predicting the future load which concluded with the value of MAPE 0.15% for Australian city and 0.002%, 0.209%, 0.077%, 0.140% respectively for four Indian cities in Maharashtra state. The combination of LSTM (Long short-term memory) and RNN (Recurrent neural network) based model has been proposed in [Agrawal et al,2018]utilizing last twelve years of data precisely from 2004 to 2015.The proposed model estimates the load demand with MAPE (Mean Absolute Percentage error) of 6.54% and confidence interval of 2.25%.In [Hosein andHosein,2017], study presented a short-term load forecasting approach that uses DNN (Deep Neural Network) applications along with other Machine learning techniques. In addition to which the input dataset was taken from periodic smart meter energy usage reports. The dataset was then trained with the proposed methods to carry forward the process in order to have forecast result. The proposed models were evaluated through MAPE (Mean Absolute Percentage Error) to find the accuracy among the methods used. On comparison, it is shown that DNN combination approach outperforms the other traditional method used. The study presented in [Danladi et al, 2016] utilizes a fuzzy logic model for long term load forecasting. A year ahead prediction is presented in which two factors such as temperature and humidity are considered as input data whereas active power demand is taken as output. The study ensures that the prediction performed through fuzzy logic model is capable of predicting future load demand. The result obtained a load with a MAPE of 6.9% and efficiency of 93.1%

The study in [Ortiz-Arroyo et al,2005] presents a simple ANN prediction that can outperform other than many complex models with the timeframe of 1997 to 1998. Of the three experiments presented in the study, the first experiment configured 5 nodes in the hidden layer with 2.59-9.91% MAPE. Further adding 5 more nodes in the hidden layer expressed the MAPE as 2.94-9.0%. Finally, MAPE obtained in the last case obtained as 2.52-12.68% where 15 nodes configured in hidden layers. Overall experiment performed with ANN model gives result as 29% in terms of Mean-Absolute Percentage Error indicating the best performance when compared with other complex models. The study followed an effective GPR approach in [Yadav et al,2021] to perform monthly load forecasting model. The proposed model was tested by different combinations of kernels and basis function to identify the best model with higher efficiency which is able to predict future load demand accurately with 1.79%MAPE. The experiment was performed with Data analysis tools of Microsoft excel (Multiple linear regression) in [Bareth et al, 2023] to forecast a load demand data for September month 2022 of last 10 days for which the historical and temperature data has been considered from September month 2022.Microsoft Excel has been utilized for prediction of load demand, but the error in load estimate is quite high. A comparative study of different machine learning methods is presented in [Bareth et al 2023] for load forecasting. The prediction is then covered by eight algorithms along with another six methods to perform the forecasting task. During testing, it has been found that the RF (random forest) and SVR (support vector regression) provides the result with accuracy of 60.78%. and 48.34% respectively. In study [Kochar et al 2023], three different models such as LSTM, ANN, and GPR were trained using historical data incorporating the other features generation and load to forecast a load demand. The results obtained were further evaluated in the form of evaluation matrices MAPE. As compared to the three models proposed in study, LSTM is found to be more accurate and efficient with MAPE of 0.6805%. Hence LSTM is considered the most suitable for LTLF prediction performed in the study. The paper presented in [Xu et al,2023] first analyzes the current state of the energy system load forecasting and finds that there are still some shortcomings in existing forecasting models. It is then proposed to construct a short and medium-term energy system load forecasting models based on hybrid deep learning. After optimization, the results show that MAPE is obtained with 2.36%. In [Farsi et al,2021],a hybrid model of long short-term memory networks (LSTM) and convolutional neural network (CNN) model named LSTM-CNN Network or PLC Net is proposed. The forecasting method is executed on Malaysia for hourly load consumption and on Germany for daily power consumption. Among the various prediction algorithms and hybrid models, the most accurate results come from the PLC network. The accuracy found in PLC Net for Malaysian dataset is between 94.16% and 98.14% and for German dataset, it is between 82.49% and 91.31%.

In view of the literature survey carried out in this paper, it has been found that there is very limited research available in literature which is proposed for Indian dataset. Thus, in this paper, the historical load of one Indian State namely Chhattisgarh is considered, and it is collected from the load dispatch center of CG state power transmission company. The AI based model utilizing ANN is explored to forecast the load demand on short term basis. This paper follows the structure as: Section 2 of the report includes a summary of prior studies on short-term load forecasting with different models. An overview and architecture of the proposed model is presented in Section 3. The detailed approach of ANN model for a day ahead prediction is presented in Section 4, and the outcome of the model performed along with its calculation performance matrices is described in Section 5. Along with this comparison of the proposed model with different datasets is presented in terms of training and testing model, finally concluding remarks of the proposed study is given in Section 6 along with the acknowledgement in Section 7.

2. Methodology

Artificial neural networks (ANNs), also abbreviated to NNs or neural networks) are a division of machine learning models that are constructed using the principles of neuronal organization determined by connectionism within the biological neural networks that make up animal brains. Artificial Neural Network (ANN) is mainly composing a large number of highly interconnected processing elements that work parallel to solve a given problem. ANN model is capable of doing several works effectively. The network creates its own representation of information during the time of learning [Chen et al, 2017]. The application of ANN has been used in a number of fields with the aim of prediction. There are number of studies available which comprises the effective use of ANN with different aspects in different fields such as the unsupervised learning based on ANN carries a potential to redefine an artificial ecosystem[Dike et al,2018], whereas many research activities in artificial neural works (ANNs) have shown that ANNs have powerful pattern classification and pattern recognition capabilities also [Zhang et al,1998].There are variations of ANN architectures such as multilayer feedforward network (MFN), radial basis function (RBF), recurrent network, feedback network, and Kohonen Self-Organizing Map (SOM) network [Huang and Yanbo.2009]. It has been mentioned that ANNs work more accurately than other various techniques used, such as statistical classifiers, especially when the feature space gets complex, and the source data covers different statistical distributions [Mas and Flores,2008]. Exploratory analysis regarding functional form can also be extended using a neural network approach [McMenamin and Monforte,1998]. The ANN architecture includes the following nodes in one or more layers [Panda et al, 2017]: Input nodes, Hidden layers and their nodes, Output nodes, Connection of nodes. An implementation of activation function is performed on the basis of available data and type of layers [Gupta and Raza, 2020]. Figure 11.1 below represents the basic structure of an artificial neural network. The approach presented in this paper follows the general *feed forward architecture*, commonly used in artificial intelligence [Teixeira and Fernandes, 2012]. Prediction in neural network helps to recognize a complex network of interconnected nodes that "learn"

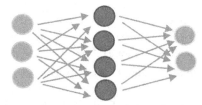

Input Nodes   Hidden Nodes   Output Nodes

Fig. 11.1 Basic structure of artificial neural network

the structure of your data. First, they analyze historical data to determine how to predict known output values using given predictor variables. The ANN model was trained using the Levenberg-Marquardt optimization algorithm.

3. Proposed Model

In this paper, development of load forecasting approach for Chhattisgarh state using the Short-Term Load Forecasting (STLF) method i.e., a day ahead. ANN approach is presented where the predicted values will be based on its past values. The general framework of the proposed model is shown in Fig. 11.2. It is used for forecasting depending upon the nature of data, geographical area and many other internal key features.

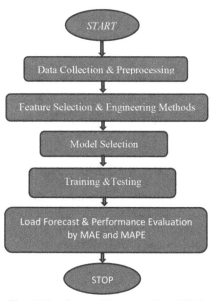

Fig. 11.2   General framework of STLF

- To accomplish the task of forecasting, a data collection step must be performed which will work as a past value for the predicted data and will also help to find the accuracy among the data values. The load data of Chhattisgarh state is obtained from Chhattisgarh State Load Dispatch Centre. The collected dataset consists of parameter of load demand which is further arranged in hourly and 15-minutes interval. Here, the timeframe of dataset varies from 25 Dec 2021 to 30 June 2023 i.e. for prediction of 1 Jan 2022, the data values from 25 Dec 2021 to 31 Dec 2021are considered as a past value. Further the same process is performed through shifting of each data ahead in a way where a week ago data work as input for target data. The dataset is then arranged to perform the proposed model on it, which includes preprocessing of data that ensures the quality of data, normalization and also helps to identify the missing values. Each day consists of 96 samples i.e. for every month a dataset may vary from 2880 (96*30) to 2976 (96*31) as time period for months includes 30 and 31 days respectively.
- In addition to this the data is further analyzed to perform the *feature selection and engineering methods* like extracting new features or enhancing the existing ones. For feature extraction process, historical data are mainly considered.
- Further, the appropriate *model selection* for the study is important, in particular the ANN architecture, no. of neurons and no of hidden layer. This paper addresses the ANN (Artificial Neural Network) approach used for day ahead forecasting. There are seven inputs representing the last week's data and one output representing the next day's load demand. There are 30 neurons in the hidden layer. Thus, the three-layered ANN is trained by Levenberg Marquardt algorithm.

- The data values are then trained and tested using the proposed model i.e., ANN. All the other parameters and features are adjusted according to the demand or need of model during the process.
- Finally, statistical matrices like MAPE and MAE are calculated using (1) and (2). Also, comparison of different time frames is shown in order to address the fitness of model used and to ensures the reliability of the study done. The evaluation matrices are calculated with the help of given formula:

$$\text{MAPE} = \frac{1}{n}\sum_{i=1}^{n}\left|\frac{Yi - Y'i}{yI}\right| * 100 \tag{1}$$

$$\text{MAE} = \frac{1}{n}\sum_{i=1}^{n}|Yi - Y'i| \tag{2}$$

where n represents length of the dataset whereas Yi and $Y'i$ are Actual and Predicted values respectively.

4. Result and Discussion

In this study, the application of ANN (Artificial Neural Network) model is proposed to predict the day ahead forecasting values particularly for Chhattisgarh State. The input dataset covers the dataset of the years 2021-2022-2023 which is then segregated as input data and target values equally for performing both training and testing. Figure 11.3 shows the actual and

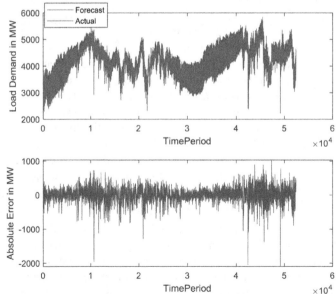

Fig. 11.3 Graph of actual and forecasted load during training of ANN along with its associated absolute error

forecasted load along with the error performed during training of ANN model. Figure 11.3 clearly indicates the complete overlapping of actual and forecasting data, which is then ensured in the corresponding absolute error plot as well.

Further the statistical matrices such as MAPE (Mean Absolute Percentage Error) and MAE (Mean Absolute Error) were evaluated which confirms the accuracy of proposed ANN model. The MAPE observed during training process was 0.228367% indicating the low percentage error, whereas MAE observed for the same was 133.2578 measuring the average magnitude of the error obtained. The observed values of evaluation matrices are clearly demonstrating the effective nature of the proposed model. After training, a model is then tested on the dataset that covers a timeframe of month July to Sep 2023 for a day ahead prediction. The below Fig. 11.4 showcases the actual and forecasted load along with its corresponding error during testing of ANN. The graph clearly showcasing the conjoining the actual and forecasted values. Evaluation matrices is then performed after ANN model testing. The MAE (Mean Absolute Error) and MAPE (Mean Absolute Percentage Error) during testing was observed as 316.5128 MW and 0.905671% respectively. The observed values ensure the lower percentage error along with its average magnitude. These observed values verify the effectiveness of the proposed model. Here, the proposed model based on ANN is shown to be effective enough to predict future values on short term basis.

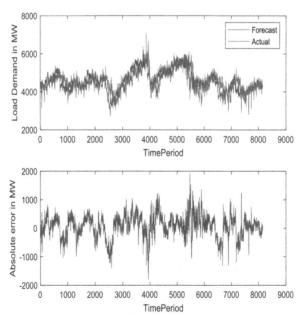

Fig. 11.4 Graph of actual and forecasted load during testing of ANN along with its associated error

The complete architecture of Neural Network performed is stored in variable called net. into which its factors affects the overall performance of the network and can be change accordingly.

The factors with its values used to trainthe network are shown in Table 11.1. In order to simplify the above prediction, the model is then tested for different input dataset of timeframe from 24 Sep 2023 to 7 Oct 2023 which includes a parameter of load demand arranged in hourly and 15 minutes' interval. Here, the dataset consists of 96 samples of each day. The dataset is

Table 11.1 Significant factors of ANN

Factors	Values
No. of Neurons in hidden layer	30
Maximum no. of epochs	2000
Learning rate	0.1
Momentum	0.9

arranged where data 96*7 is considered as input data and the next day data is taken as target. The arrangement of dataset includes shifting process where every time a dataset is shifted each day ahead. The model is then tested for future forecasting in particular one week ahead load demand forecasting is performed. The result obtained in Fig. 11.5, is evaluated with the help of evaluation matrices i.e., through MAPE and MAE. The values of MAPE and MAE are observed as 0.117137% and 141.8906 MW respectively. Further the below Table 11.2 specifies a tabular representation of evaluation matrices observed during training and testing of a model. It is clear that the ANN model gives accurate results for short term forecasting.

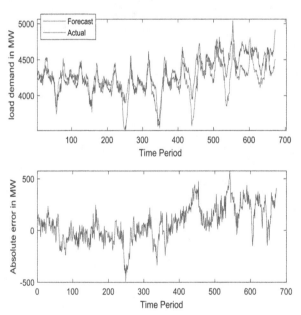

Fig. 11.5 Plot of actual and predicted load with testing error for 1week period.

Table 11.2 Evaluation matrices

Stages	MAPE (in %)	MAE (in MW)
During Training of dataset from 1st Jan 2021 to 30 June 2023	0.228367	133.2578
During Testing of dataset from July to Sep 2023	0.905671	316.5128
During Testing of ANN for 1 week of dataset Sep- Oct 2023	0.117137	141.8906

5. Conclusion

In this paper, a simple model for STLF using ANN approach is presented where a day ahead load demand prediction of Chhattisgarh state is performed. The model was trained and tested with the chosen timeframe of historical data from 1st Jan 2022 to Sep 2023 and finally the model is used to predict the future load demand. To account the efficiency of the model and to understand the model completely, the evaluation matrices such as MAPE in % and MAE in MW were used. With the same process the model is then tested on different input sets of datasets in order to showcase the simplification of method and also to ensure the accuracy, fitness and efficiency of the proposed model. In the end, the forecasted values along with the error obtained is illustrated to showcase the test and forecasted results. Therefore, with different timeframe of dataset, day ahead prediction through ANN (Artificial Neural Network) approach is found to be efficient to predict data accurately.

Acknowledgement

This work is completed under a project sponsored by the Central Power Research Institute (CPRI), Bangalore under the Research Scheme on Power (RSOP) approved vide Order No. RSOP/21|TR/17 dated 13 January 2023.

REFERENCE

1. Agrawal,Rahul, Muchahary,Frankle AND Tripathi, Madanmohan,2018. "Long term load forecasting with hourly predictions based on long-short-term-memory networks". 1-6. 10.1109/TPEC.2018.8312088.
2. Akhtar, Saima, Sulman Shahzad, Asad Zaheer, Hafiz Sami Ullah, Heybet Kilic, Radomir Gono, Michał Jasiński, and Zbigniew Leonowicz. 2023. "Short-Term Load Forecasting Models: A Review of Challenges, Progress, and the Road Ahead" *Energies* 16, no. 10: 4060. https://doi.org/10.3390/en16104060.
3. Bareth, Rashmi and Kochar, Matushree and Yadav, Anamika, 2023. "Electrical Load Forecasting based on Multiple Regression using Excel Tools," *2023 Third International Conference on Advances in Electrical, Computing, Communication and Sustainable Technologies (ICAECT)*, Bhilai, India, pp. 1-4, doi: 10.1109/ICAECT57570.2023.10117618.
4. Bareth, Rashmi, Matushree Kochar and Anamika Yadav. 2023. "Comparative Analysis of different Machine Learning Models for Load Forecasting." 2023 IEEE IAS Global Conference on Renewable Energy and Hydrogen Technologies (GlobConHT): 1-5.https://doi.org/10.1109/GlobConHT56829.2023.10087406.
5. Chen,Mingzhe & Challita, Ursula and Saad, Walid & Yin, Changchuan and Debbah, Mérouane. 2017. "Artificial Neural Networks-Based Machine Learning for Wireless Networks: A Tutorial," in *IEEE Communications Surveys&Tutorials*,vol.21,no.4,pp.3033071,Fourthquarter201doi:10.1109/COMST.2019.2926625.
6. Danladi, Ali & Yohanna, Michel & Puwu, M.I. & Garkida, B.M, 2016. Long-term load forecast modelling using a fuzzy logic approach. Pacific Science Review A: Natural Science and Engineering. 10.1016/j.psra.2016.09.011.
7. Dhaval,Bhatti,& Deshpande, Anuradha,2020."Short-term load forecasting with using multiple linear regression." *International Journal of Electrical & Computer Engineering* 10,no.4:10.11591/ijece.v 10i4.pp3911-3917.

8. Dike, Happine.ss Ugochi &Zhou,Yimin & Deveerasetty, Kranthi Kumar & Wu,Qingtian. 2018. "Unsupervised learning based on artificial neural network: A review." In *2018 IEEE International Conference on Cyborg and Bionic Systems (CBS)*, pp. 322-327. IEEE,doi: 10.1109/ CBS.2018.8612259.

9. Farsi, Behnam & Amayri, Manar & Bouguila, Nizar & Eicker, Ursula, 2021. On Short-Term Load Forecasting Using Machine Learning Techniques and a Novel Parallel Deep LSTM-CNN Approach. IEEE Access. PP. 1-1. 10.1109/ACCESS.2021.3060290.

10. Gupta, Tarun Kumar, and Raza,Khalid, 2020 "Optimizing deep feedforward neural network architecture: A tabu search based approach." *Neural Processing Letters* 51: 2855-2870. https://doi. org/10.1007/s11063-020-10234-7.

11. Hosein, Stefan and Hosein, Patrick, 2017. "Load forecasting using deep neural networks," *2017 IEEE Power & Energy Society Innovative Smart Grid Technologies Conference (ISGT)*, Washington, DC, USA, pp. 1-5, doi: 10.1109/ISGT.2017.8085971.

12. Huang, Yanbo. 2009. "Advances in Artificial Neural Networks – Methodological Development and Application" *Algorithms* 2, no. 3: 973-1007. https://doi.org/10.3390/algor2030973.

13. Kochar, Matushree; Shrivastava, Suruchi Yadav, Anamika and Gupta, Shubhrata. 2023. Long Term Load Forecasting Model for Chhattisgarh State. 1-5. 10.1109/WCONF58270.2023.10235091.

14. Mas, Jean & Flores, Juan. 2008. The application of artificial neural networks to the analysis of remotely sensed data. International Journal of Remote Sensing - Int J Remote Sens. 29. 617-663. 10.1080/01431160701352154.

15. McMenamin, J. Stuart & Monforte, Frank A. 1998. "Short Term Energy Forecasting with Neural Networks" The Energy Journal, International Association for Energy Economics, vol. 0(Number 4), pages 43-61. https://doi.org/10.5547/ISSN0195-6574-EJ-Vol19-No4-2.

16. Ortiz-Arroyo, D. and Skov, M.K. and Quang Huynh.2005."Accurate Electricity Load Forecasting with Artificial Neural Networks," International Conference on Computational Intelligence for Modelling, Control and Automation and International Conference on Intelligent Agents, Web Technologies and Internet Commerce (CIMCA-IAWTIC'06)*, Vienna, Austria, pp. 94-99, doi: 10.1109/CIMCA.2005.1631248.*

17. Panda, Sujit & Mohanty, Sachi & Jagadev, Alok. 2017. Long Term Electrical Load Forecasting: An Empirical Study across Techniques and Domains. Indian Journal of Science and Technology.10. 1-16. https://dx.doi.org/10.17485/ijst/2017/v10i26/115372.

18. Teixeira, Joao Paulo and Fernandes, Paula Odete. 2012."Tourism time series forecast-different ANN architectures with time index input." *Procedia Technology* 5: 445-454. https://doi.org/10.1016/j. protcy.2012.09.049.

19. Xu, Hongsheng and Fan, Ganglong and Kuang, Guofang and Song, Yanping,2023."Construction and Application of Short-Term and Mid-Term Power System Load Forecasting Model Based on Hybrid Deep Learning," in *IEEE Access*, vol. 11, pp. 37494-37507, doi: 10.1109/ACCESS.2023.3266783. process regression-based load forecasting model". IET Generation, Transmission & Distribution. n/a-n/a. 10.1049/gtd2.12926.

20. Yadav, Anamika, Ayush Kumar, Rudra Pratap Singh Rana, Maya Chandrakar, Mohammad Pazoki, and Ragab A. El Sehiemy.2021. "An Efficient Monthly Load Forecasting Model Using Gaussian Process Regression." In *2021 IEEE 4th International Conference on Computing, Power and Communication Technologies (GUCON)*, pp. 1-8. IEEE, 2021.10.1109/GUCON50781.2021.9574008.

21. Zhang, Guoqiang, B. Eddy Patuwo, & Michael Y. Hu,1998."Forecasting with artificial neural networks: The state of the art." International journal of forecasting 14, no. 1: 35-62. https://doi. org/10.1016/S0169 2070(97)00044-7.

Emerging Technologies and Applications in Electrical Engineering –
Prof. Dr. Anamika Yadav et al. (eds)
© *2024 Taylor & Francis Group, London, ISBN 978-1-032-82568-7*

Automatic Fault Detection of Faulty Bypass Diodes in Photovoltaic Array using Machine Learning for Residential/Small Scale Systems

Poorva Sharma[1]
Assistant Professor, Electrical & Electronics Engineering Department,
Government Engineering College Raipur

Anamika Yadav[2]
Professor, Electrical Engineering Department,
National Institute of Technology, Raipur, CG

ABSTRACT: Identification of faulty bypass diodes is a necessary factor under the fault detection and mitigation system as bypass diodes in photovoltaic systems play a significant role in improving the performance of the system by reducing the losses and providing safety to the system from the occurrence of local hotspots whenever there is an event of temporary or permanent partial shading. This paper demonstrates a method for classification of no-fault system operation, partial shading without faulty bypass diode fault condition and detection of specific faulty bypass diode under partial shading condition in a photovoltaic array using machine learning technique. Labelled dataset for the learning algorithm has been generated by using a simulated model of 2.9kW photovoltaic system under MATLAB-Simulink platform at variable irradiance and temperature values. Using output voltage, output current, power output and string currents of the photovoltaic array as the inputs, weighted K-Nearest Neighbors (KNN) algorithm is able to identify the faulty bypass diodes with an accuracy of 91.3%.

KEYWORDS: Bypass diode fault, Fault detection, Machine learning techniques (MLT), Photovoltaic systems (PVs)

[1]poorvasharma27@gmail.com, [2]ayadav.ele@nitrr.ac.in

DOI: 10.1201/9781003505181-12

1. Introduction

Due to the enormous growth in the energy and power sector, the demand for energy is rising at a very fast pace. The extraction of energy through fossil fuels has repercussions and the effects can be witnessed through climate change. To provide for the requirements of energy demand and also maintain a healthy environment, renewable energy has proved its worth. There has been a tremendous growth in the renewable energy market and India has targeted to install 500 GW of renewable energy by the year 2030 (Bureau 2022) amongst which it is estimated to meet for about 280GW of energy through solar sector. For any system, faults are inevitable and for a reliable and safe operation of a system, it is crucial to maintain a well-planned fault diagnosis and mitigation system. It has been mentioned in previous literatures of PVs and has also been observed that the safety devices like overcurrent relays, distance relays etc. (Amit Dhoke 2019) are not reliable enough to detect fault under varied temperature and irradiance conditions and various faults may happen to remain undetected which may cause deterioration of the system and can also cause fire catastrophes. Detection of faults in PVs can be done analytically. However, the algorithms used for detection are usually very complex. This paves the way for machine learning techniques (Adel Mellit 2018).

In the previous literatures, various techniques have been demonstrated to identify different faults occurring in PV panels. (Tyrrell 2023) proposed a method to detect the number of short and open Photovoltaic bypass diode fault conditions using artificial neural networks (ANN) but the actual defective module could not be identified. (Sherif S. M. Ghoneim 2021) demonstrated detection of string fault, string to ground fault, and string to string fault in PVs for which thirty features were used for fault diagnosis using ANN. Similarly, (A. Eskandari 2021) identified line - line (LL fault) and line to ground (LG fault) and (W. Miao 2021) identified LL faults using hierarchical classification & machine learning technique and algorithm based on string-current behaviour respectively.

There is a variety of faults to which the PVs may be subjected to, which can be environmental or physical in nature (Rajasekar 2018). Due to environmental conditions, PVs can have shading phenomenon (temporary or permanent) that causes solar panels to lose the power generated by the shaded panel. By the use of bypass diodes, such shaded panels can be isolated from the rest of the panels by providing a bypass path and they reroute the current path. However, if these bypass diodes become faulty i.e., short or open circuited, then the performance of the entire system will get affected. Thus, an effort has been made in this paper to automatically detect the faulty bypass diodes so that they can be effectively removed/repaired for the reliable operation of the entire PVs by using machine learning technique.

2. PV System Model and its Parameters

To diagnose and classify different conditions in a PVs, a model has been designed in MATLAB/ Simulink platform with a total capacity of 2.9 kW consisting of nine panels as a matrix of 3x3. Each panel has a bypass diode connected across it which works under the condition of partial shading and provides a bypass route to the current path. Under normal operating condition i.e., when there is no shading effect and all the bypass diodes are in healthy condition, each panel gets same insolation and temperature is same for all the panels. During the case of partial shading, the irradiance values of the panels is changed. The model has been simulated under the condition that maximum power is being extracted from the panels. In the model under consideration, the simulation has been done by taking the parameters under standard test conditions (temperature: 25°C, irradiance: 1000 W/m^2) as mentioned in Table 12.1.

Table 12.1 Simulation parameters

Parameters	Values
Maximum Power of individual solar panel (in Watts):	327.106
Open circuit voltage V_{OC} (in Volts):	65.1
Short Circuit current I_{SC} (in Amperes):	6.46
Voltage at maximum power point (in Volts):	54.7
Current at maximum power point (in Amperes):	5.98

Figure 12.1 shows the configuration of PVs wherein it is demonstrated that the bypass diode number 4 and 8 are faulty due to accidental short circuit across the diodes.

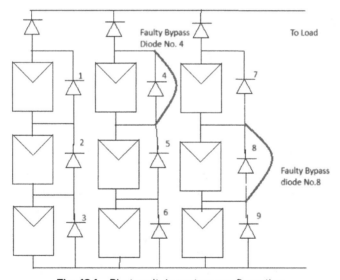

Fig. 12.1 Photovoltaic system configuration

3. Methodology

In this paper, different scenarios have been investigated under the condition of changing the temperature from 25°C to 58°C, and a dataset has been created by acquiring the readings for the parameters mentioned in Table 12.1. These parameters are the inputs to the machine learning algorithm. Figure 12.2. depicts the simulated model of the PV system modeled using MATLAB/Simulink platform which has been used for generation of dataset under different scenarios.

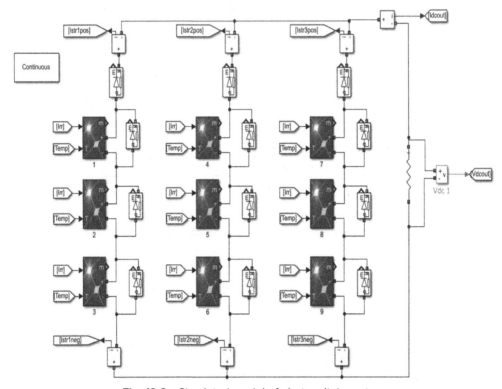

Fig. 12.2 Simulated model of photovoltaic system

Under the normal conditions, all the panels receive equal solar irradiance and the systems contains no faulty bypass diodes. Dataset for normal condition has been generated by varying the irradiance values as 100, 500 and 1000 W/m^2. Under the condition of partial shading, cases have been considered by taking one panel as being partially shaded at a time while keeping all the rest of the panels receiving same irradiance value and all the bypass diodes are working under healthy condition. For identifying the faulty bypass diode, 9 different conditions have been investigated taking each of the connected bypass diode at a time as being short circuited when shading occurs in the panel and dataset has been generated for the same. Thus 11 different operations have been performed and a dataset of 49,374 readings has been generated. Table 12.2 specifies the operations performed, their associated labels and sample size for each operation. Weighted KNN algorithm is used for fault classification using the complete data set for which the details are given in Table 12.3.

Table 12.2 Input parameters

S.No.	Parameters
1.	Temperature
2.	Top current of string 1 (instantaneous value)
3.	Bottom current of string 1 (instantaneous value)
4.	Top current of string 2 (instantaneous value)
5.	Bottom current of string 2 (instantaneous value)
6.	Top current of string 3 (instantaneous value)
7.	Bottom current of string 3 (instantaneous value)
8.	Total current out of the panel (instantaneous value)
9.	Voltage across the panel (instantaneous value)
10.	Power output of the panel (instantaneous value)

Table 12.3 Operations and labels

S.No.	Operation performed	Label	Sample Size of Dataset
1.	Normal	1	3798
2.	Partial shading	2	22797
3.	Bypass Diode '1' faulty	3	2531
4.	Bypass Diode '2' faulty	4	2531
5.	Bypass Diode '3' faulty	5	2531
6.	Bypass Diode '4' faulty	6	2531
7.	Bypass Diode '5' faulty	7	2531
8.	Bypass Diode '6' faulty	8	2531
9.	Bypass Diode '7' faulty	9	2531
10.	Bypass Diode '8' faulty	10	2531
11.	Bypass Diode '9' faulty	11	2531

4. Results

The results have been obtained by applying weighted KNN algorithm to the dataset generated from the simulated model. The trained model can classify normal operation (no fault and no partial shading, Label 1), partial shading operation (partial shading without bypass diode fault, Label 2) and bypass diode fault condition (Label 3 to Label 11) with 100 % accuracy as shown in Figure 12.3.

Table 12.4 Weighted KNN model description

Parameters	Details
Number of neighbours	20
Distance metric	Euclidean
Distance weight	Inverse

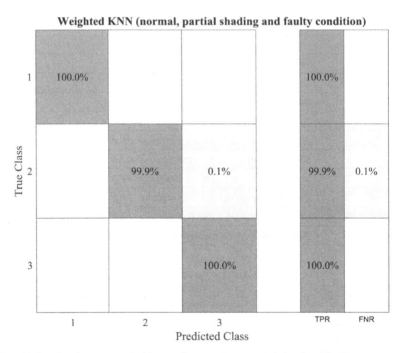

Fig. 12.3 Confusion matrix 1(normal operation, partial shading, faulty operation)

However, when classification is done by taking the cases of faulty bypass diodes separately (panel wise identification of faulty bypass diodes), then the algorithm is able to classify the 11 operations with an accuracy of 91.3 % as can be seen from Fig. 12.4. It can be seen from Fig. 12.4 that the identification of faulty bypass diodes in the same string of the PV array have major misclassification, whereas the model can accurately distinguish the faulty bypass diodes lying in different strings of the PV array. The obtained dataset was applied on different ANN algorithms namely fine tree, weighted KNN, Naïve Bayes, and Quadratic SVM. Amongst all these algorithms, highest accuracy was obtained for weighted KNN algorithm.

5. Conclusions

In this paper, an automatic fault detection system based on machine learning technique has been demonstrated. A small-scale photovoltaic model of 2.9kW was used for the dataset generation purpose. Amongst the various machine learning algorithms, weighted KNN algorithm provided the highest accuracy. It has been demonstrated that the proposed method for fault classification can predict the faulty bypass diodes with a reasonable accuracy. The classification is 100 % in terms of identifying normal operation, partial shading operation and faulty operation. For the panel wise classification of faulty bypass diodes, the machine learning model can predict with an accuracy of 91.3%. This model can be used for small scale PVs by varying the input parameters as per the number of strings (parallel paths) in the photovoltaic array.

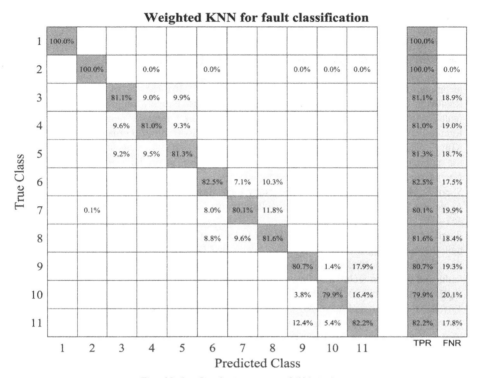

Fig. 12.4   Confusion matrix 2(11 labels)

REFERENCES

1. A. Eskandari, J. Milimonfared and M. Aghaei. 2021. "Fault Detection and Classification for Photovoltaic Systems Based on Hierarchical Classification and Machine Learning Technique." *IEEE Transactions on Industrial Electronics, vol. 68* 12750-12759.
2. A. Mellit, G. M. Tina and S. A. Kalogirou. 2018. "Fault detection and diagnosis methods for photovoltaic systems: A review." *Renewable and Sustainable Energy Reviews, vol. 91* 1-17.
3. Adel Mellit, Soteris A. Kalogirou. 2018. *Chapter II-1-D - A Survey on the Application of Artificial Intelligence Techniques for Photovoltaic Systems,.* McEvoy's Handbook of Photovoltaics (Third Edition), Academic Press,.
4. Amit Dhoke, Rahul Sharma, Tapan Kumar Saha. 2019. "An approach for fault detection and location in solar PV systems." *Solar Energy, Volume 194* 197-208.
5. B. P. Kumar, D. S. Pillai, N. Rajasekar, M. Chakkarapani, and G. S. Ilango. 2021. "Identification and localization of array faults with optimized placement of voltage sensors in a PV system." *IEEE Trans. Ind. Electron., vol. 68* 5921–5931.
6. Bureau, Snapshot of Press Information. 2022. "Press Information Bureau." *Ministry of New and Renewable Energy,Renewable Energy in India.* 09 September. https://pib.gov.in/FeaturesDeatils. aspx?NoteId=151141&ModuleId%20=%202:.

7. Chen, M. Dhimish and Z. 2019. "Novel open-circuit photovoltaic bypass diode fault detection algorithm." *IEEE J. Photovolt., vol. 9* 1819–1827.

8. Chine, W., Mellit, A., Pavan, A. M., & Kalogirou, S. A. 2014. " Fault detection method for grid-connected photovoltaic plants." *Renewable Energy* 99-110.

9. Madeti, S.R., Singh, S.N. 2017. "A comprehensive study on different types of faults and detection techniques for solar photovoltaic system." *Sol. Energy 158* 161-185.

10. Rajasekar, Dhanup S. Pillai and N. 2018. "A comprehensive review on protection challenges and fault diagnosis in PV systems." *Renewable and Sustainable Energy Reviews, Volume 91* 18-40.

11. Ramón Fernando Colmenares-Quintero, Eyberth R. Rojas-Martinez, Fernando Macho-Hernantes, Kim E. Stansfield & Juan Carlos Colmenares-Quintero | (2021) , Cogent. 2021. "Methodology for automatic fault detection in photovoltaic arrays from artificial neural networks." *Cogent Engineering* 1-21.

12. Sherif S. M. Ghoneim, Amr E. Rashed, Nagy I. Elkalashy. 2021. "Fault Detection Algorithms for Achieving Service Continuity in Photovoltaic Farms." *Intelligent Automation & Soft Computing* 467-479.

13. Singh, M. S. Ramakrishna and S. N. 2018. "Modeling of PV system based on experimental data for fault detection using kNN method." *Solar Energy, vol 173* 139-151.

14. Siva Ramakrishna Madeti, S.N. Singh. 2017. "A comprehensive study on different types of faults and detection techniques for solar photovoltaic system." *Solar Energy* 161-185. https://doi.org/10.1016/j.solener.2017.08.069.

15. Tyrrell, M. Dhimish and A. M. 2023. "Photovoltaic Bypass Diode Fault Detection Using Artificial Neural Networks." *IEEE Transactions on Instrumentation and Measurement, vol. 72* 1-10.

16. W. Miao, K. H. Lam and P. W. T. Pong. 2021. "A String-Current Behavior and Current Sensing-Based Technique for Line–Line Fault Detection in Photovoltaic Systems." *IEEE Transactions on Magnetics, vol. 57* 1-6.

Emerging Technologies and Applications in Electrical Engineering –
Prof. Dr. Anamika Yadav et al. (eds)
© 2024 Taylor & Francis Group, London, ISBN 978-1-032-82568-7

Effect of Wave Rise and Tail Times of Standard Lightning Current on Transient Response of the Horizontal Buried Conductor

Prasad Chongala
[1]ResearchScholar, Dept.of EEE, GIET University, Gunupur,Odisha

Gandi Ramarao*
Assistant Professor, Dept.of EEE,
Aditya Institute of Technologyand Management Tekkali, A.P.

Pratap Kumar Panigrahi
Professor, Dept. of EEE,
GIET University, Gunupur, Odisha

ABSTRACT: This paper presents the TC is analysed through the horizontal conductor buried in lossy grounding conditions (i.e., $\sigma = 0.01$ s/m, 0.001 s/m, 0.0001 s/m, 0.00001 s/m) due to the variations in rise and tail time of applied standard lightning current (LC) of 100kA, 1.2/50 μs. This is the first of its kind attempt that the intermittence nature of rise and tail time of standard LC is applied for the analysis of transient current (TC) along the horizontal buried conductor (BC). The results are reported and observed that the conductivity of ground as well as the variation in rise and tail time of standard LC showing much impact on peak of TC along the BC. The rise time (t_r) variation (0.84 μs $\leq t_r \leq$ 1.56 μs) while keeping tail time (t_{tail}) constant (50 μs) is much effecting on peak current of TC flowing through the BC at different distances along the conductor and at different grounding conductivities are identified and reported. Similarly, the variation in t_{tail} (40 μs $\leq t_{tail} \leq$ 60 μs) while keeping the t_r constant (1.2 μs) is least effecting on peak current of TC flowing through the BC at different distances along the conductor and at different grounding conductivities are also evaluated and reported in this paper. Finally, it is observed from the results that Peak current of TC along the conductor depends upon the t_r and t_{tail} variations. Further, this approach will be needful for designing of appropriate grounding system to lowering the TC through the horizontal BC.

KEYWORDS: Ground conductivity, Rise time, Standard LC, Tail time

*Corresponding author: grr231@gmail.com

DOI: 10.1201/9781003505181-13

1. Introduction

Generally, the lightning and switching impulse voltages portrays a vital role in order to analyse the TC along the underground and overhead transmission lines. The aforementioned objective is significant for proper economical designing of insulation for high voltage (HV) equipments used in power system network. The whole process needs to be analysed for providing un interrupted power supply to end users. The transient behaviour of lightning current (LC) along such equipments has been analysed by the prominent researchers since past decades. Numerous uses of antenna theory and electromagnetic compatibility (EMC) in underground power conductors and telecommunication lines can be seen in these types of investigations (Jin and Ali, 2010), (Poljak and Doric, 2006). The primary goal of a grounding system is to switch over voltages and currents while also shielding underground cables from lightning. If the ground is not ideal, a significant amount of lightning impulse current (LIC) flows into the buried conductor (BC) whenever a lightning discharge occurs on it, developing transient current (TC) and voltages into it. The insulation of these buried power conductors may be damaged by these transient produced voltages, making it impossible to maintain the continuity of the power supply. Therefore, the research of coupling between electromagnetic fields created by LIC and the horizontal grounding conductor, as well as the analysis of transient behaviour of currents and voltages, are the fundamental goals significant in the design of a correct grounding system (Leonid and Menter, 1996, Liu et.al., 2001, Ala and Silvestre, 2002). The transient analysis of LC along a grounding conductor that has been triggered by impulsive currents is illustrated in (Suflis et al., 1998, Gonos et al., 2003, Masanobu et al., 2006). To analyse the transient behaviour of voltages and currents in grounding conductors caused by transient magnetic field coupling, there are basically two models. One is a transmission line (TL) model, and the other is an antenna theory model (Tesche et al., 1997). The aforementioned models can be solved either in the time domain or the infrequency domain. In comparison to the TL approach, the antenna theory approach is more complex and time-consuming to implement. The transient response of the modified transmission line model (MTLM) and the wire antenna theory model (WAT) in the frequency domain for the horizontal grounding conductor are illustrated in (Poljak et al., 2008). Based on the simulation of an unconventional LIC waveform, the transient analysis of a BC was reported in (Ramarao and Chandrasekaran, 2018).

In general, standard and abnormal lightning impulse waveforms mostly impact on overhead and underground lines. Thus, it is possible to use the TL model to analyse the transient behaviour of the current along the horizontal BC by employing the conventional LIC wave shape as the exciting current. In this paper, the finite difference time domain (FDTD) method is used to first solve the telegrapher's equations before applying the standard LIC as an exciting current to BC at various ground conductivities. The analysis of the results is then computed and reported.

2. Electrical Stresses on the Outer Insulation of Buried Cables Due to a Nearby Lightning Strike to a Ground

Cables buried in the ground, whether directly or inside an insulating pipe, are nonetheless susceptible to disruptions and damage from lightning. Accounts of cable pairs fusing, sheath holes, and furrows longer than 100 feet caused by lightning strikes arcing directly to the buried cable date back to 1945. A frequency domain study employing wave theory for transmission line was presented, presuming that the outer jacket will eventually come into touch with the soil. For several scenarios, including direct arcing to the cable, strokes to ground without arcing to the cable, two-layer soil model, cables with insulated sheaths, soil ionisation, and discussion of corrective actions, voltage between the sheath and the cable conductors was thoroughly analysed. The Sunde's work was incorporated into the majority of the papers that followed on lightning stroke arcing to cable sheaths. For instance, it was expanded to cover insulating jackets with a certain number of defined punctures (Poljak and Doric, 2006). For the over voltage prediction in (Leonid and Menter, 1996), the non-linear circuit model for switching the break down channel to the cable, the transmission line model for wave response, and earlier efforts on the generation of break down channels in 2D geometry are combined. A more hazardous scenario for buried tele-communication wires is considered in (Liu et.al., 2001). A strike to the top headlines can result from an arcing to the underground cable. The arcing to the wire will continue if the strike results in power follow current, causing significantly more damage than the lightning itself. The topic was investigated in (Liu et.al., 2001), where safe distance guidelines for preventing lightning-caused power frequency arcs were offered. The protective zone of buried shield wires is described in (Ala and Silvestre, 2002) for the purpose of shielding buried cables from direct impacts. Also taken into account is the spark over between the cable and the buried shield wire. A nearby object striking the subterranean electrical lines causes damage (Suflis et al., 1998). This was thoroughly examined in the triggered lightning tests at the Camp Blanding, Florida-based International Centre for Lightning Research and Testing (ICLRT) (Suflis et al., 1998, Gonos et al., 2003, Masanobu et al., 2006). Damages seen in these studies range from modest neutral wire melting and cable jacket punctures to significant jacket punctures and nearly complete neutral wire melting (Tesche et al., 1997, Poljak et al., 2008). After drilling a hole in the conduit and melting the insulating jacket for the cable contained in PVC conduit, the stroke ended on the concentric neutral wires. The corrosion of the neutral caused by the water leak caused by the ruptured jacket causes a delayed failure. The runway lighting system could also sustain damage (Poljak et al., 2016, Paul, 2008, Sunde, 1968, Naidu and Kamaraju, 1995, Qin et al., 2016, Camp and Garbe, 2004).

At the ICLRT, an experiment with triggered lightning was carried out on a test runway lighting system (Gonos et al., 2003, Tesche et al., 1997). Measurements were made of the currents flowing through the driven rods, single core lighting wire, and counter poise as well as the voltage differential between the two. The damages to the terminal equipment are also looked into in the two experimental experiments mentioned above. There aren't many studies on induced currents caused by a strike to the earth in the buried cable. For instance, the 3 ICLRT evaluated the induced currents in the buried cable using both theoretical and experimental methods (Gonos et al., 2003, Poljak et al., 2008). These don't really matter for

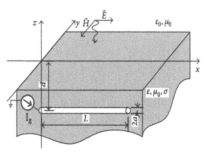

Fig. 13.1 Conductor of 10m length underlying in the ground

the current project, which deals with tension on the outside insulation. Most, if not all, of the earlier theoretical investigations focused on direct arcing to the cable as a result of a close contact with the ground. The following presumptions are either implicitly or expressly made in the immediate area. The FDTD method is created and used to solve the telegrapher problem in (Poljak et al., 2016).

3. Use FDTD Technique to Solve Telegrapher's Equations

Using the FDTD approach, Telegrapher's equations are solved (Nagel, 2010, Gedney, 1996, Sullivan, 2013). YEE developed this approach as an effective numerical solution methodology in 1966. Additionally, a common technique for assessing the transient response LC down a transmission line or underground conductor is the FDTD approach.

The derivatives stated in equation (1) are invalid, and numerous Finite Differences are used to approximate them. The position and time parameters are discretized as Δx and Δt.

$$\frac{\partial v(x,t)}{\partial x} + Ri(x,t) + L\frac{\partial i(x,t)}{\partial t} = 0 \tag{1}$$

$$\frac{\partial i(x,t)}{\partial x} + Gv(x,t) + C\frac{\partial v(x,t)}{\partial t} = 0 \tag{2}$$

Typically, the differentiation is considered as a finite difference in the Maxwell's equations (Song et al., 2012), where the differentiation of time is nothing but the variation between future and past values in time and the differentiation of space is nothing but the variation between the next and previous values in space. Both time and space differentiation are represented by these Maxwell's equations. Thus, there is a relationship between the magnetic and electric fields. Accordingly, current magnetic field values must be known in order to determine the next step's electric field values, and vice versa (Nagel, 2010).

Figure 13.2 shows how a BC is discretized in space and time. The feasibility of using the finite differences of voltages and current as described in (Gedney, 1996) are readily seen in Fig. 13.2, which is demonstrated by equation (3)

$$\frac{\partial v(x,t)}{\partial x} = \frac{v_{k+1}^n - v_k^n}{\Delta x}$$

$$\frac{\partial i(x,t)}{\partial t} = \frac{i_{k+1/2}^{n+1/2} - i_{k+1/2}^{n-1/2}}{\Delta t} \tag{3}$$

By substituting the equation (3) in equation (1), then the equation (1) becomes as equation (4) is illustrated below.

$$\frac{v_{k+1}^n - v_k^n}{\Delta x} + R\frac{i_{k+1/2}^{n+1/2} + i_{k+1/2}^{n-1/2}}{2} + L\frac{i_{k+1/2}^{n+1/2} - i_{k+1/2}^{n-1/2}}{\Delta t} = 0 \tag{4}$$

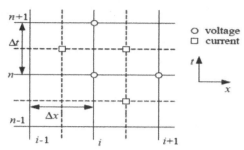

Fig. 13.2 Discretization of the conductor in space and time

By simplifying the equation (4), eventually, the current through the BCs is given by equation (5) as follows

$$i_k^{n+3/2} = A \cdot \left[B \cdot i_k^{n+\frac{1}{2}} - \left(v_{k+1}^{n+1} - v_k^{n+1} \right) \right] \text{ for k = 2, 3, ... , N} \tag{5}$$

Where in (5), $A = \left(\dfrac{\Delta x}{\Delta t} l + \Delta x \dfrac{r}{2} \right)^{-1}$, $B = \left(\dfrac{\Delta x}{\Delta t} l - \Delta x \dfrac{r}{2} \right)^{-1}$,

The equation (6) can also formed from Fig. 13.2 and is given as follows

$$\frac{\partial i(x,t)}{\partial x} = \frac{i_{k+1/2}^{n+1/2} - i_{i-1/2}^{n+1/2}}{\Delta x}$$

$$\frac{\partial v(x,t)}{\partial t} = \frac{v_k^{n+1} - v_k^n}{\Delta t} \tag{6}$$

By substituting the equation (6) in equation (2), then the equation (2) is modified as equation (7) as follows

$$\frac{i_{k+1/2}^{n+1/2} - i_{i-1/2}^{n+1/2}}{\Delta x} + G \frac{v_k^{n+1} + v_k^n}{2} + C \frac{v_k^{n+1} - v_k^n}{\Delta t} = 0 \tag{7}$$

Where, $L = \dfrac{\mu_0}{2\pi} \ln\left(\dfrac{2l}{\sqrt{2ad}} - 1 \right)$; $G = \dfrac{2\pi\sigma}{\ln\left(\dfrac{2l}{\sqrt{2ad}} - 1 \right)}$; $C = \dfrac{2\pi\varepsilon_0\varepsilon_r}{\ln\left(\dfrac{2l}{\sqrt{2ad}} - 1 \right)}$

By simplifying the equation (7), then Equation (8) illustrates the final formula for the voltage variation at different distances along the BC in terms of space and time.

$$v_k^{n+1} = D \cdot \left[E \cdot v_k^n - \left(i_k^{n+\frac{1}{2}} - i_{k-1}^{n+\frac{1}{2}} \right) \right] \text{ for k = 2, 3, ..., N} \tag{8}$$

Where in (8), $D = \left(\dfrac{\Delta x}{\Delta t} c + \Delta x \dfrac{g}{2} \right)^{-1}$; $E = \left(\dfrac{\Delta x}{\Delta t} c - \Delta x \dfrac{g}{2} \right)^{-1}$

As indicated in Fig. 13.1, an excitation current identical to a conventional lightning impulse is injected at the conductor's beginning. Using equations (5) and (8), it is then possible to detect TC at various places through the conductor.

4. Results and Discussion

The TC through a conductor underlying at a depth of d = 0.5m and a radius of a = 5mm with a ε_0 = 10 has been taken for the analyzing the concept of transient response of BC when it is excited with standard LC. The BC is assumed to be 10m long.

LIC waveform of 100 kA, 1.2/50 μs produced by using double exponential function (DEXP) as given in equation (9), which is excited to this BC.

$$I_g(t) = I_0 . K . (e^{-\alpha t} - e^{-\beta t}) \tag{9}$$

Equation (7) illustrates how the TC from the telegrapher's equations ((1) & (2)) was solved using FDTD. These equations were implemented in MATLAB. The horizontal conductor that is buried and the standard LIC with 100 kA peak, t_r and t_{tail} of 1.2/50 μs (Fig. 13.3) is applied in order to analyse the transient response in BC.

For the purpose of keen observation of the TC along the BC, the waveform in Fig. 13.3 is applied to a horizontal BC. The data of the equation (9) are provided below in order to provide exciting current with the necessary wave shape (i.e., 1.2/50 μs). These values are taken from (Ramarao and Chandrasekaran, 2018, Ramarao and Chandrasekaran, 2017) for generating the LIC waveform with required waveshape.

$$\alpha = 1.4912 * 10^4 \ s^{-1}$$
$$\beta = 1.6335 * 10^6 \ s^{-1}$$

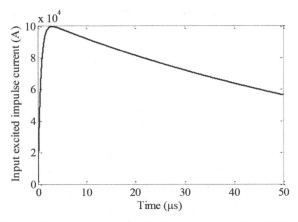

Fig. 13.3 1.2/50 μs input triggered LIC waveform with a 100 kA peak

The effect of ground conductance, wave t_r and t_{tail} of standard LIC on TC in BC is analyzed and illustrated clearly in below subsections. The variation of ground conductances (σ = 0.01

s/m, 0.001 s/m, 0.0001 s/m, and 0.00001 s/m) affecting affecting on TC in BC is analyzed in subsection 4.A. The variation of rise time (t_r) $(0.84\,\mu s \leq t_r \leq 1.56\,\mu s)$ by keeping the t_{tail} constant at 50 μs of standard LIC affecting on TC in BC is analyzed in subsection 4.B. Moreover, the variation of t_{tail} (40 μs $\leq t_r \leq$ 60 μs) by keeping the t_r constant at 1.2 μs of standard LIC affecting on TC in BC is analyzed in subsection 4.C.

4.1 Effect due to Variation of Ground Conductances

The analytical findings were obtained for various ground conductivities (σ = 0.01 s/m, 0.001 s/m, 0.0001 s/m, and 0.00001 s/m). The findings for the transient LC peak were collected at several locations along the buried horizontal conductor (at 2 m, 4 m, 6 m, 8 m, and 10 m), and they are shown in Table 13.1. Also, the transient response of LC is drawn at 2 m along the BC for different ground conductivities and illustrated in Fig. 13.4.

Table 13.1 Transient behavior of current when 100 kA, 1.2/50 μs LIC is triggered at various distances and ground conductivities

Conductivity of the ground (s/m)	Peak current at 2 m (A)	Peak current at 4 m (A)	Peak current at 6 m (A)	Peak current at 8 m (μA)	Peak current at 10 m (μA)
0.01	153.3	0.9221	0.005545	33.65	0.2377
0.001	1298	10.34	0.08245	657.1	5.75
0.0001	4754	39.05	0.3209	2636	23.82
0.00001	7857	64.75	0.5336	4395	39.86

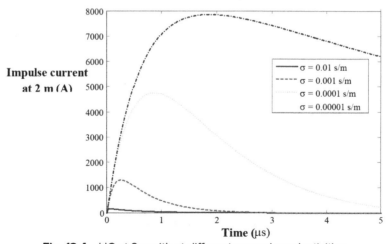

Fig. 13.4 LIC at 2m with at different ground conductivities

It is observed from the Fig. 13.4 that the magnitude of TC along the BC (moving from exciting point towards end point of BC) increases with decreasing of ground conductivities (from σ = 0.01 s/m to σ = 0.00001 s/m).

4.2 Effect Due to Variation in Rise Time by Keeping Tail Time Constant

The effect of TC along BC at different grounding conditions (σ = 0.01 s/m, 0.001 s/m, 0.0001 s/m, 0.00001 s/m) due to the application of standard LIC with the variation of t_r (0.84 µs $\leq t_r \leq$ 1.56 µs) by keeping the t_{tail} constant at 50 µs is analyzed and reported in this section. The TC along BC at the conductance of ground of σ = 0.01 s/m at different distances (i.e., 2m, 4m, 6m, 8m, 10m) are calculated by the application of standard LIC with the variation of t_r (0.84 µs $\leq t_r \leq$ 1.56 µs) by keeping the t_{tail} constant at 50 µs and are reported in Table 13.2.

Table 13.2 TC at different distances at the ground conductivity of 0.01 s/m when t_r of LIC is varied from 0.84 µs to 1.56 µs

t_r and t_{tail} of injected IC	At 2 m		At 4 m		At 6 m		At 8 m		At 10 m	
	Peak current	Time to peak	Peak current	Time to peak	Peak current	Time to peak	Peak current	Time to peak	Peak current	Time to peak
0.84/50 µs	214.6 A	0.06 µs	1.292 A	0.07 µs	7.785 mA	0.09 µs	47.13 µA	0.1 µs	0.341 µA	0.12 µs
1.2/50 µs	153.3 A	0.06 µs	0.922 A	0.08 µs	5.545 mA	0.1 µs	33.33 µA	0.12 µs	0.2195 µA	0.12 µs
1.56/50 µs	118.4 A	0.06 µs	0.712A	0.08 µs	4.289 mA	0.1 µs	25.80 µA	0.12 µs	0.1686 µA	0.12 µs

Figure 13.5 plots and illustrates the transient response of LC along the BC at a distance of 2m with respect to time. At different t_r of standard lightning impulse waveform (such as 0.84/50 µs, 1.2/50 µs, and 1.56/50 µs by keeping the t_{tail} constant) and at the ground conductance of 0.00001 s/m, this response is produced. The peak current and its peak time are seen to decrease as the t_r increases, as can be seen in Fig. 13.5.

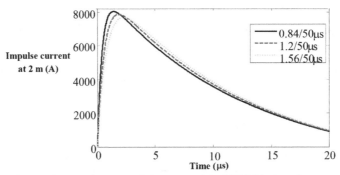

Fig. 13.5 LIC at 2m at the conductivity of the ground is 0.00001 s/m when t_r varies from 0.84 to 1.56 µs

Figure 13.6 plots and illustrates the transient response of LC along the BC at a distance of 6 m with respect to time. At different t_r of standard lightning impulse waveform (such as 0.84/50 µs, 1.2/50 µs, and 1.56/50 µs by keeping the t_{tail} constant) and at the ground conductance of

0.001 s/m, this response is produced. The peak current and its peak time are seen to decrease as the t_r increases, as can be seen in Fig. 13.6.

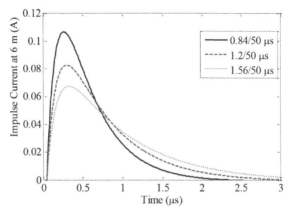

Fig. 13.6 LIC at 6 m at the conductivity of the ground is 0.001 s/m when t_r varies from 0.84 to 1.56 μs

It is noted from Fig. 13.5 and Fig. 13.6 that the peak of TC flowing along the BC is much affected by the variation of t_r of applied LC at different grounding conditions.

4.3 Effect due to Variation in Tail Time by Keeping Rise Time Constant

The effect of TC along BC at different grounding conditions ($\sigma = 0.01$ s/m, 0.001 s/m, 0.0001 s/m, 0.00001 s/m) due to the application of standard LIC with the variation of t_{tail} (40 μs ≤ t_r ≤ 60 μs) by keeping the t_r constant at 1.2 μs is analyzed and reported in this section. The TC along BC at the conductance of ground of $\sigma = 0.01$ s/m at different distances (i.e., 2m, 4m, 6m, 8m, 10m) are calculated by the application of standard LIC with the variation of t_{tail} (40 μs ≤ t_r ≤ 60 μs) by keeping the t_r constant at 1.2 μs and are reported in Table 13.3.

Table 13.3 TC at different distances at the ground conductivity of 0.01 s/m when t_{tail} of LIC is varied from 40 μs to 60 μs

t_r and t_{tail} of injected IC	At 2 m		At 4 m		At 6 m		At 8 m		At 10 m	
	Peak current	Time to peak	Peak current	Time to peak	Peak current	Time to peak	Peak current	Time to peak	Peak current	Time to peak
1.2/40 μs	151.6 A	0.06 μs	0.9118 A	0.08 μs	5.483 mA	0.1 μs	32.97 μA	0.1 μs	0.2348 μA	0.1 μs
1.2/50 μs	153.3 A	0.06 μs	0.9221 A	0.08 μs	5.545 mA	0.1 μs	33.28 μA	0.1 μs	0.2377 μA	0.1 μs
1.2/60 μs	154.5 A	0.06 μs	0.9294 A	0.08 μs	5.588 mA	0.1 μs	33.56 μA	0.1 μs	0.2398 μA	0.1 μs

Figure 13.7 plots and illustrates the TC along the BC at a distance of 2 m with respect to time. At t he ground conductivity of 0.01 s/m, and this response is recorded for different t_{tail} of standard LIC waveform (such as 40 μs, 50 μs, and 60 μs by keeping the t_r constant at 1.2 μs).

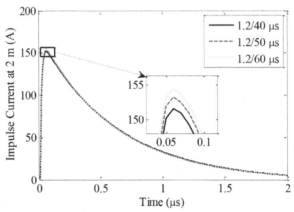

Fig. 13.7   LIC at 2m at the conductivity of the ground is 0.01 s/m when t_{tail} varies from 40 to 60 µs

From Fig. 13.7, it can be seen that when the t_{tail} increases, there is a slight increase in the peak current and its peak time.

Figure 13.8 plots and illustrates the TC along the BC at a distance of 6m with respect to time. At the ground conductivity of 0.01 s/m, and this response is recorded for different t_{tail} of standard LIC waveform (such as 40 µs, 50 µs, and 60 µs by keeping the t_r constant at 1.2 µs). From Fig. 13.8, it can be seen that when the t_{tail} increases, there is a slight increase in the peak current and its peak time.

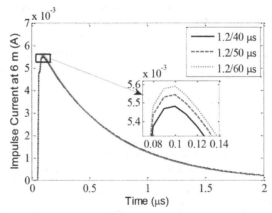

Fig. 13.8   LIC at 6m at the conductivity of the ground is 0.01 s/m when t_{tail} varies from 40 to 60 µs

It is observed from Fig. 13.7 and Fig. 13.8 that the peak of TC flowing along the BC is least affected by the variation of t_{tail} of applied LC at different grounding conditions.

5. Conclusion

This paper discusses and analyses the transient behaviour of the LIC through the horizontal conductor buried in lossy medium using the transmission line (TL) approach. The Finite Difference Time Domain (FDTD) approach is used to solve the telegrapher's equations for BCs and analyse the transient LC at various locations along the conductor at varying conductivities of lossy medium. Peak currents are calculated and reported at distances of 2 m, 4 m, 6 m, 8 m, and 10 m along the BC and at the ground conductivities of $\sigma = 0.01$ s/m, 0.001 s/m, 0.0001 s/m, 0.00001 s/m. The results show that, as the observation point travels away from the initial or exciting point toward the end, the peak of the current falls. Moreover, when the conductivity of lossy ground is reduced, a significant change in peak current has been seen. This indicates that the ground conductivity has an inverse relationship with the peak current that is achieved. In order to create a correct grounding system and reduce the peak current along the horizontal BC, this kind of study will be useful.

REFERENCES

1. Ala, G., Silvestre, M. L. D. (2002). A simulation model for electromagnetic transients in lightning protection systems. IEEE Transactions on Electromagnetic Compatibility. 44 (4): 539–554.
2. Camp, M., Garbe, H. (2004). Parameter Estimation of Double Exponential Pulse (EMP, UWB) With Least Squares and Nelder Mead Algorithm. IEEE Trans. Electromagnetic. Compat. 46 (4): 675–678.
3. Gedney, S. D. (1996). The application of the finite-difference time domain method to EMC analysis. Proceedings of Symposium on Electromagnetic Compatibility. IEEE.
4. Gonos, I. F., et al. (2003). Transient behavior of a horizontal electrode under LIC, XIIIth International Symposium on High Voltage Engineering.
5. Jin, X., Ali, M. (2010). Embedded antennas in dry and saturated concrete for application in wireless sensors," Progress In Electromagnetic Research. 102: 197–211.
6. Leonid G. D., Menter. F. E. (1996). Transient electromagnetic fields near large earthing systems, IEEE Transactions on Magnetics. 32 (3): 1525–1528.
7. Liu, Y., Zitnik, M., Thottappillil, R. (2001). An improved transmission-line model of grounding system, IEEE Transactions on Electromagnetic Compatibility. 43 (3): 348-355.
8. Masanobu, T., et al. (2006). FDTD simulation of a horizontal grounding electrode and modeling of its equivalent circuit. IEEE transactions on electromagnetic compatibility. 48 (4): 817–825.
9. Nagel, J. R. (2010). The Finite-Difference Time-Domain (FDTD) Algorithm. Department of Electrical and Computer Engineering. University of Utah.
10. Naidu, M. S., Kamaraju, V. (1995). High Voltage Engineering, New Delhi, Tata McGraw-Hill.
11. Paul, C. R. (2008). Analysis of Multi conductor Transmission Lines. John Wiley & Sons. New York. NY. USA.
12. Poljak, D., et al. (2016). Transient electromagnetic field coupling to buried thin wire configurations: antenna model versus transmission line approach in the time domain. International Journal of Antennas and Propagation.
13. Poljak, D., et al. (2008). Comparison of wire antenna and modified transmission line approach to the assessment of frequency response of horizontal grounding electrodes. Engineering analysis with boundary elements. 32 (8): 676–681.

14. Poljak, D., Doric, V. (2006). Wire antenna model for transient analysis of simple grounding systems, part II: The horizontal grounding electrode. Progress In Electromagnetic Research. 64: 167–189.
15. Qin, F., Mao, C., Wu, G., Zhou, H. (2016). Characteristic Parameter Estimation of EMP Energy Spectrum, 7th Asia Pacific International Symposium on Electromagnetic Compatibility. pp. 11–13.
16. Ramarao, G., & Chandrasekaran, K. (2018). Evaluating Lightning Channel-Base-Current Function Parameters for Identifying Interdependence of Wavefront and Tail by PSO Method. IEEE Transactions on Electromagnetic Compatibility. 62(1): 183–180.
17. Ramarao, G., Chandrasekaran, K. (2018). Calculation of Multistage Impulse Circuit and Its Analytical Function Parameters. International Journal of Pure and Applied Mathematics. 114 (12): 583–592.
18. Ramarao, G., Chandrasekaran, K. (2017). Evaluation of circuit and its analytical function parameters for lightning and switching impulse. 2017 International Conference on Innovations in Electrical, Electronics, Instrumentation and Media Technology (ICEEIMT), Coimbatore. India. 3-4.
19. Song, J., Liu, Y., Yu, Y. (2012). Numerical analysis of transmission line telegraph equation based on FDTD method. JCIT. 7 (20): 258–265.
20. Suflis, S. A., et al. (1998). Transient behavior of a horizontal grounding rod under LIC, Recent Advances in Circuits and Systems. Word Scientific Publishing Company. Singapore. pp. 61–64.
21. Sullivan, D. M. (2013). Electromagnetic simulation using the FDTD method, John Wiley & Sons.
22. Sunde, E. D. (1968). Earth Conduction Effects in Transmission Systems, Dover Publications. New York. NY. USA.
23. Tesche, Frederick M., et al. (1997). EMC analysis methods and computational models, John Wiley & Sons.

Emerging Technologies and Applications in Electrical Engineering –
Prof. Dr. Anamika Yadav et al. (eds)
© 2024 Taylor & Francis Group, London, ISBN 978-1-032-82568-7

Analysis and Protection of Different Fault in Transmission line using MHO Digital Distance Protection Relay

Dewashri Pansari[1]

Assistant Professor, Electrical & Electronics Engineering Department,
Government Engineering College, Raipur

Anamika Yadav[2]

Professor, Electrical Engineering Department,
National Institute of Technology, Raipur

ABSTRACT: Transmission lines are vital components of electrical power systems and their reliable operation is critical for maintaining the integrity of power grids. To ensure the continuous supply of electricity it is imperative to analyze and protect transmission lines from various faults promptly. This research paper explores the novel applications of MHO digital distance protection relay for the analysis and protection of different types of faults in transmission line and also provides backup protection. This paper discusses innovative techniques to enhance the fault detection, classification and response capabilities of the Distance protection relay system. The Digital MHO relay is implemented in a model consisting of sources 31800 MVA, 25 kA connected by two 230 kV distributed transmission lines of 200 km length each. The relay measures various impedances and as per the impedance seen by the MHO digital distance relay impedance trajectory for fault in zone 1and zone 2 of the relay is depicted.

KEYWORDS: MHO digital distance protection relay, Impedance trajectory, MATLAB.

1. Introduction

The stability and reliability of electrical power systems are paramount necessitating the need for robust protection schemes. Distance protection relays play a crucial role in safeguarding

[1]pansaridewashri@gmail.com, [2]ayadav.ele@nitrr.ac.in

DOI: 10.1201/9781003505181-14

transmission lines, transformers and other critical assets. Therefore, it is imperative to rigorously validate and verify the operation of these relays in various scenarios. Traditional relay testing methods often fall short of replicating real-time conditions and cannot comprehensively assess the relay behavior under dynamic conditions (M. Sreeram and P. Raja, 2013). The fundamental concept in a distance protection scheme involves assessing the line impedance from the relay installation point to the point of fault occurrence. If the impedance seen by the relay falls below a specified threshold (predefined value) the relay will initiate its operation. In contrast, if the impedance exceeds this threshold the relay remains inactive. Zone 1 provides primary high-speed protection to a substantial section of transmission line. Zone 2 on the other hand protects remaining part of line and also serves as backup protection (D. Pal, 1 Dec.1, 2017).

2. Power System Modelling

The first step involves creating an accurate real-time model of the power system under protection. This model includes representations of generators, transformers, transmission lines and loads. The test setup has two perfectly balanced power sources operating at 31800 MVA connected by 230 kV transmission lines spanning a distance of 200 kilometres. The configuration of the transmission line allows for the deliberate introduction of faults at various locations along its path (as distributed transmission line is used). Two parallel transmission lines namely line 1 and line 2 is placed between two bus bars B1 and B2. Line 1 is protected by MHO distance relays placed at each end of line (Brunelle, 2020). Each relay comprises of four subsystems i.e. Measurement unit, Impedance computation unit, Fault detection unit, Tripping unit. The disturbance or fault is created in line1 (protected line) by fault programming system and system is analysed at various fault applied at t =0.1s and cleared at t =0.2s. Load of 300MW and 200MW is connected near the bus bar B1and B2 respectively. Figure 14.1 shows the MATLAB Simulink design of the modelled power system. The distance relay protection is a widely used for safeguarding transmission lines and it serves as the primary protective devices for overhead lines. In this paper a MHO distance protection relay estimates the impedance between fault and relay to identify faults by utilizing measurements of current and voltage up-to the fault point. Zone 1is adjusted to protect 80% of the protected line and zone 2 is adjusted to protect 120% of the line's impedance to intentionally extend protection further down the line. Zone 3 is deactivated. Each relay measures the voltages and currents at the input of the line with the help of some mathematical equations.

The relay calculates six different impedances (D.Tholomier, 2004). This computation is carried out by assessing the voltage (V) and current (I) phasors using discrete fourier analysis. When measured impedance falls in the protection zone(zone1and zone 2)the relay gets activated . Consequently the trip signal is generated by the relay switch. The trip signal is then sent to the line breaker and the start permissie trip signal is transmitted to the relay present at the other end of the line. If relay B2 detects the fault within the second zone and after receiving the SPT signal, it immediately get activated without any time delay.

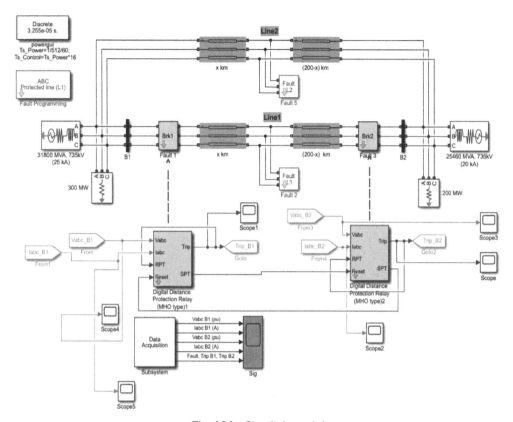

Fig. 14.1 Simulink model

3. Methodology

In this paper different scenarios have been investigated under various faulty conditions. A distributed line is used to create fault at different distances in transmission line. Line to ground, line to line and LLL-G fault is created to understand different fault condition and the behavior of system was analyzed. The inputs to relay are three phase voltage and current. Further, discrete fourier is used to convert it in fundamental form in the measurement unit (J. R. Camarillo-Peñaranda, 2020). Output from measurement unit is given to the impedance computation block to measure impedance using some mathematical equations for different fault conditions. Further fault detection is made by comparing the values with the radius of zone 1and zone 2, then as per the output of fault detection unit tripping command is activated. Fig. 14.2 shows different units inside MHO digital distance relay.

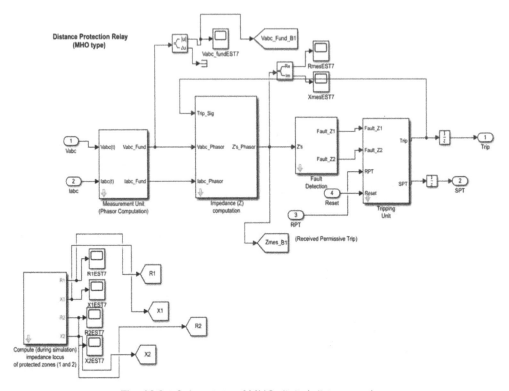

Fig. 14.2 Subsystem of MHO digital distance relay

4. Results

The model was tested by performing various faulty conditions as mentioned in Table 14.1 and operation of MHO digital distance relay was analyzed. Line-to-ground fault is created at a distance of 100 km length i.e. at the middle of protected zone and as this fault is equidistant from the two breakers, both the breaker are equally responsible for protecting the transmission line 1. This test confirms that relay1 and relay2 senses the fault and trip simultaneously. Figure 14.3 illustrates the trajectory of impedance observed by the relay 1. Figure 14.4 and 14.5 shows current and voltage waveform at bus 1 and 2 respectively. In the second case, a fault is created between line A and B taking x=20 km in model. As a result of this relay B1 activate first and transmits the Permissive Trip (SPT) signal to relay 2. Impedance locus, current and voltage waveforms are as shown below. In the third case fault is created between all three lines and ground at a distance x=20 km in first section of line 1 and relay 1 fails to operate while relay 2 get activated. In the fourth case, a fault is created on line 2 which lies outside the protection zones of both relays so no tripping takes place. This test confirms that neither relay1 nor relay2 initiates any protective action.

Table 14.1   Relay performance validation

Test	Fault Location	Fault Type	Observation
1	L_1, x=100km	A-G	Line to ground fault: Both relay trips
2	L_1, x=20km	AB	Line to Line fault:B_1 trips faster than relay B_2
3	L_1, x=20km	ABC	Trip in zone 2 of relay B2
4	L_2, x=20km	ABC	No Tripping

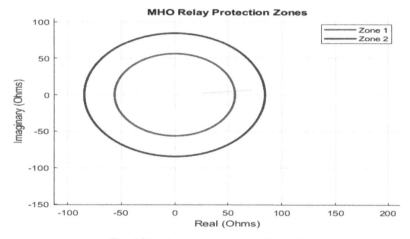

Fig. 14.3   Impedance locus of relay 1

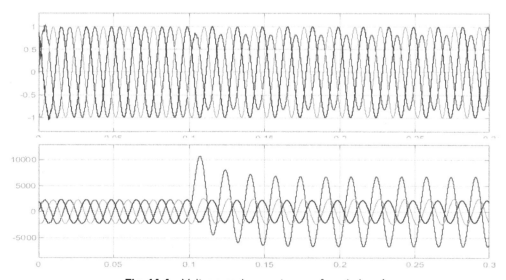

Fig. 14.4   Voltage and currents waveform in bus 1

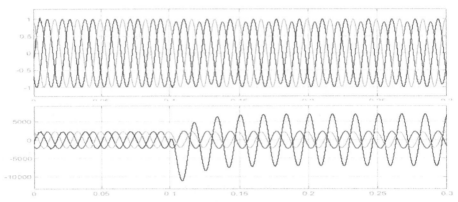

Fig. 14.5   Voltage and currents waveform in bus 2

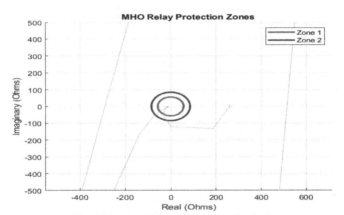

Fig. 14.6   Impedance locus for LL fault

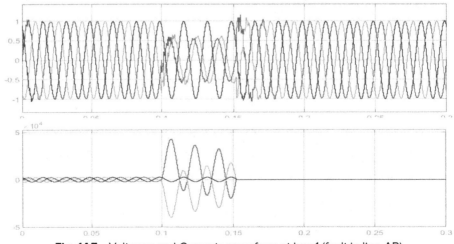

Fig. 14.7   Voltages and Currents waveform at bus 1 (fault in line AB)

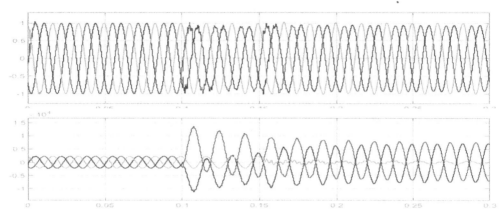

Fig. 14.8 Voltages and currents waveform at bus 2 (fault in line AB)

The proposed methodology is validated through a series of case studies. The results demonstrate the effectiveness of real-time modelling and testing of MHO distance protection relay by identifying potential issues such as mal-operation, incorrect settings and sensitivity to system transients.

5. Conclusion

This research paper presents an innovative method for the real-time modelling and validation of distance protection relays. By utilizing digital simulation techniques for MHO distance relay, it offers a precise and controlled testing environment that faithfully emulates real-world power system conditions. In the future, potential research avenues could explore the integration of machine learning and artificial intelligence for automated relay testing and fault diagnosis. Furthermore, the establishment of standardized testing procedures and performance benchmarks for distance protection relays can further propel advancements in this field.

REFERENCES

1. Brunelle, A. D. (2020). Real Time Modeling and Testing of Distance Protection Relay based on IEC 61850 Protocol. *canadian journal of electrical and computer engineering*, 157-162.
2. D. Pal, B. M. (1 Dec.1, 2017). Synchrophasor Assisted Adaptive Relaying Methodology to Prevent Zone-3 Mal-Operation During Load Encroachment, *IEEE Sensors Journal*, 7713-7722.
3. D.V.Coury, D. C. (1998). Artificial Neural Network Approach to Distance Protection of Transmission Lines . *IEEE Transactions on Power Delivery*, 102-108.
4. D.Tholomier.(2004).Distance Protection and Dynamic Loading of Transmission Lines. *Power Engineering Society General Meeting*, 100-105.
5. Erezzaghi M. E., C. P. (2003). The Effect Of High Resistance Faults On A distance relay. *Power Engineering Society General Meeting* , 2128-2133.
6. Fitiwi, D. K. (2009). Assessment of ANN based Auto-Reclosing scheme Developed on Single MachineInfinite Bus Model with IEEE 14-Bus System Model Data. *TENCON Publication*, 1-6.

7. H. Khorashadi-Zadeh, M. R. (2004). An ANN based Approach to Improve the Distance Relaying Algorithm . *IEEE Conference on Cybernetics and Intelligent Systems Singapore* , 1374-1379.

8. J.R.Camarillo-Peñaranda,M.A.(2020).Hardware-in-theloop testing of virtual distance protection relay. *IEEE/IAS 56th Ind. Commercial Power Syst. Tech. Conf* , 1–6.

9. M.Sreeram, P. Raja. (2013). Implementation of DSP based numerical three-step distance protection scheme for transmission lines, *INDICON*, 1-6.

10. S. M. Hashemi,M.Sanaye-Pasand.(Apr. 2019).A new predictive approach to wide-area out-of-step protection. *IEEE Trans. Ind. Informat*,1890–1898.

11. Sherwali, A.A.(2009).Modelling of Numerical Distance Relays Using MATLAB. *2009 IEEE Symposium on Industrial Electronics and Applications (ISIEA 2009)*, 389-393.

Emerging Technologies and Applications in Electrical Engineering –
Prof. Dr. Anamika Yadav et al. (eds)
© 2024 Taylor & Francis Group, London, ISBN 978-1-032-82568-7

A Comprehensive Review on Switched Capacitor Based DC-DC Converters

15

Jeetender Vemula[1]

Assistant Professor, EEE Dept.,
Vignana Bharathi institute of technology, Hyderabad

E. Vidya Sagar[2]

Professor, Head,
EEE Dept., Osmania University, Hyderabad

B. Sirisha[3]

Assosiate Professor,
EEE Dept., Osmania University, Hyderabad

Vadthya Jagan[4]

Professor, EEE Dept.,
Vignana Bharathi institute of technology, Hyderabad

ABSTRACT: DC-DC converters with voltage buck and boost capabilities are extensively used in a great number of applications. The purpose of the paper is to present detailed analysis of DC-DC converters and their classifications. This analysis shows that popular Switched-Capacitor (SC) converters achieve the limits of utilization for reactive components. The primary configurations of switched capacitors (SCs) utilised in DC-DC power conversion are reviewed in this document. The design of efficient switched-capacitor based converters is the key subject of this literature. The main topologies of SCBC and their performance under steady state are discussed. This review compares the cost, applications, reliability, advantages, disadvantages and efficiency of multistage converters such as; M-SCBC, M-SIBC, Transformer and coupled inductor-based family, Luo converters, Multilevel configuration, X-Y configuration family. The SC converter is clearly a good choice for future integrated DC-DC converters with great power density.

KEYWORDS: High voltage gain, DC-DC converter, Switched capacitors, Multi-stage converters

[1]jeetender.eee@gmail.com, [2]evsuceou@gmail.com, [3]sirishab2007@yahoo.com, [4]jagan.iitr@gmail.com

DOI: 10.1201/9781003505181-15

1. Introduction

Automobile manufacturers are now concentrating their efforts on new advancements such as hydrogen production and Fuel Cell (FC) technology to boost efficiency. And we can see that these days everyone are facing difficulties on petroleum because of insufficiency of resources and also the growth in production of injurious gases. The energy conversion is handled by the power electronic circuits. The types of fuel cells, the number of stacks, and the size of the fuel cells all contribute to the creation of electrical power fuel cells.

DC-DC topologies are used to buck and boost the output voltages. Non-isolated, isolated and Coupled Inductor based DC-DC converters are the three primary types of DC-DC converters. A non-isolated type configurations generally shares a common ground between input and output. While isolated topology, input and output terminal are electrically isolated. Isolated and Non-isolated are divided into two types; Unidirectional and Bidirectional configuration. Unidirectional converters are further categorized into three types, Single-stage Power Converter (SPC), Multistage Power Converters (MPC), DC-DC Multiphase Converter.

Single stage Power Converters constitutes of two basic topologies among which one is buck converter and the other one is boost converter. A cascaded assembly of these two is buck-boost configuration. These converters perform step-up and step-down operations on the input voltage [1]. Multistage Power converters are designed by the hybridisation of conventional converters. CUK, SEPIC and ZETA are the multistage power converters. Other important Multistage power converters are Cascaded and Quadratic boost converters, Switched Inductor Based Converters family (SIBC), Switched Capacitor Based Converter family (SCBC), Transformers and Coupled Inductor topologies, Luo Converters family, Multilevel configuration, X-Y configuration family. DC-DC Multiphase Converters further classification is beyond the scope of the article. The purpose of the literature is to extend the idea of introducing a switched-capacitor circuit into a multistage converter; Cuk, Zeta and SEPIC. Figure 15.1 illustrates a flowchart that describes the further classification of power converters.

2. Single Stage Power Converters

Single stage Power Converters are the conventional converters which step up, step down the terminal voltages. A Buck Conventional Converter steps down the terminal voltage. The Boost Conventional Converter operates to step up the input voltage. Buck-Boost Conventional Converter performs both step-up and step-down operations on the applied voltage.

2.1 Buck Conventional Configuration

Buck converter constitutes of an active switch, a diode, inductor and a capacitor. The switch is used to control the power flow in the circuit. Figure 15.2(a) depicts the circuit diagram of buck conventional converter.

Duty Ratio is the proportion of time during which a switch is in the on position over the total time period. The duty ratio D value can range from 0 to 1. When D is equal to 0, there is no voltage, but when D is equal to 1, the full input voltage is applied across the load. The buck

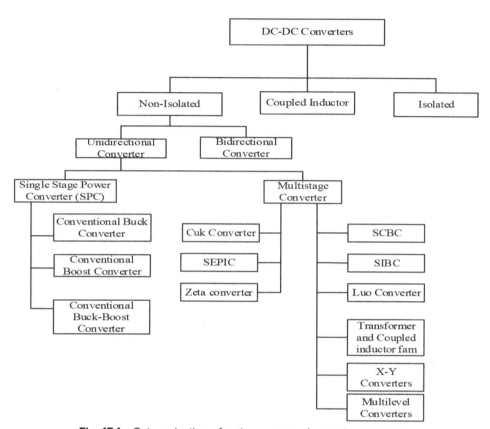

Fig. 15.1 Categorization of various power electronic converters

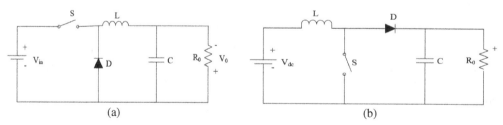

Fig. 15.2 Circuit configurations of (a) buck conventional converter and (b) boost conventional converter

converter operates when the duty cycle (D) is between 0 and 1. Equation (1) specifies the voltage gain.

$$M(D) = D \qquad (1)$$

2.2 Boost Conventional Converter

A fundamental type of switch-mode converter is a boost converter, also mentioned to as a step-up circuit. The converter increases the input voltage. It steps up DC voltage from a low level to a high level while reducing current from large to small, thus maintaining the constant supplied power. Boost converters are easy to design and require only a limited amount of components to be integrated into the system. An inductor, a diode, a capacitor, and a semiconductor switch are the components that make up the system. The illustration of the circuit diagram of a traditional boost converter may be found in Fig. 15.2(b).

The major benefit of this converter is, it poses high efficiency and in some of the boost converters, we can gain efficiency of 99%. The voltage gain is mentioned in equation (2).

$$M(D) = \frac{1}{1 - D} \tag{2}$$

2.3 Buck-Boost Conventional Converter

The two basic topologies available are buck circuit and the boost circuit. A buck-boost configuration is the cascade connection of these two converters. In this converter the output voltage can be either higher or lower than the input voltage. Hence, it is called buck-boost converter. This converter is also called as an inverting converter as it provides a reversal in the output voltage polarity without a transformer. Figure 15.3(a) displays a schematic topology of a buck-boost converter.

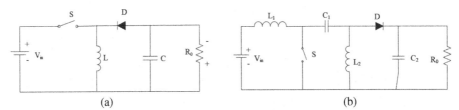

(a)   (b)

Fig. 15.3 Circuit configurations of (a) Buck-boost Conventional topology and (b) Cuk topology

The buck-boost network uses unregulated DC voltage as input to produce a controlled DC output voltage with negative polarity relative to the input voltage's common terminal. The DC output voltage of a buck-boost Conventional converter is controlled by adjusting the duty ratio. Equation (3) specifies the voltage gain.

$$M(D) = -\frac{D}{1 - D} \tag{3}$$

3. Multistage Power Converters

The Multistage Power Converters are derived by the hybridisation of conventional converters. The Cuk, SEPIC, Zeta converters. These DC-DC multistage power converters are grouped into three primary groups derived on the basis of boosting stages and conversion ratio.

3.1 Low Voltage Step-up MPC

These are SEPIC, Zeta and Cuk converters. These are also known as derived or two stage converters as the converters derived by hybridising two conventional converters [2-4].

Cuk Converter

Cuk converter is a DC-DC MPC with a step-up/down inverting output created by combining popular Boost and Buck converters. The basic structure of Cuk converter can be seen in Figure 15.3(b). In a Cuk Converter the front-end acts a typical boost converter, while the back-end acts as a typical buck converter.

$$M(D) = -\frac{D}{1-D} \tag{4}$$

SEPIC

Single Ended Primary Inductance Converter (SEPIC) is a multistage power converting DC-DC MPC derived by hybridisation of conventional boost and buck-boost converters. The circuit diagram of SEPIC is designed as in Fig. 15.4(a).

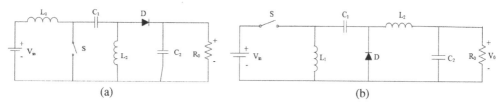

(a)   (b)

Fig. 15.4   Circuit topologies of (a) SEPIC topology and (b) Zeta topology

A SEPIC converter's front-end circuit is a typical boost circuit, while the back end is a typical buck-boost circuit. In equation (5), the voltage gain is mentioned.

$$M(D) = \frac{D}{1-D} \tag{5}$$

Zeta Converter

The Zeta is a non-inverting DC-DC converter that combines features of the original buck-boost and buck converters to step up or step-down voltage levels. Figure 15.4(b) shows the fundamental circuit diagram of the Zeta converter.

A Zeta converter's front-end is a typical buck-boost converter, while the back end is a typical buck converter. In equation (6), the voltage gain is mentioned.

$$M(D) = \frac{D}{1-D} \tag{6}$$

3.2 Medium Voltage Step-up MPC

These are Quadratic or Cascaded boost converters. To achieve a moderate voltage, they are connected in cascaded form [7]. The strategies employed by QBC are using a single active switch and also utilising the number of passives switching devices. The main drawbacks of the QBC are voltage stresses across the controlled switch, complexity and low efficiency [4].

3.3 High Voltage Step-up MPC

They hold high voltage gain intended by hybridisation of DC-DC configurations. Of different switched capacitor and inductor family Switched Capacitor Based Converters (SCBC) [9] offer new possibilities for attaining high DC voltage ratios.

4. Switched-Capacitor Cells

Now-a-days, the Switched-Capacitor (SC) based configurations gained more attractions to generate a high voltage gain in many step-up applications. The gain ratio is determined by the number of capacitors and their configuration in the converter [1]. In a SC cell, all the capacitors are charged parallelly and discharges its energy in series. Aside from that, some of the SC circuits use the charge pumping idea, which involves transferring energy from one capacitor to another, and is hence referred to as a charge pump network. Figure 15.5 depicts various SC cell circuitries.

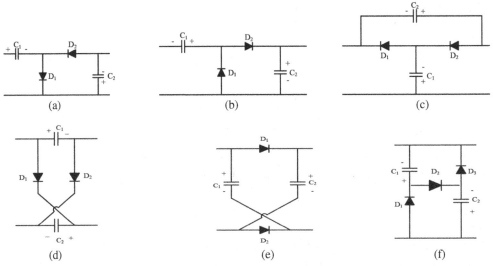

Fig. 15.5 Circuit configurations of switched capacitor cells

5. Converters based on Switched Capacitors

Switched Capacitors (SC) circuits have previously been imposed for different voltage step-up operations in order to obtain high voltage gain. Using SC cells in boost configuration, a number of high step-up DC-DC MPCs have been explored in this literature [5]. A general SCBC circuit diagram is depicted Fig. 15.6(a).

It's an SCBC, with a diode and SC in the Zeta configuration. The FEB-BC, SC, and C filter stages are included in this three-stage step-up converter. In comparison to conventional boost topologies with minimal switching losses, this converter has high negative voltage gain.

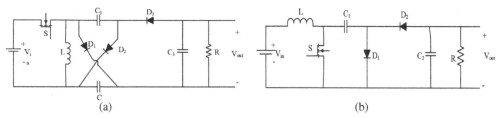

Fig. 15.6 Circuit configurations of (a) Inverting SC with FEB-BC and C-filter and (b) TSHV Front-end boost converter

5.1 Switched-Capacitor and Hybrid Switched-Inductor

Using SC stages in traditional boost converters, a number of high step-up DC-DC MPCs have been investigated [1]. The M-SCBC (Multistage Switched Capacitor Based Converter) is praised for its simple structure and modular approach. With an intermediate duty cycle, the suggested ultra-converter achieves a high voltage gain. In boost converter [4], HSI-SC is put in the middle with a voltage multiplier. As a result, transmission losses are decreased, and it is appropriate for high-frequency operation. The structure of converter can indubitably be expanded to N-stages by using more-number of multiplier stages. To reduce output ripples, the C-filter is used, and intermediary HIS-SC stages are used to increase the voltage with a high value.

5.2 Conventional Converter with Switched-Capacitor and Hybrid Switched-Inductor

Switched capacitor converters are not the same as traditional capacitor converters. The resistance of a power converter dictates its performance, and it is typically significantly more than the impedance of a converter that stores energy via inductance. Controlling the output voltage can be done in a variety of ways. SI-C circuits have been constructed from switched-capacitor cells. New topologies have been proposed based on the duality of a SC cell, SI cell, duality of current-driven converters and voltage-driven converters. The steady state behaviour of a SC converter can be represented as an ideal transformer with omitting frequency dependent parasitic losses , and series output impedance [2]. The conversion ratio is dictated by the converter topology and is represented by the turns ratio n.

5.3 Front-end Boost and Buck Boost M-SCBC Converter

A number of M-SCBC topologies have been developed with frond-end boost and buck-boost structure. The FEB and switched capacitor are combined to create the TH-SVBC. The inverting power converter are made use for medium to high boost. It comprises of one inductor, two capacitors, and two diodes. Equations (7) shows the voltage gain of Converter in CCM mode. Figure 6(b) comprises the circuit diagram of M-SCBC converter [3].

$$M(D) = \frac{-1}{1-D} \tag{7}$$

Figure 15.7(a) shows the power circuit for an inverted switched capacitor converter with an inductor on the input side. This circuit includes LC filter, a switching capacitor stage and an input inductor stage. A negative voltage gain is provided by the converter. This converter requires two inductors, two diodes, three capacitors, as well as a single active switch. The converter's voltage gain is specified in equation (8).

The power circuit for a switched-capacitor converter with an inductor at the input side is shown in Fig. 15.7(b). The three-stage converter structure is made by the LC filter, switched capacitor stage and input inductor stage. A positive voltage conversion ratio is provided by the converter. This converter requires three capacitors, two inductors and two diodes, as well as a single control switch. Equation (9) presents the voltage gain of the converter [6].

$$M(D) = \frac{-(1+D)}{1-D} \tag{8}$$

$$M(D) = \frac{1+D}{1-D} \tag{9}$$

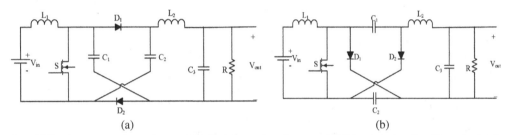

(a)   (b)

Fig. 15.7 Circuit configurations of (a) non-inverting front-end SC boost circuit with output LC filter and (b) front-end inverting SC boost converter with lc filter.

6. Comparison Graph

Figure 15.8 shows the voltage conversion ratios of different converters in which it elaborates the converter performance at different duty ratios.

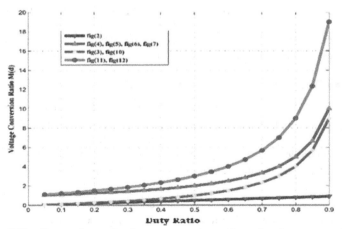

Fig. 15.8 Comparison of voltage conversion ratios of various converters

7. Conclusion

In this paper there is a brief explanation done on DC-DC converters classifications and spell out about the DC-DC converter family. Concisely described about conventional converters-Buck, Boost, and Buck-Boost converters and addressed their voltage conversion ratios. And have shortly explained about the hybridization of conventional Converters-Cuk, SEPIC, Zeta converters with the required voltage conversion ratio equations. A brief note on switched capacitor cells with circuit diagrams and their applications in converters been clearly signified. This literature speaks about conventional converters with switched capacitors and Hybrid switched inductor capacitor cell. Conclusively, these high-efficiency configurations with high step-up gain can cover a wide variety of voltages are in great requirement in the automobile industry.

8. Comparison Table

Table 15.1   Cost, applications, reliability, advantages, disadvantages, efficiency

DC-DC Converter	Cost	Applications	Reli-ability	Advantages	Disadvantages	Effi-ciency
M-SIBC	High	1. The low-duty vehicle's powertrain 2. Loads with high voltage	Very High	1. High voltage gain 2. merges with any configuration.	1. More number of Inductors 2. Weight	High
M-SCBC	Avg	1. The low-duty vehicle's powertrain 2. Loads with high voltage	High	1. Small in size 2. Less weight 3. Modular	1. Low voltage regulation 2. High starting current 3. More number of capacitors	Very High
Luo Converter Family	High	1. Large power 2. Loads with high voltage	High	1. ultra step-up gain 2. Size is average	1. More number of inductors, diodes and capacitors 2. Complexity in control	Average
Multilevel Converter	Avg	1. high duty vehicles power train 2. Large power 3. Loads with high voltage	Average	1. Self Balanced 2. Less number of switches	1. More number of capacitors and diodes	Very High
Coupled Inductor and transformer-based converters	High	1. The power train of low duty vehicles 2. Large power 3. Loads with high voltage	Average	1. High power 2. soft switching 3. Large step-up gain	1. Bulky 2. More losses because magnetic coupling is high	High
X-Y converter Family	Avg	1. Large power 2. Loads with high voltage	High	1. Single switch 2. High conversion ratio 3. Extendable structure	1. More number of diodes 2. Discontinuous current.	High

Table 15.2 Comparisons between modified and traditional converters

S. No.	Converter type	Fig. No.	No. of Switches	No. of Inductors	No. of Capacitors	No. of Diodes
1	Single Stage Converter (Buck)	2(a)	1	1	1	1
2	Single Stage Converter (Boost)	2(b)	1	1	1	1
3	Single Stage Converter (Buck-Boost)	3(a)	1	1	1	1
4	Two Stage Converter (Cuk)	3(b)	1	2	2	1
5	Two Stage Converter (SEPIC)	4(a)	1	2	2	1
6	Two Stage Converter (Zeta)	4(b)	1	2	2	1
7	Multi Stage with C Filter	5	1	1	3	3
8	TS-HVBC With LC filter	6(a)	1	1	2	2
9	Inverting SCBC with input side inductor and LC filter.	6(b)	1	2	3	2
10	Non-inverting SCBC with FEBC and LC filter.	7	1	2	3	2

REFERENCES

1. Bhaskar, Mahajan Sagar, Vigna K. Ramachandaramurthy, Sanjeevikumar Padmanaban, Frede Blaabjerg, Dan M. Ionel, Massimo Mitolo, and Dhafer Almakhles. "Survey of DC-DC non-isolated topologies for unidirectional power flow in fuel cell vehicles." IEEE Access 8 (2020): 178130-178166.

2. Bhaskar, Mahajan Sagar, Pandav Kiran Maroti, and Draxe Kaustubh Prabhakar. "Novel topological derivations for DC-DC converters." International Journal of Computational Engineering & Management 16, no. 6 (2013): 49-53.

3. Divya Navamani, J. "High step-up DC-DC converter by switched inductor and voltage multiplier cell for automotive applications." Journal of Electrical Engineering and Technology 12, no. 1 (2017): 189-197.

4. Li, Shouxiang, Yifei Zheng, Bin Wu, and Keyue Ma Smedley. "A family of resonant two-switch boosting switched-capacitor converter with ZVS operation and a wide line regulation range." IEEE Transactions on Power Electronics 33, no. 1 (2017): 448-459.

5. Marangalu, Milad Ghavipanjeh, Seyed Hossein Hosseini, Naser Vosoughi Kurdkandi, and Arash Khoshkbar-Sadigh. "A new five-level switched-capacitor-based transformer-less common-grounded grid-tied inverter." IEEE Journal of Emerging and Selected Topics in Power Electronics (2022).

6. Maroti, Pandav Kiran, Mahajan Sagar Bhaskar Ranjana, and B. Sri Revathi. "A high gain DC-DC converter using voltage multiplier." In 2014 International Conference on Advances in Electrical Engineering (ICAEE), pp. 1-4. IEEE, 2014.

7. Nagi Reddy, B., Chandra Sekhar, O., Ramamoorty, M.. "Implementation of zero current switch turn on based buck-boost-buck type rectifier for low power applications" International Journal of Electronics, 106(8), pp. 1164–1183.

8. Nagi Reddy, B., M. Bharathi, M. Pratyusha, K. S. Bhargavi, and B. Srikanth Goud. "Design of a novel isolated single switch AC/DC integrated converter for SMPS applications." International Journal of Emerging Trends in Engineering Research 8, no. 4 (2020): 1111-1119.

9. Sivani, L. Sri, L. Nagi Reddy, B. K. SubbaRao, and A. Pandian. "A new single switch AC/DC converter with extended voltage conversion ratio for SMPS applications." Int. Journal of Innovative Technology and Exploring Engineering 8, no. 3 (2019): 68-72.

Emerging Technologies and Applications in Electrical Engineering –
Prof. Dr. Anamika Yadav et al. (eds)
© 2024 Taylor & Francis Group, London, ISBN 978-1-032-82568-7

Optimizing Solar Power: ANFIS-Based MPPT for Photovoltaic Systems

16

MD Mujahid Irfan[1]
SMIEEE & Assistant Professor, Dept. of EEE,
SR University – Warangal

M. Raghavendra[2]
UG Student, Dept. of EEE, SR University – Warangal

Chidurala Saiprakash[3]
Assistant Professor, Dept. of EEE, SR University – Warangal

Md. Madhar Ahmed[4]
UG Student, Dept. of EEE, SR University – Warangal

ABSTRACT: Solar energy is an abundant and sustainable source of power, but harnessing its full potential requires efficient management of PV systems. This paper presents a novel approach for maximizing the energy output of PV systems through the integration of ANFIS into a MPPT controller. By combining the strengths of artificial neural networks and fuzzy logic, the ANFIS controller provides an effective and computationally efficient solution to the MPPT challenge. The presented system ensures that the PV system operates at its maximum power point under varying conditions, offering higher efficiency and power quality while meeting grid requirements. The results of simulations demonstrate the effectiveness of the ANFIS-based MPPT system, showcasing improved performance in extracting energy from PV arrays compared to traditional MPPT techniques. This research contributes to the advancement of solar energy generation systems, enabling them to deliver higher energy yields with greater precision, thereby promoting the integration of renewable energy sources into the power grid.

KEYWORDS: Photovoltaic (PV) Systems, Adaptive Neuro Fuzzy Inference System (ANFIS), Maximum Power Point Tracking (MPPT)

[1]mm.irfan@sru.edu.in, [2]margamraghavendra2112@gmail.com, [3]saiprakash.ace@gmail.com, [4]madharahmed521@gmail.com

DOI: 10.1201/9781003505181-16

1. Introduction

The global dependence on traditional fossil fuels has resulted in a complex network of interrelated problems, including the greenhouse gas emissions, degradation of the environment, and the quick depletion of conventional energy supplies. The burning of fossil fuels releases carbon dioxide into the atmosphere which in turn causes global warming and alters our climate [1][2]. This clean, renewable energy source offers a possible path ahead in the search for sustainable energy by lowering greenhouse gas emissions and reducing our need on finite fossil fuels. The wind energy offers a scalable and environmentally responsible method of producing power [5][6]. With so many benefits, solar energy is a sustainable energy source that has developed into growing in popularity as a solution to meet modern energy demands. Among the MPPT techniques that are most frequently employed is Perturb & Observe (P&O) [3][4]. It makes periodic adjustments to the operating voltage of solar panels and tracks the resulting change in power output. Another MPPT method is Incremental Conductance (IncCond). It tracks the highest power point using the rate of change of conductance (di/dV) [8][9]. The mostly used technique is Fuzzy Logic-based MPPT which is based on linguistic principles, this research proposes the ANFIS [7], which blends the fuzzy logic and artificial neural networks to maximize the performance of renewable energy systems [9][10], which can be problematic in applications where transparency and interpretability are essential [11][12]. Given changes in system behaviour or environmental conditions [13], ongoing maintenance and upgrades are also essential to guarantee the ANFIS model's continued efficacy [14][15].

2. ANFIS

The adaptive neuro-fuzzy interference system integrates fuzzy logic and ANN. Based on the interconnection of the variables; an ANN can be used to train the fuzzy controller to determine

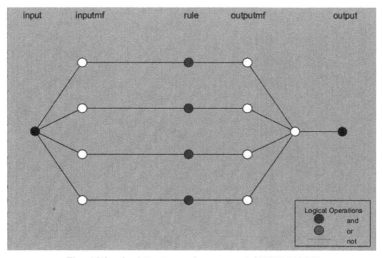

Fig. 16.1 Architecture of proposed ANFIS MPPT

the exact membership functions for the variables. Through continuous learning and adaptation made possible by this integration, the photovoltaic system is able to maintain its maximum power output even in the face of variations in the temperature and sunshine intensity [16][17]. The fuzzy logic part of ANFIS uses a language rule set to represent the relationship between the desired output and the input variables. Real-time data from the photovoltaic array, including factors like solar irradiance and panel temperature, truly has collected via ANFIS-based MPPT [18].

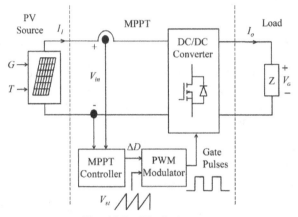

Fig. 16.2 Block diagram

The ANFIS has adaptive learning capabilities, it can adjust to shifting environmental circumstances and changes in the properties of the PV system with efficiency. The flexibility guarantees excellent precision, allowing for precise control of the PV system to run at or close to its maximum power point, improving the efficiency of energy conversion [19].

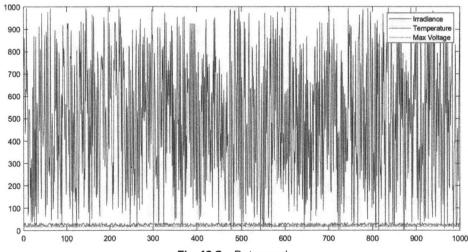

Fig. 16.3 Data graph

Because solar systems are inherently nonlinear, it can effectively track the fluctuating maximum power point and optimize energy production. It also improves performance even in harsh environments, lowers energy loss, and increases efficiency. Due to its versatility across various photovoltaic technologies, ease of parameter adjustments, and quick reaction to environmental changes, ANFIS-based MPPT is a flexible and useful instrument for maximizing solar energy production, enhancing its sustainability and efficiency [20].

Comparing the conventional MPPT techniques like P&O, IncCond, Hill Climbing algorithms, ANFIS offers a number of advantages. The technique struggle to adjust to shifting environmental conditions because they frequently rely on heuristic and trail-and-error approaches for parameter tweaking. Because of ANFIS' superior accuracy, the PV system will run close to its maximum power point, increasing its efficiency of energy conversion. In the Fig. 16.4, It has FIS output of the trained data, where the Output is plotted on the Y-axis, and Index is plotted on X-axis. This trained data is in the FIS file format which is uploaded in the fuzzy controller block for the MPPT technique operation [10][11].

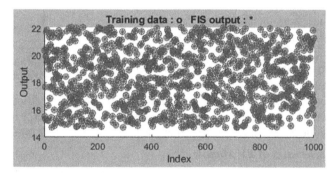

Fig. 16.4 FIS output

In the Fig. 16.5, the Training data is plotted as Output on the Y-axis, and data set index on X-axis. In the Fig. 16.6, the training error of the data which is trained of value 2.264×10^{-7}, it is nearly zero. The data trained and monitored so that the error is minimum.

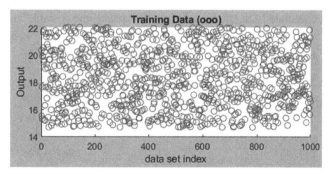

Fig. 16.5 Training data

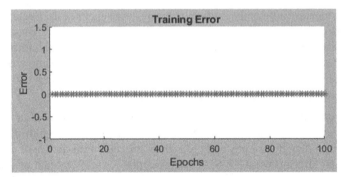

Fig. 16.6 Training error

2.1 ANFIS based MPPT

Solar PV

PV Module is the necessary tool made to collect sunlight and transform it into electrical power. These panels, which are culminated with photovoltaic cells that are usually composed of semiconductor materials like silicon, use the photovoltaic effect to produce electricity

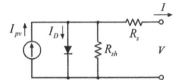

Fig. 16.7 Internal circuit of solar PV module

when sunlight touches them. They have a junction box for secure electrical connections, a strong aluminium frame, tempered glass for protection, and a back sheet. Light absorption is improved by anti-reflective coatings on the front surface.

P-I and V-I Characteristics of PV Module

The solar PV module's Power-Voltage (P-V) and Voltage-Current(V-I) characteristics are essential to comprehending its efficiency and performance. The link between electrical power output and voltage across the module terminals is depicted by the P-V Curve.

Boost Converter:

The Boost Converter represents the power conditioning unit that is responsible for converting the variable and unregulated DC output from the PV panel into a stable and controlled DC

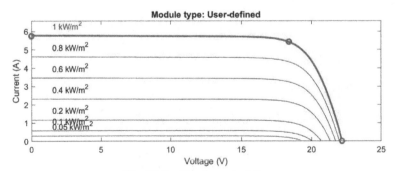

Fig. 16.8 V-I characteristics

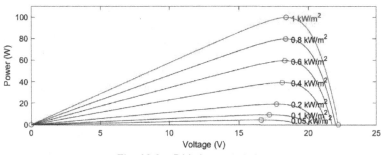

Fig. 16.9 P-V characteristics

voltage. This component typically includes a power electronic circuit that regulates the voltage to match the load or battery requirements. In your context, it ensures that the power extracted from the PV panel is efficiently delivered to the load or grid.

PWM Generator:

The PWM Generator in Simulink generates pulse-width modulated signals based on the control signals from the MPPT Fuzzy Controller. PWM is commonly used in power electronics to control the duty cycle of switches in the Boost Converter. The PWM signals determine the on/off states of the switches, regulating the output voltage and power flow to the load.

3. Methodology

Simulink Model:

When the Sun rays falls on the solar panel, the solar panel converts the solar energy into electrical energy, low voltage DC and then the voltage is fed to the boost converter, which boosts the voltage to a higher value. The boosted voltage is fed to the load or battery for utilizing the electric energy. So, the boosted voltage might be variable sometimes as the weather conditions

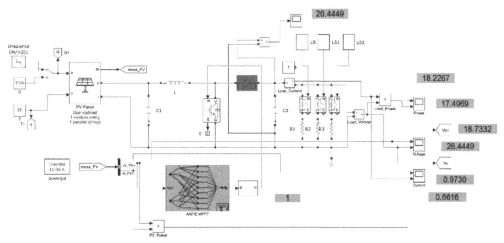

Fig. 16.10 ANFIS-MPPT simulink model

are cloudy, and during night, the voltage might not be maximum. So, to increase the voltage to its maximum, we need to use MPPT Techniques, the MPPT is Maximum Power Point Tracking, there are many techniques for improving the power output of the PV system. Such as Perturb & Observe Method, Incremental Conductance Method, Artificial Neural Network, etc. The technique we are using in this model is, ANFIS which is a hybrid system having Artificial Neural Network and Fuzzy System. The voltage, current and power data are taken from the solar panel and the model is trained with the data. ANFIS output is fed to the PWM Converter to control the switching operation of the voltage fed to the PWM Converter as shown in Fig. 16.10. The boost converter gate pulse is taken from the PWM Converter, and the Voltage gets boosted and the boosted voltage is fed to the load. The ANFIS algorithm will make sure that at any point of time, the voltage delivered to the load to be maximum regardless of the environmental conditions.

4. Results

In the Fig. 16.11, the Load Power is the Power coming from the ANFIS controller which is tuned and boosted, the PV Power is the Power coming directly from the Solar Panel. In the Fig. 16.12, The Load Voltage is the Voltage coming from the ANFIS controller which is tuned and boosted, the PV Voltage is the Voltage coming directly from the Solar Panel. The voltage from the ANFIS model is 40% greater than the voltage coming from the Solar panel.

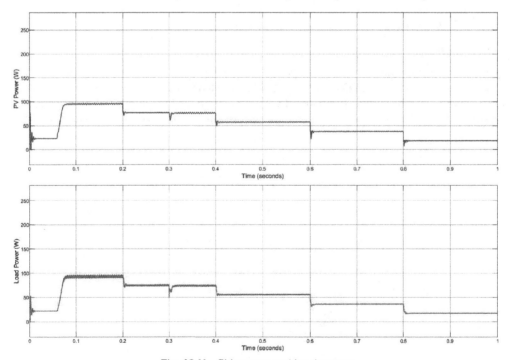

Fig. 16.11 PV power and load power

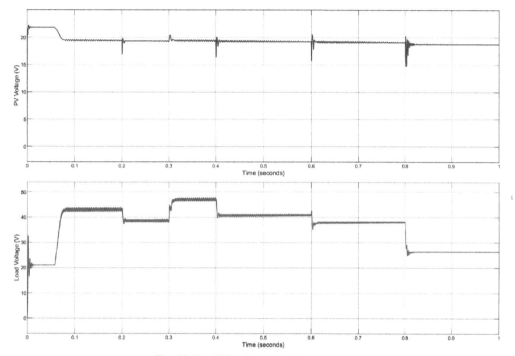

Fig. 16.12 PV voltage and load voltage

In the Fig. 16.13, The Load Current is the Current coming from the ANFIS controller which is tuned, the PV Current is the Current coming directly from the Solar Panel. The variation in the graphs describes that the Current generated is varying with the environmental conditions. The Current from the ANFIS model is 40% lesser than the Current coming from the Solar panel.

Figure 16.14 projects the graph of Duty Ratio on Y-axis and Time on X-axis from the PWM Converter. The switching frequency of the PWM Converter is set to 10000 Hz, where the IGBT switch turns ON and OFF 10000 times during the operation of the Boost converter.

5. Conclusion

The paper introduces an ANFIS based MPPT controller for solar photovoltaic modules. This study evaluates the ANFIS based MPPT controller's performance under two different output situations and provides a thorough analysis of the control structure. The results unambiguously show that the suggested controller is effective in maximizing the solar PV module's available power output. Furthermore, the study shows notable gains in output voltage and power, especially when the PV module runs under less-than-ideal load circumstances. The simulation findings confirm that the controller operates reliably and efficiently in a variety of load scenarios and solar irradiation levels, further underscoring its suitability for real-world applications. This research offers valuable insights into the practical implementation of ANFIS-based MPPT controllers, contributing to the advancement of solar energy harvesting technology.

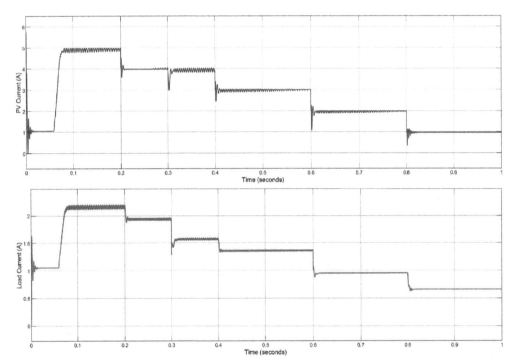

Fig. 16.13 PV current and load current

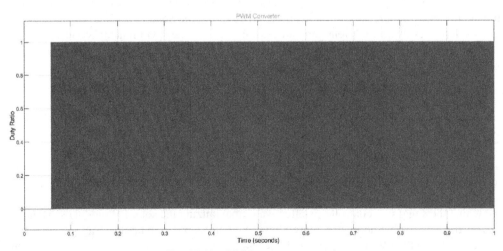

Fig. 16.14 PWM converter graph

REFERENCES

1. A. Iqbal, (2010), "Adaptive Neuro-Fuzzy Inference System based Maximum Power Point Tracking of a Solar PV Module".
2. Abd El Hakim Ali Elagori and M. Emin Tacer, (2017), "Implementation and Evaluation of Maximum Power Point Tracking (MPPT) Based on Adaptive Neuro-Fuzzy Interference System for Photovoltaic PV System".
3. Adilson Puna, and A.B.K. Bhattacharya, (2021), "Modelling and Simulation of PV Connected with Step-up Converter based on ANFIS MPPT Controller".
4. Epyk Sunarno, Indhana Sudiharto, and Dian Yolanita, (2023), "Design MPPT with ANFIS Method on Zeta Converter with DC Load".
5. Gerome I. Pagatpatan, Jessa P. Pagkaliwangan, Diether Kyle A. Torcuato, Glenn C. Virrey, John Peter, M. Ramos, and Ira C. Valenzuela, (2020), "Neuro-Fuzzy based MPPT for Solar PV Panel".
6. I. Made Ari Nratha, I. Made Ginarsa, Agung Budi Mulijono, Ida Ayu Sri Adyani and Sultan Sultan, (2023), "ANFIS based MPPT for Rooftop Solar Panels Connected to Single Phase Power Grid".
7. Mohak Jain, and Bharat Bhushan, (2019, "Performance Analysis of FIS and ANFIS based MPPT for Solar PV System with Boost, SEPIC and CUK Converter Topologies".
8. Muhammad Shahid and Naghma Noorani, (2016), "Efficiency Improvement of Grid Connected PV using ANFIS based MPPT".
9. Muhammed Y. Worku and M.A. Abido, (2016), "Grid Connected PV system using ANFIS based MPPT Controller in Real Time".
10. N. Sai kumar, V. Tharun Yadav, S. Sai krishna, M.A. Sameer and G. Upendra Rao, (2022), "An Experimental Estimation of Hybrid ANFIS-PSO based MPPT for PV Grid Integration under Fluctuation Sun Irradiance".
11. Needhu Varghese, and Reji P., (2019), "ANFIS based Maximum Power Extraction from Standalone Photovoltaic System for a DC House".
12. Pascal Kuate Nkounhawa, Dieunedort Ndapeu, and Bienvenu Kenmeugne, (2021), "MPPT Based on Adaptive Neuro-Fuzzy Interference System (ANFIS) for a Photovoltaic System under Unstable Environmental Conditions".
13. Pradeep Hunoor and SR Savanur, (2015), "Design and Analysis of ANFIS Controller to Control Modulation Index of VSI Connected to PV Array".
14. Rati Wongsathan, (2018), "Optimized Fuzzy and Neuro-Fuzzy Controller based MPPT using MOHGA applied for a Solar Photovoltaic Module".
15. S. R. Revathy, V. Kirubakaran, M. Rajeshwaran, T. Balasundaram, and V. S. Chandra Sekar, (2022), "Design and Analysis of ANFIS – Based MPPT Method for Solar Photovoltaic Applications".
16. S. Saranya, Bibhuti Bhusan Rath, and P. Guruvulu Naidu, (2021), "Design of an ANFIS based MPPT controller for a solar photovoltaic system under partial shading condition".
17. Shashank Pareek and Tarlochan Kaur, (2021), "Hybrid ANFIS-PID based MPPT Controller for a Solar PV System with Electrical Vehicle Load".
18. Shiwani Singh, and Dr. Pratibha Tewari, (2021) "Design & Simulation of an ANFIS MPPT Controller for Solar Power Application in MATLAB".
19. Siddaraj, Udaykumar R. Yaragatti and Nagendrappa Harishchandrappa, (2023), "Coordinated PSO-ANFIS-Based 2 MPPT Control of Microgrid with Solar Photovoltaic and Battery Energy Storage System".
20. Wahyu Setyo Pambudi, Riza Agung Firmansyah, Yuliyanto Agung Prabowo, Titiek Suheta, and Fathammubina, (2022), "Designing ANFIS Controller for MPPT on Photovoltaic System".

Emerging Technologies and Applications in Electrical Engineering –
Prof. Dr. Anamika Yadav et al. (eds)
© 2024 Taylor & Francis Group, London, ISBN 978-1-032-82568-7

An Adaptive DC Link Voltage Three-Phase Photovoltaic Structure for CPI Voltage Fluctuations

17

K. Neelima*

Electrical and Electronic Engineering Department,
Vignana Bharathi Institute of Technology, Hyderabad, India

A. Manoj Kumar

PG scholar, Electrical and Electronic Engineering Department,
Vignana Bharathi Institute of Technology, Hyderabad, India.

Vadthya Jagan

Electrical and Electronic Engineering Department,
Vignana Bharathi Institute of Technology, Hyderabad, India

ABSTRACT: This paper discusses a solar photovoltaic (SPV) system with two stage 3-φ, and is grid-connected. The first stage consists of a boost converter, which is responsible for maximum power point tracking and feeding the extracted solar energy to the DC link of the PV inverter. The second stage consists of a two-level VSC (voltage source converter), which serves as a PV inverter and feeds power into the grid from a boost converter. In the system that is being proposed, an adaptive DC link voltage is used. This voltage is made adaptive by altering the reference DC link voltage in accordance with the voltage at the CPI (common point of interconnection). Switching power losses can be reduced with the use of adaptive DC link voltage regulation. To get a more desirable dynamic response, a feed forward term for the solar contribution is applied. The suggested approach offers benefits not only in scenarios in which there is frequent and prolonged under voltage, but also in scenarios in which there is normal voltage at CPI. The THD of the grid current was determined to be significantly lower than the limit set by the IEEE-519 standard.

KEYWORDS: Solar photovoltaic, Two-level VSC, DC link, Grid integration

*Corresponding author: neelimarakesh@gmail.com

DOI: 10.1201/9781003505181-17

1. Introduction

Last century's progress was fueled by electricity. Global energy imbalance has resulted from traditional power source loss. Solar, wind, and tidal energy can alleviate energy shortages. Technology must be profitable. Solar panels' high cost delayed SPV (Solar Photovoltaic) systems by years (M. Pavan, 2013, M. Delfanti, 2013). Solar energy systems are grid-connected or stand-alone. A solitary system's energy storage management usually batteries are most important. Storing energy with batteries independent solar energy converters are problematic (D. Debnath, 2015). If the grid is present, grid-interfaced technologies are better for energy storage. All electricity can go to the energy buffer grid. Grid-interfaced SPV systems handle isolation, volatility, modelling and more (W. Xiao, 2013). As the electricity grid grows, so does distributed generation. Distributed production lowers losses, optimizes distribution assets, and flattens load profiles (A Yadav, 2014). Rooftop solar, modular power converters, and static energy conversion suit distributed generating SPV systems. Solar panels make SPV systems pricey (B. N. Reddy, 2023). Thus, to save plant startup costs, increase energy output from available capacity. DC link voltage depends on filter reactive power. The proposed system adjusts VSC DC link voltage to CPI voltage changes. Circuit topologies vary. It varies from peak three-phase line voltages above the DC link voltage reference control VSC currents. The reference DC link voltage will always exceed the Hub of All Connections (CPI) maximum voltage. Thus, VSC DC link voltage regulation maintains a worst-case DC link voltage. This study provides an adaptive DC link voltage architecture for grid-connected PV systems to account for CPI voltage changes. Boost converter, then two-stage voltage regulator. This study provides a variable DC connection voltage for VSC, unlike earlier studies. The adaptive DC connection voltage considerably minimizes ohmic losses at high frequencies in the interface inductor, as well as switching losses in all power devices. Remote radial endpoints and nominal grid voltage conditions benefit most from the recommended DC link voltage arrangement.

Several researchers have built systems with two stages and three phases that are grid-interfaced, but none have shown stability throughout such a wide CPI voltage range (350 V to 480 V for a nominal 415 V). The VSC's rating and cost increased due to its CPI voltage response. A little VSC cost increase may be justified due to PV array costs. Grid current and voltage total harmonic distortions must always be below the IEEE-519 standard of 5%. Both techniques employ the worst-case scenario; therefore, the suggested method does not lose power device ratings. The proposed control method creates more power with the same ageing infrastructure.

2. Configuration of the System

Figure 17.1 shows the suggested system's configuration. A grid-connected SPV system requires a two-stage method. The two-stage, three-phase voltage source converter and maximum power point tracking DC-DC boost converter are stages. The boost converter may maximize PV array power with the array hooked into its input. The boost converter powers the VSC's DC connector. Grid-tied VSC may adjust real-time to the DC link CPI voltage. Three-phase VSC legs are IGBTs. The interface inductors' opposite ends are connected to the CPI and the VSC's

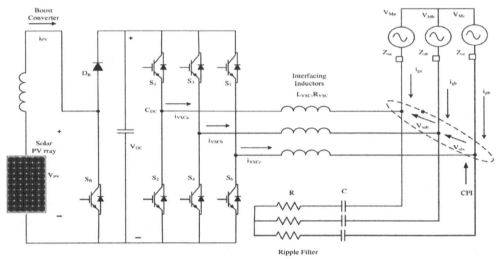

Fig. 17.1 Configuration of the system

output terminals. The CPI dampens VSC high-frequency switching ripples via a ripple filter. Simulation variables and settings are displayed.

3. Control Approach

Figure 17.2 shows SPV's principal control approach. Boost converter control and VSC control with grid connection make up the system's control. Maximum power point tracking (MPPT) regulates the boost converter's input voltage, while the CPI voltage condition maintains the adaptive output voltage, which acts as the VSC's DC link voltage. These strategies work together. The suggested system modifies the boost converter's input and output voltages based on circuit characteristics. After being fed into the boost converter, energy from the VSC's DC connection enters the grid that operates in three phases and has a power factor of one relative to the CPI, VSC does this. An MPPT method based on composite *InC* devices estimate the reference PV array voltage, and a PLL-less control regulates the VSC. a PV feedforward (PVFF) term as well as an inaccuracy in the PI controller's DC link voltage can approximate reference grid currents. Grid voltages are needed to estimate VSC current synchronization unit vectors. A hysteresis current controller compares observed and computed reference grid currents for VSC switching logic.

4. Control Algorithm for VSC

Figure 17.2 shows VSC control. VSC's control algorithm feeds the grid PV energy at a power factor of one relative to CPI and regulates the voltage on the DC connection to a value that has been predetermined as a reference. To regulate VSC output currents or currents on the grid, including reference currents on the grid, are calculated. In order to preserve a symmetrical

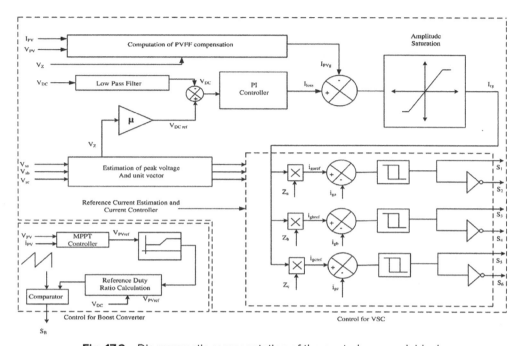

Fig. 17.2 Diagrammatic representation of the control approach block

and sinusoidal distribution of grid currents, calculate the grid's electrical current intensity and multiply the predicted amplitude by in-phase unit vectors (synchronization signals). For control, CPI line voltages, DC link voltage and grid currents are measured. Line voltages are used to figure out phase voltages. Calculate in-phase unit vectors from estimated phase voltages. Estimate unit vectors (Z_a, Z_b, Z_c) by dividing phase voltages by Vz. Fig. 17.3 shows how PV power and grid voltage determine the system's reference current. The loss factor includes nominal grid voltage (10%), interface inductor (5%) and switch (5%) variation. To Calculate CPI voltage amplitude.

$$V_z = \sqrt{\frac{2(v^2{}_{sa} + v^2{}_{sb} + v^2{}_{sc})}{3}} \tag{1}$$

The estimated unit vectors for each of the three stages are:

$$z_a = \frac{v_{sa}}{V_z}, z_b = \frac{v_{sb}}{V_z}, z_c = \frac{v_{sc}}{V_z} \tag{2}$$

The reference DC link voltage should be

$$V_{DCref} = \mu\sqrt{3}V_z, \text{ where } \mu > 1 \tag{3}$$

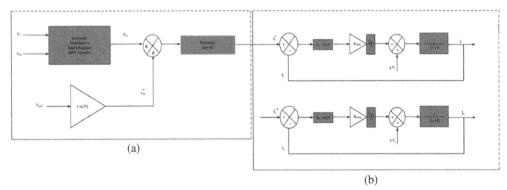

(a)

(b)

Fig. 17.3 The control design with a forward-feeding control and a current feedback loop

5. Simulation Results

5.1 Case (i) Change in Solar Insolation without PV Feedforward

The statistics below indicate the performance of the proposed system with and without feed forward adjustment under a quick fall in insolation from 1000W/m2 to 500W/m2. The system is in a steady condition with no changes expected 1000W/m2 SPV insolation before 0.3 s. Well-balanced grid currents are sinusoidal, 1000W/m2 drops to 500W/m2 at 0.3 s. Insolation decreases PV array power and current. The suggested system reacts better to rapid insolation changes. The technology progresses quickly and provides the reduced amount of power to be fed back into the system using the suggested control mechanism. VSC DC link voltage remains unchanged. Figure 17.4(a) demonstrates the PV array output voltage with a magnitude of 450V. The Fig. 17.4(b) illustrates the PV array output current with a magnitude of 47A up to 0.3 sec and is 30A after 0.3 sec. Figure 17.4(c) shows the power from PV array, which is of 2.2 kW up to 0.3 sec and then shifted to 1.4 kW. Figure 17.4(d) illustrates the dc output voltage Vdc is 630V. Figure 17.5 demonstrates the grid current which is 38A till 0.3 sec and changed to 22A

5.2 Case (ii) Change in Solar Insolation with Feedforward for PV

Figure 17.6(a) demonstrates the PV array output voltage with a magnitude of 450V. Figure 17.6(b) illustrates PV-array output current with an approximate magnitude of 52A up to 0.3 sec and shifted to 32A after 0.3 sec. Figure 17.6(c) shows the power calculated from PV array, a value of 2.4kW up to 0.3 sec and then 1.4kW after 0.3 sec. Figure 17.6(d) illustrates the dc output voltage Vdc is 630V. Figure 17.7(a) shows the grid side voltage with a constant magnitude of 400V. Figure 17.7(b) demonstrates the grid current with 40A till 0.3 sec and changed 22A.

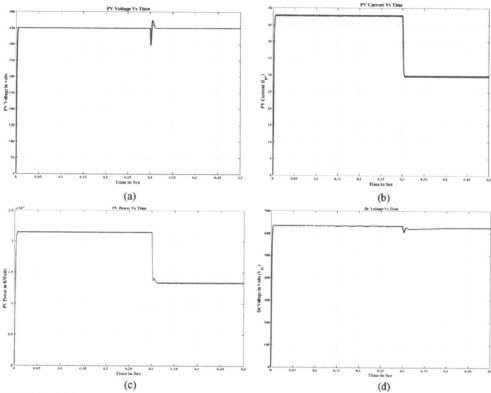

Fig. 17.4 (a) Output voltage from PV array, (b) Output current from PV array, (c) Output power of PV array, (d) Output voltage from DC-DC converter

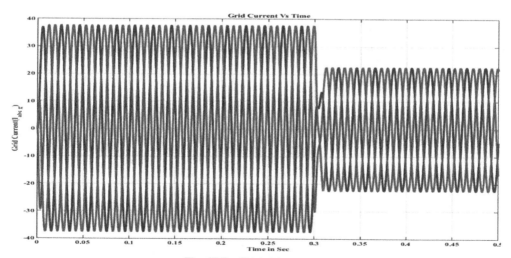

Fig. 17.5 Grid side current

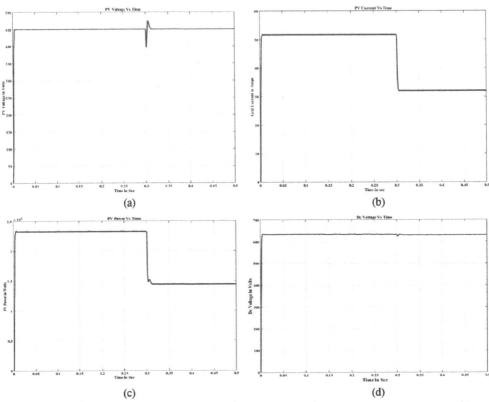

Fig. 17.6 (a) Output voltage from PV array, (b) PV array output current, (c) Output power calculated from PV array, (d) Output voltage from DC-DC converter

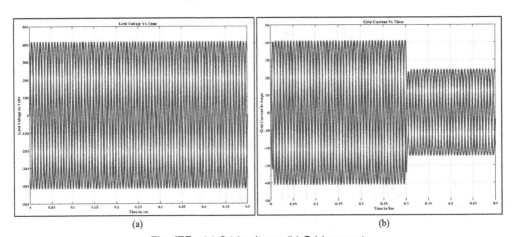

Fig. 17.7 (a) Grid voltage, (b) Grid current

6. Conclusion

This paper discussed a SPV with two stage 3-φ, grid- connected system. The effectiveness of the suggested approach has been shown using a wide range of CPI voltage adjustments. A simple and novel adaptive DC link voltage control approach has been proposed for the regulation of grid-tied VSC. In addition, a feed forward term is used for the PV array, which allows for instantaneous dynamic response. Adjustments in PV power and CPI voltage are taken into consideration while selecting the word used for the feed forward function of the PV array. The total harmonic distortion (THD) of the grid currents and voltages is less than 5%, as specified by the IEEE-519 standard. The simulation findings support the effectiveness of the proposed control method.

REFERENCE

1. Yadav, Alka, and Laxmi Srivastava. "Optimal placement of distributed generation: An overview and key issues." In 2014 International Conference on Power Signals Control and Computations (EPSCICON), pp. 1-6. IEEE, 2014.
2. B. N. Reddy., GOUD, SRIKANTH, B. Sai Kumar, B. Sai Sri Vindhya, D. Bharath Kumar, E. Laxman Kumar, O. Chandra Shekar, A. Ajitha, Kambhampati Venkata Govardhan Rao, and Thalanki Venkata Sai Kalyani. "Switched Quasi Impedance-Source DC-DC Network for Photovoltaic Systems." International Journal of Renewable Energy Research (IJRER) 13, no. 2 (2023): 681–698.
3. Debnath, Dipankar, and Kishore Chatterjee. "Two-stage solar photovoltaic-based stand-alone scheme having battery as energy storage element for rural deployment." IEEE Transactions on Industrial Electronics 62, no. 7 (2014): 4148–4157.
4. Delfanti, Maurizio, Valeria Olivieri, Batuhan Erkut, and Gaston A. Turturro. "Reaching PV grid parity: LCOE analysis for the Italian framework." (2013): 1506–1506.
5. Pavan, Alessandro Massi, and Vanni Lughi. "Grid parity in the Italian commercial and industrial electricity market." In 2013 International Conference on Clean Electrical Power (ICCEP), pp. 332–335. IEEE, 2013.
6. Xiao, Weidong, Fonkwe Fongang Edwin, Giovanni Spagnuolo, and Juri Jatskevich. "Efficient approaches for modeling and simulating photovoltaic power systems." IEEE journal of photovoltaics 3, no. 1 (2012): 500–508.

Emerging Technologies and Applications in Electrical Engineering –
Prof. Dr. Anamika Yadav et al. (eds)
© 2024 Taylor & Francis Group, London, ISBN 978-1-032-82568-7

Development and Analysis of IoT-based Smart Healthcare Monitoring System

18

Subhrat Tripathi[1], Kaushal Pratap Sengar[2], Praveen Bansal[3]
Madhav Institute of Technology and Science, Gwalior, M.P

ABSTRACT: The healthcare monitoring system is growing significantly in the medical field, including hospitals and various healthcare developments in the medical field. In this paper, an IoT-based smart healthcare system has been put forward to monitor the basic health symptoms of the patient along with regular monitoring of its room parameters. This proposed model uses several sensors to gather the data parameters of the patient as well as from its environment, with the help of a variety of sensors. Obtained results are compared with the actual data of standard devices and have been analyzed and represented in different forms. The difference in error values between the proposed model and the standard devices is under satisfactory limits for every case study. The condition of the healthcare consumer can be communicated with health professionals and they act accordingly in departments. It is a significant technical parameter to monitor health using effective healthcare systems with recent.

KEYWORDS: Co2 sensor, ESP32, Healthcare monitoring system, Internet of things, Pulse rate sensors

1. Introduction

Healthcare is a key component of a person's need to live a good life. Healthiness is comprised of a condition encompassing physical, mental, and social aspects, rather than simply the lack of diseases. Unfortunately, the problems faced by the global health industry are complicated by conditions portraying ailing health services, the larger gap among the cities and villages, and the poor availability of doctors along with supporting staff over the past few decades as given in Elija Perrier (2015). Internet of Things (IoT) has arisen as a transformative technological

---
[1]subhrattripathimits@gmail.com, [2]kaushalsengar@mitsgwalior.in, [3]pbansal444@mitsgwalior.in

DOI: 10.1201/9781003505181-18

revolution, enabling the interconnection of various devices as mentioned by (Prosanta and Hwang, 2016). IoT applications have found their place in numerous domains, including intelligent health tracking, intelligent parking, smart homes, smart cities, environmental conditions control, operations in industries, and agriculture (Ashok, 2020 and Kumari, 2022). One of the most prominent applications of IoT is in the management of healthcare that offers various facilities including the monitoring of the health conditions along with the surrounding environment of the patient. IoT essentially involves connecting different types of computers to the internet using various sensors and different networks, offering a straightforward, energy-efficient, intelligent, reliable, scalable as well as interoperable approach to healthcare optimization (Sarkar and Misra, 2016). Nowadays, modern systems are evolving with flexible interfaces, assisting devices, and mental health management to enhance the quality of human life. Among the vital health indicators, human heart rate and body temperature play an important role. Heart rate, often referred to as pulse rate, is the count of heartbeats per minute. The normal range for a healthy individual typically falls ranging from 60 to 100 beats per minute. Resting heart rhythms tend to be at approximately 70 beats per minute for adult males and approximately 75 bpm for adult females. Females aged 12 and beyond typically exhibit higher heart rates compared to males of the same age by Reddy(2015)and Islam (2020) . Body temperature, on the other hand, represents the heat within the human body and is influenced by variables like ambient temperature, gender, and dietary habits. Among healthy adults, body temperature usually falls between 97.8 °F (36.5 °C) and 99 °F (37.2 °C) according to Santoso & Dalu (2015). Numerous methods have been developed for both invasive and non-invasive assessment of pulse rate and temperature of the body by Yin (2016) and Islam (2015). Ensuring that healthcare facilities maintain optimal room conditions is essential for patient comfort, considering factors such as humidity levels and the presence of gases like CO and CO2. High levels of toxic gases and inappropriate humidity can be detrimental to patients' well-being. In the present scenario a multitude of critical health conditions, encompassing heart disease, diabetes, breast cancer, liver disorders, and various others, pose significant health challenges (Ayon & Hossain, 2020). The main objective of the proposed model is to observe key healthcare variables related to patients and the environment within patient rooms. This paper introduces a custom healthcare monitoring model that tracks patient pulse rate, body temperature, ambient humidity, and the concentration of CO and CO2 gases using sensors. The Internet of Things (IoT) is widely recognized as a potential solution to alleviate the pressures on healthcare systems (Al-Ali, 2017). This model is also capable of maintaining a unified patient database in hospitals while providing personalization for essential health criteria. The proposed model is a suitable choice for a health monitoring system, as evidenced by the system's demonstrated effectiveness.

2. System Design and Working

2.1 Hardware Components

The elements which are used to design the proposed model are mentioned below:

(i) Processor: ESP32 plays a crucial role in the IoT learning tools. It is based completely on a fully-fledged Linux system in a minimized platform at an economical cost. The connections of ESP32 device sensors and actuators are achieved by using GPIO pins. The design of ESP32 is achieved by the integration of switches of antenna, RF-balun, amplification controllers, and amplifiers to lower the noise, and the addition of filters along with the modules involved in power management. EPS32 has the ability to communicate with outside devices of Bluetooth Wi-Fi by using of the interfaces like SPI/SDIO, or by using I2C/UART.

(ii) Pulse Rate Sensor: The pulse rate sensor is designed according to the plethysmography theory. It counts the changes that happen in the quantity of blood flowing through the organs of anyone resulting in passing of the light intensity through that organ. This becomes more critical in systems where the timing of the pulses or the heart's pulse rate needs to be monitored. The rate at which heartbeats decides how the blood volume is distributed and the pulse of the signal gets equalized to the heartbeat pulse when blood consumes the light.

(iii) Body Temperature Sensors (LM35): The series of LM35 temperature sensors offer precise and temperature-optimized measurement circuits, providing a voltage output that linearly corresponds with the temperature which is in degrees Celsius. LM35 has an advantage in that it provides a temperature reading in degrees Celsius, making it convenient for consumers without the need to account for a significant constant voltage offset in the display. The LM35 is designed to function within a temperature range of -55°C to 150°C.

(iv) Room temperature and humidity sensor (DHT11): The DHT11 is one of the widely utilized sensors designed for measuring both temperature and humidity. The operating range of voltage is between 3.3V and 5.5V and provides a temperature resolution of 1°C with ±2°C of accuracy, covering a range of temperature from 0°C to 50°C. For humidity measurement, it offers a resolution of 1%RH and maintains an accuracy of ±5%RH within the same temperature range.

(v) CO Sensor (MQ-9): The MQ-9 sensor is used for the identification of carbon monoxide (CO) along with flammable gases. It functions by relying on a chemical reaction that modifies the electrical resistance of its sensing element. This sensor is equipped with an internal heater to maintain temperature control and generates analog output. It demands a preheat time of a few minutes and exhibits different response times. MQ-9 sensors can also be used for various applications including gas leak detectors and fire detection systems.

(vi) CO2 Sensor (MQ-135): MQ-135 sensors are employed to identify and quantify various gases, including NH3, Nicotine, Benzene, Smoke, and carbon dioxide. The sensor module of MQ-135 features a dedicated digital pin which helps it to function independently, eliminating the need for a microcontroller, and making it particularly useful for specific gas detection.

2.2 Working

Figure 18.1 represents the construction diagram of a monitoring system based on IoT with important components. The core concept of the proposed model is online monitoring of both patients and their room conditions. This healthcare monitoring system is the combination of three main design components (1) Module containing sensors, (2) Module processing the data, and (3) Serial monitor of the Arduino IDE. The wired sensors play an important role in gathering the necessary data of healthcare from both the body of patients and from the

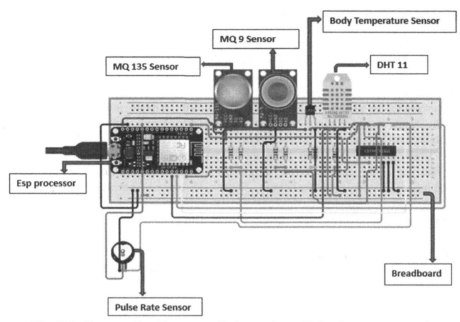

Fig. 18.1 Proposed healthcare monitoring system with hardware components

environment of their room. Collected data from sensors are subsequently processed with the help of the ESP32 module and transmitted to the serial monitor of Arduino IDE via data cable which also provides power to the ESP32 microcontroller. The Arduino IDE provides a user-friendly interface and real-time monitoring with the help of an inbuilt serial monitor. Figure 18.2 represents the working model of the system that has been proposed. The sensor senses different types of data for each test case and sends these data to the ESP32. ESP 32 process the obtained data according to the program code given to it and display it on the Arduino IDE (integrated development environment) serial monitor.

Fig. 18.2 Working model of proposed system

3. Results and Discussion

The effectiveness and performance of the suggested model were tested among various patients of different age groups in various climatic and environmental conditions. It measured the rate of pulses, the temperature of the human body, and patient room or environmental humidity for different test cases. The upcoming section provides the outcomes and analysis of measurements for various health parameters of all the patients involved in the study. To assess the system's performance, we conducted real-life testing with different test subjects by involving five humans. The measurements were acquired by using the developed system for this study, and the outcomes from these real-life tests are sequentially presented in Tables 18.1, Table 18.2, and Table 18.3. Table 18.1 shows the heart rate data (in bpm) obtained from the proposed model with the help of a pulse rate sensor by keeping the index finger on

Table 18.1 Heart rate data (in bpm) collected by measured device (actual data) and proposed model

No. of Patient	Heart Rate (bpm)	Heart Rate (Proposed Model)	% Error
P1	69	66	4.34
P2	79	75	5.06
P3	78	76	2.56
P4	67	70	-4.48
P5	73	76	-4.10

Table 18.2 Body temperature data (F) collected by measured device (actual data) and proposed model

No. of Patient	Body Temperature F [10]	Body Temperature (Proposed Model)	% Error
P1	96.4	93.1	3.42
P2	96.1	92.4	3.85
P3	97.6	95..6	2.05
P4	97.3	94.8	2.57
P5	94.0	92.9	1.12

Table 18.3 Humidity data collected by measured device (actual data) and proposed model

Experiments	Humidity	Humidity (Proposed Model)	% Error
E1	66	65	1.51
E2	70	68	2.85
E3	65	63	3.07
E4	72	70	2.78
E5	65	62	4.61

the sensor module and compared with measured data collected from standard devices. The model exhibited a 4.48% error, which is considered satisfactory and significant. Figure 18.3(a) illustrates the deviation between data obtained from the proposed model and the actual data collected from the standard medical device. Table 18.2 depicts patient body temperature data (in degrees Fahrenheit) for different test cases, obtained from the proposed model with the help of the lm35 sensor by keeping the body part on the sensor module and compared with actual measured data. A maximum error of 3.85% was found, which represents the effectiveness of the developed model. Figure 18.3(b) depicts the deviation between the measured data using the proposed model and the actual data. It shows that the measured data closely matches with actual data measured using a standard device. Table 18.3 presents the patient's room humidity data for different test cases in percentage which is measured as a relative percentage (rh). The model exhibited a maximum error of 4.61%, which is an acceptable limit. Variations may occur due to differences in patient age groups, the nature of the disease, and the geographical and climatical locations of the patient. Figure 18.4 (a) and Fig. 18.4(b) represents a diagram of the difference in patient body temperature and patient's room humidity in percentage.

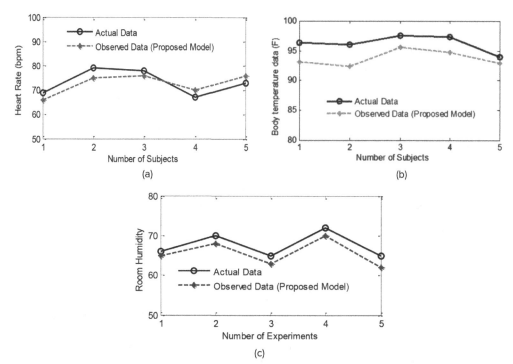

Fig. 18.3 Results of actual and proposed model (a) Heart rate in bpm (b) Body temperature in F, (b) Room humidity in %.

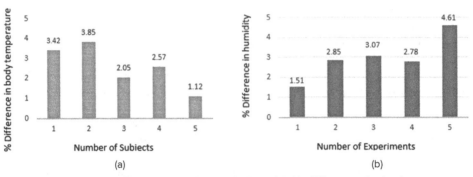

Fig. 18.4 Percentage difference in observed data (a) % difference in body temperature, (b) % difference in room humidity

4. Conclusion

The research work in this manuscript comprised of, an IoT-based smart healthcare monitor model that collects the essential and fundamental data of the subject such as pulse rate, and temperature of the subject's body, along with basic measures of the patient's environmental condition such as humidity present in the surroundings, and the level of different carbon gases present in the surroundings. Observed data is compared with the actual data collected from the published literature and not more than 5% error is found. The device can communicate data so the medical team will be able to observe and keep an eye on the real-time sensed data even in conditions where the patient's location is outside of the hospital. The developed prototype poses no complications in both creation and utilization. The system is designed in a way that it can play a pivotal role in case of major disease treatment. The suggested model will advance the present-time system of healthcare with the proper application of the Internet of Things. By using the video feature, face-to-face cases can be consulted by doctors for patients in need. Subsequently, sensor readings are transmitted through a wire to a medical server and accessed by authorized healthcare professionals. Based on the sensed data, the doctors can diagnose the patient's overall health condition and recommend appropriate treatment, thereby promoting the patient's well-being.

References

1. Al-Ali, A.R., et al. (2017). A smart home energy management system using IoT and big data analytics approach. IEEE Trans Consum Electron. https://doi.org/10.1109/TCE.2017.015014.
2. Ashok, D., Tiwari, A., & Jirge, V. (2020). Smart Parking System using IoT Technology. In 2020 International Conference on Emerging Trends in Information Technology and Engineering (ic-ETITE) (pp. 1–7). Vellore, India.
3. Ayon, S., Islam, M.M., & Hossain, M.R. (2020). Coronary artery heart disease prediction: a comparative study of computational intelligence techniques. IETE J Res. https://doi.org/10.1080/03772063.2020.1713916.

4. Chang, S.-H., Chiang, R.-D., Wu, S.-J., & Chang, W.-T. (2016). A context-aware, interactive M-health system for diabetics. IT Prof., 18(3), 14–22.
5. Foysal, M. R., et al. (2021). IoT Based Temperature Control System of Home by using an Android Device. In 2021 1st International Conference on Emerging Smart Technologies and Applications (eSmarTA) (pp. 1–8). Sana'a, Yemen.
6. Gope, P., & Hwang, T. (2016). BSN-care: A secure IoT-based modern healthcare system using body sensor network. IEEE Sensors J., 16(5), 1368–1376.
7. Islam, M.M., et al. (2020). Development of a smart healthcare monitoring system in the IoT environment. SN Computer Science, 1, pp. 1–11.
8. Islam, S. M. R., et al. (2015). The Internet of Things for healthcare: A comprehensive survey. IEEE Access, 3, 678–708.
9. Kumari, K.S., et al. (2022). Agriculture monitoring system based on the internet of things by deep learning feature fusion with classification. Computers and Electrical Engineering, 102, p. 108197.
10. Perrier, E. (2015). Positive Disruption: Healthcare, Ageing and Participation in the Age of Technology. Sydney, NSW, Australia: The McKell Institute.
11. Reddy, G.K., & Achari, K.L. (2015). A non-invasive method for calculating calories burned during exercise using heartbeat. In 2015 IEEE 9th International Conference on Intelligent Systems and Control (ISCO) (pp. 1–5).
12. Santoso, D., & Dalu Setiaji, F. (2015). Non-contact portable infrared thermometer for rapid influenza screening. In 2015 International Conference on Automation, Cognitive Science, Optics, Micro Electro Mechanical System, and Information Technology (ICACOMIT) (pp. 18–23).
13. Sarkar, S., & Misra, S. (2016). From micro to nano: The evolution of wireless sensor-based healthcare. IEEE Pulse, 7(1), 21–25.
14. Yin, Y., et al. (2016). The Internet of Things in Healthcare: An overview. J. Ind. Inf. Integr., 1, 313.
15. Zhu, N., et al. (2015). Bridging e-health and the Internet of Things: The SPHERE project. IEEE Intell. Syst., 30(4), 39–46.

Emerging Technologies and Applications in Electrical Engineering –
Prof. Dr. Anamika Yadav et al. (eds)
© 2024 Taylor & Francis Group, London, ISBN 978-1-032-82568-7

Performance Investigation of Three Phase Induction Machine with Scalar and Vector Control Method

19

Manish Bisoi[1], Rada Kalyan[2], Ramya Selvaraj[3], Hari Priya Vemuganti[4]
Department of Electrical Engineering, NIT Raipur, India

ABSTRACT: The induction motor, which is the most widely used motor type in the industry, has been favoured for its good self-starting capability, simple and rugged structure, low cost, and reliability, among other advantages. This paper analyses the performance of a three-phase induction machine using closed-loop scalar control and indirect vector control. The entire analysis is conducted using the popular simulation software, MATLAB SIMULINK. There are two commonly used methods for controlling induction motors: scalar control and vector control. In the case of scalar control, the closed-loop scalar method is often preferred due to its simplicity and cost-effectiveness, but it cannot be used of dynamically changing speed and loads, and has less precise control. The popularly used vector control method is the Indirect Vector Control method, which is somewhat more complex and expensive, but it offers better dynamic and precise control of the induction machine.

KEYWORDS: Field oriented control (FOC), PI controller, Squirrel cage induction motor (SCIM), Indirect vector control, Closed loop V/f control

1. Introduction

The utilization of three-phase induction machines has witnessed a significant surge in industrial applications, owing to the availability of various methods for controlling motor speed and torque. Control strategies for induction motors can be broadly categorized into two approaches: scalar control and vector control. Scalar control, being a relatively simpler method, focuses on regulating the magnitude of selected quantities. In the case of Induction

---

[1]mani789bisoi@gmail.com, [2]kalyanrada206@gmail.com, [3]rselvaraj.ee@nitrr.ac.in, [4]hpvemuganti.ee@nitrr.ac.in

DOI: 10.1201/9781003505181-19

Motors (IM), this technique involves the application of Volts/Hertz constants. On the other hand, vector control is a more intricate strategy compared to scalar control. Scalar control proves inadequate for systems with dynamic behaviour, prompting the need for vector control. Also termed field-oriented control, vector control necessitates the identification of the motor's field flux for implementation (Sarde, Auti Gadhave, 2014).

This paper presents for speed control of an Induction Motor (IM) using both the closed-loop V/f method and indirect vector control method, which has been developed in Simulink and analysed in detail. The simulation employs a squirrel-cage Induction Machine for its robust performance, mechanical durability, simple construction, and low maintenance (Divyasree and Binojkumar, 2017). The Induction Machine (IM) is controlled using scalar and vector control technology. Scalar control employs a simple strategy based on the steady-state model, using magnitude V/f control to regulate synchronous speed. Closed-loop V/f control enhances control by incorporating a PI controller to compensate for slip speed, improving rotor speed control (Habbi, Ajeel, Ali, 2016). Vector Control, or Field-oriented Control, for the Induction Machine relies on its dynamic model. This method can be realized through different approaches, depending on how the rotor field is measured or estimated. Implementation options include sensor control or sensor less control (Kumar, Das, Syam, & Chattopadhyay, 2015) (Rajesh, Babu, & Tagore. 2012). In this analysis, we utilize the indirect vector control scheme due to its popularity and advantages over other methods.

2. Speed Control Methods

Two popular methods of speed control for IM Drives are Scalar control and Vector control.

2.1 Scalar Control Method

Volts/Hz control, based on the steady-state model of an induction machine, adjusts supply frequency so to change synchronous speed for speed control with a constant voltage-to-field ratio (OTKUN, 2020).

Open loop V/f control: Open-loop V/f control regulates the frequency while keeping a constant voltage-to-frequency (V/f) thus controlling speed. A small voltage boost is necessary under low-speed conditions due to significant stator resistance-induced voltage drop (Zhang, Liu, & Bazzi, 2017).

Closed loop V/f control: To achieve better performance under loaded condition, a PI based slip compensation is provided via speed encoder and reference speed, thus to get better control through PWM based inverter (Meghana, Cherukupalli, Sravani, & Naidu, 2021).

2.2 Vector Control Method

Vector control overcomes the drawbacks of scalar control, providing decoupled comparable to that of separately excited DC machines (Thool, & Wakhare). The primary objective of vector control was to achieve superior performance in case of speed and torque variations. Similar to separately excited DC machines, Field-Oriented Control (FOC) enables the inverter-fed induction machine drive to independently control the flux and torque components of the stator current (Aziz, Abdelaziz, Ali, & Diab, 2023). The basic principle of vector control is to separate the components of the stator current responsible for producing flux and torque through coordinated changes in the magnitude, frequency, and phase of the stator voltage. In induction machines it involves aligning the current vector with the rotor flux axis for natural decoupling, known as field-oriented control, optimizing performance compared to alignment with stator or air gap flux axes that require additional compensation for decoupling(Ahmad, 2010).

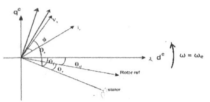

Fig. 19.1 Showing relationship between different frames

Below given the Torque expression for Induction machine,

$$T_e = KT\lambda_r i_{qs} = Ki_{ds}i_{qs} \tag{1}$$

For the control scheme, certain parameters are required, such as θ_f or θ_e, which represent the angle between the field axis and the stationary reference axis and will vary with time.

$$\theta_f = \theta_r + \theta_{sl} = \int(\omega_r + \omega_{sl})dt \tag{2}$$

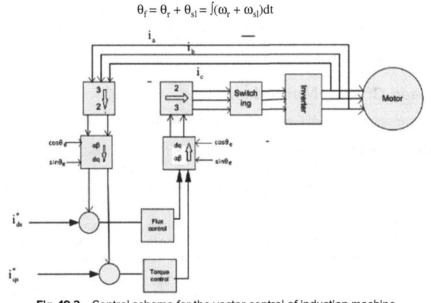

Fig. 19.2 Control scheme for the vector control of induction machine

Two vector control methods are as follows:

1. *Direct method* - This method acquires flux linkage directly through sensors or models, using line voltages/currents or stator flux/induced EMF.

2. *Indirect method* - Indirect vector control is a method of vector control that overcomes drawbacks of direct methods. In this method angle between stationary axis and Rotor field axis estimated or measured thus to transform stator currents for decoupled control (Ahmad, 2010).

The equations from the dynamic model which are useful deriving control system for indirect vector control given below.

Rotor equations can be given for SCIM in the form of Flux linkage as below,

$$d\lambda^e_{dr}/dt + R_r i^e_{dr} - \omega_{sl}\lambda^e_{qr} = 0 \tag{3}$$

$$d\lambda^e_{qr}/dt + R_r i^e_{qr} - \omega_{sl}\lambda^e_{dr} = 0 \tag{4}$$

From FOC we know rotor flux is aligned with d^e thus $\lambda_{qr} = 0$ and $\lambda_r = \lambda_{dr}$

Flux Component (i_{ds}) - Oriented in phase with rotor flux vector

Torque component (i_{qs}) – Quadrature with rotor flux vector

From those equation we have following conclusions;

$$i_f = i_{ds} = (1/L_m) * (\lambda_r + (L_r R_r * (d\lambda_r/dt))) \tag{5}$$

$$i_T = i_{qs} = (\omega_{sl} L_r\lambda_r)/(L_mR) \tag{6}$$

$$\omega_{sl} = R_rL_m i_T/(L_r \lambda_r)$$

$$\omega_e = (R_rL_m i_T)/(L_r \lambda_r) + (P/2) * \omega_m \tag{7}$$

$$\lambda_r = (L_mi_f)/(1 + T_r) \tag{8}$$

and torque can be given as

$$T_e = (3/2) * (P/2) * (L_m L_r \lambda_r i_{qs}) \tag{9}$$

3. Simulation

The simulation is performed using an induction machine with the following parameters as shown in the table.

Table 19.1   Parameters used in the in the simulation

S. No.	Parameter of Induction machine	Values
1.	Voltage Rating	460 V
2.	Frequency	50 Hz
3.	Power	37.3kWor 50 Hp
4.	Number of poles (P)	4
5.	Stator resistance (R_s)	0.087 Ω
6.	Stator Leakage Inductance (L_s)	0.0001 H
7.	Rotor resistance (R_r)	0.0347 Ω
8.	Rotor leakage inductance (Lr)	0.0008 H
9.	Mutual Inductance (Lm)	0.0347 H
10.	Inertia of Rotor (J)	0.662kg.m^2
11.	Friction constant (Bw)	0.1 N.m.sec.rad^{-1}

Three Simulink models were simulated respectively closed-loop V/F control and indirect vector control with two inverter PWM topology i.e., SPWM and SVPWM. The simulation is carried for 10 sec with changing speed and load torque for every 2 sec intervals.

A. *Closed loop V/f control with SPWM Inverter:* In this model closed loop control is obtained via taking feedback from speed encoder and compared with reference value though PI controller. In the model we can see control strategy (Bharti, Kumar, & Prasad, 2019)

B. *Indirect vector control with SPWM and SVPWM Inverter:* In this model all blocks are realized form of equations as shown in previously. First of stator current is transformed

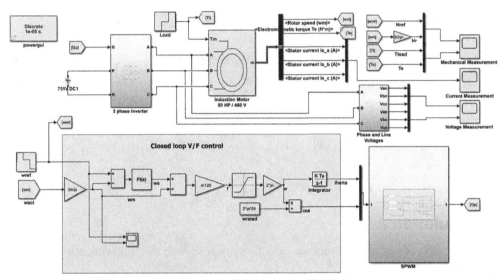

Fig. 19.3 Closed loop v/f with SPWM inverter MATLAB simulink model

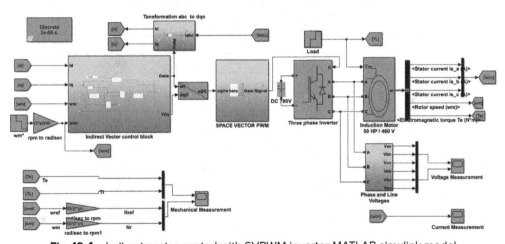

Fig. 19.4 Indirect vector control with SVPWM inverter MATLAB simulink model

from abc to d-q through combined Clark and Park with theta (angle between stationary and d-q axis) as input (Park, 1933). For outer speed control PI loop and inner current control PI loops, K_p and K_i are respectively 50 & 20 and 100 & 26. All value are estimated and calculated in control block.

4. Simulation Analysis and Results

A) Closed loop v/f control with SPWM Inverter

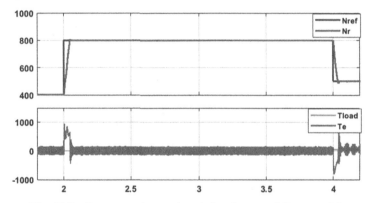

Fig. 19.5   Torque and speed variation from t = 3.8s to t = 4.2s

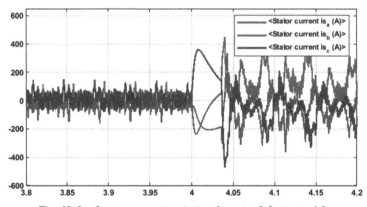

Fig. 19.6   Stator current variation from t = 3.8s to t = 4.2s

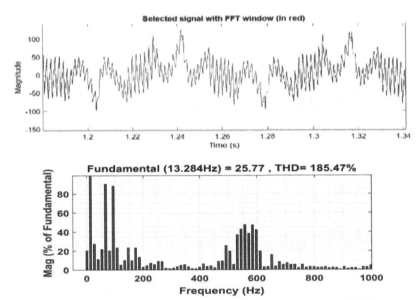

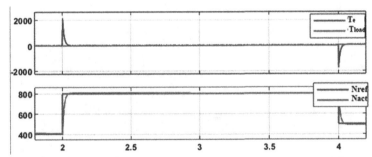

Fig. 19.7   FFT analysis for the stator current for V/f control with SPWM inverter

B) Indirect vector control with SPWM Inverter

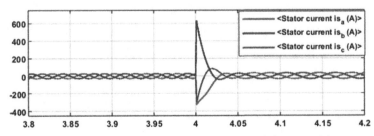

Fig. 19.8   Torque and speed variation from t = 1.8 s to t = 4.2s

Fig. 19.9   Stator current variation from t = 3.8s to t = 4.2s

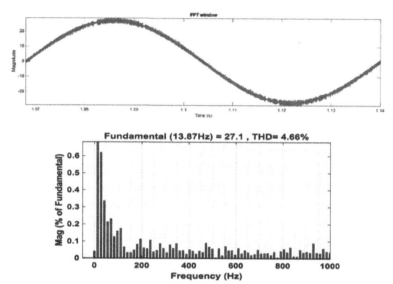

Fig. 19.10   FFT analysis for the stator current for indirect vector control with SPWM inverter

C) Indirect vector control with SVPWM Inverter

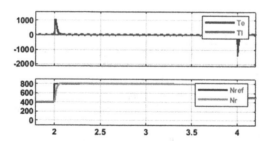

Fig. 19.11   Torque and speed variation form t= 1.8 s to t = 4.2s

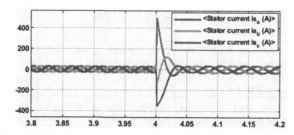

Fig. 19.12   Stator current variation from t = 3.8s to t = 4.2s

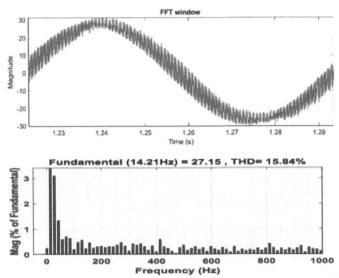

Fig. 19.13 FFT analysis for the stator current for indirect vector control with SVPWM inverter

Result of the simulation is tabulated below.

Table 19.2 Comparison of simulation results of all simulated models

S. No.	Comparison between with different methods of speed control of three phase induction machine			
	Parameters	Closed loop v/f control with SPWM inverter	Indirect vector control with SPWM inverter	Indirect vector control with SVPWM inverter
1	Speed Response	Moderate and accurate contains ripple but in shorter range of magnitude.	Faster and smooth, with little discrepancy.	Faster and smooth, at starting little overshoot but then smoother response.
2	Torque	Tracking the load torque but having large ripples more than 80%.	Tracking the load torque, very smooth operation ripple content less than 20%.	Tracking the load torque with ripple content 30-80% and depend upon speed.
3	Stator current	Little distortion from sinusoidal current.	Nearly sinusoidal.	Nearly sinusoidal.
4	Steady state response	For steady state point having good speed response but having large torque ripples.	Very smooth speed and torque response.	Very smooth speed and torque response.
5	Dynamic Response	During load changing, the speed is tracking but torque response is not satisfactory	During changing loads and speed, it has very smooth response	During changing loads and speed, it has quick and smooth response
6	Total harmonic distortion in stator Current	High THD (175.86% for observed time) content due distorted sinusoidal stator current	Low THD (4.6% for observed one cycle), nearly sinusoidal stator current.	Low THD (16.02% for observed one cycle), nearly sinusoidal stator current.

From the above table it can be seen that indirect vector control has better response to dyanmic changing load than the scalar control method as well torque variation is smooth during stable speed and stator current is more sinusoidal in case of the Indirect vector control method.

5. Conclusion

This paper presented the comparative analysis of closed-loop v/f control and indirect vector control for an Induction machine. The indirect vector control demonstrated superior performance across various speed regions and under changing loads. The performance was examined under various dynamic conditions, leading to the conclusion that indirect vector control has better performance than scalar control. In the low-speed region, the torque response is better with the indirect vector control method. The indirect vector control was investigated using both SPWM and SVPWM methods, while the v/f method was only investigated with the SPWM method. The results showed that in the case of indirect control, the THD level in the stator current is lower compared to the closed-loop v/f control.

New advancements in technology have led to the development of superior methods, such as direct torque control, which exhibits superior performance compared to field-oriented control (FOC).

REFERENCES

1. Ahmad, M. (2010). High performance AC drives: modelling analysis and control. Springer Science & Business Media.
2. Aziz, A. G. M. A., Abdelaziz, A. Y., Ali, Z. M., & Diab, A. A. Z. (2023). A Comprehensive Examination of Vector-Controlled Induction Motor Drive Techniques. Energies, 16(6), 2854.
3. Bharti, R., Kumar, M., & Prasad, B. M. (2019, March). V/f control of three phase induction motor. In 2019 International Conference on Vision Towards Emerging Trends in Communication and Networking (ViTECoN) (pp. 1-4). IEEE.
4. Divyasree, P., & Binojkumar, A. C. (2017, August). Vector control of voltage source inverter fed induction motor drive using space vector PWM technique. In 2017 International Conference on Energy, Communication, Data Analytics and Soft Computing (ICECDS) (pp. 2946-2951). IEEE.
5. Habbi, H. M. D., Ajeel, H. J., & Ali, I. I. (2016). Speed control of induction motor using PI and V/F scalar vector controllers. International Journal of Computer Applications, 151(7), 36-43.
6. Kumar, R., Das, S., Syam, P., & Chattopadhyay, A. K. (2015). Review on model reference adaptive system for sensorless vector control of induction motor drives. IET Electric Power Applications, 9(7), 496-511.
7. Meghana, I., Cherukupalli, K., Sravani, M., & Naidu, P. C. B. (2021, November). Simulation of Slip Compensation for Induction Motor Drive Using MATLAB. In 2021 Innovations in Power and Advanced Computing Technologies (i-PACT) (pp. 1-7). IEEE.
8. OTKUN, O. (2020). Scalar speed control of induction motors with difference frequency. Politeknik Dergisi.
9. Park, R. H. (1933). Two-reaction theory of synchronous machines-II. Transactions of the American Institute of Electrical Engineers, 52(2), 352-354.

10. Rajesh, T., Babu, Y. S. K., & Tagore, Y. R. (2012, November). Performance Evaluation of Indirect Vector Controlled Induction Motor Sensorless Drive. In 2012 Fourth International Conference on Computational Intelligence and Communication Networks (pp. 656-660). IEEE.
11. Sarde, H., Auti, A., & Gadhave, V. (2014). Speed control of induction motor using vector control technique. International Journal of Engineering Research, 3(4).
12. Thool, M. B., & Wakhare, M. K. C. Induction motor control by vector control method. International Refereed Journal of Engineering and Science (IRJES).
13. Zhang, Z., Liu, Y., & Bazzi, A. M. (2017, March). An improved high-performance open-loop V/f control method for induction machines. In 2017 IEEE Applied Power Electronics Conference and Exposition (APEC) (pp. 615-619). IEEE.

Emerging Technologies and Applications in Electrical Engineering –
Prof. Dr. Anamika Yadav et al. (eds)
© 2024 Taylor & Francis Group, London, ISBN 978-1-032-82568-7

Analysis of Single-Phase H6-Clamping Transformerless Inverter Topology for Constant Common Mode Voltage

20

Ahmad Syed[1]
Asst. Prof, Chaitanya Bharathi Institute of Technology(A), Hyderabad, India

Nagaraju Budidha[2]
Asst. Prof, Vaagdevi College of Engineering Bollikunta, Warangal, India

N. Santosh Kumar[3], P. Hemeshwar Chary[4]
Asst. Prof, Chaitanya Bharathi Institute of Technology(A), Hyderabad, India

G. Bhanu Prasad[5]
Scholar, Chaitanya Bharathi Institute of Technology(A), Hyderabad, India

Freddy Tan Kheng Suan[6]
Assoc. Prof University of Nottingham Malaysia, Selangor

ABSTRACT: Grid-connected photovoltaic systems typically employ transformers to ensure isolation between the panels and the grid, but these components increase costs and reduce efficiency. In response, transformerless inverters (TI) have gained traction due to their lower costs and enhanced efficiency, although they lack this isolation, leading to potential leakage current issues. Here a modified H6 topology is proposed with rectifier bridge such as consist of one switch ad four diodes to facilitate a direct current path during active modes, effectively reducing conduction losses and maintaining a constant common mode voltage to eradicate leakage current. Lastly, the performance parameters in terms of components, structure, common mode voltage, leakage current and total harmonic distortion (%THD) are validated through the simulation results. This innovation shows promise in addressing leakage current and enhancing efficiency in transformerless grid-connected photovoltaic systems.

KEYWORDS: Transformerless inverter, Photovoltaic system, Common mode voltage, Common mode leakage current, %THD

[1]ahmadsyed_eee@cbit.ac.in, [2]nagaraju_b@vaagdevi.edu.in, [3]santoshkumar_eee@cbit.ac.in, [4]hemeshwar_eee@cbit.ac.in, [5]ugs206166_eee.prasad@cbit.org.in, [6]freddy.tan@nottingham.edu.my

DOI: 10.1201/9781003505181-20

1. Introduction

Amid the escalating global energy demand, photovoltaic (PV) systems have emerged as a vital solution, harnessing abundant, free solar energy as a clean renewable source. With advancements in PV technology, the mass production cost has significantly reduced, bolstered by government incentives and subsidies, propelling its widespread adoption. PV systems convert solar energy into DC electrical energy, necessitating a well-designed DC to AC converter when connecting to the grid for safe and efficient operation. Grid-connected PV systems are categorized depends on the existence or absence of a transformer. While a transformer provides galvanic isolation among the PV system and the grid, preventing leakage current and DC injection into the grid, it also adds weight, bulk, and cost, thereby reducing overall efficiency. Transformerless grid-connected PV systems have gained attention for being lighter, smaller, and more cost-effective, offering enhanced efficiency compared to transformer-based counterparts (Khan, M.N et al.2019). However, their primary drawback lies in the absence of galvanic isolation, posing safety concerns due to potential high leakage currents from the PV to the grid. Therefore, in transformerless inverters used in grid-connected PV systems, strict adherence to safety standards such as IEEE 1547.1, VDE0126-1-1, IEC61727, EN50106, and AS/NZS5033 (G.N.2005) is crucial to mitigate these potential hazards and ensure the system's safety and performance (Mathi, A.A.D et al. 2018). Therefore, a multitude of researchers have suggested several approaches to eliminate leakage current in transformerless PV systems. These methods encompass AC and DC coupling, the utilization of common mode grounds configurations, and the application of voltage clamping techniques (Kumar, K.S,2020).

Several studies have contributed innovative transformerless PV inverter designs aimed at improving efficiency, safety, and performance in grid-connected photovoltaic systems. A novel topology is introduced with six switches and two diodes, emphasizing constant common mode voltage and high efficiency across different power factors (Mathi, A.A.D et al, 2018) the HB-ZVR topology to address bidirectional switch requirements. Multiple researchers have reviewed and proposed various transformerless inverter topologies for single-phase grid-connected PV systems. These include diverse designs such as positive and negative neutral point clamped cells, improved H6 topologies, and strategies for reactive power control in H5((D. Schmidt et.al,2003) and HERIC inverters (M. Victor et.al,2008)). For instance, another introduced a novel transformerless PV inverter, emphasizing its ability to maintain constant common mode voltage, operate efficiently across power factors, and leverage six switches and two diodes (Zhang, L et.al,2013).

The paper is structured into several sections. Section 2 introduces the proposed topology, offering a detailed explanation of its structure and operating modes. Following this, Section 3 presents and scrutinizes the simulation results related to this new topology, Section 4 encapsulates the conclusion, summarizing key findings, highlighting the proposed topology and finally Section 5 gives references.

2. Proposed Configuration

This paper introduces a new modified H6 transformerless inverter design aimed at eradicating common mode leakage current with constant common mode voltage. Illustrated in Fig. 20.1, the inverter's circuit represents a modification of the H6 inverter, incorporating an extra switch (S7) including with four diodes (D1-D4) known as H6-M topology. In the presented configuration, a rectified bridge is strategically positioned between the negative terminal of the DC supply and terminal A.

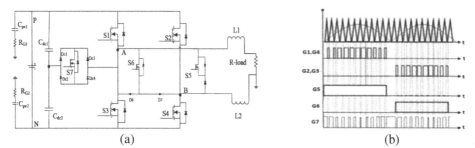

(a)   (b)

Fig. 20.1 Proposed topology (a) H6-M, (b) Control scheme

The proposed H6-M method, employing low-loss MOSFETs, adopts the common mode leakage current elimination of the neutral point clamped (NPC) method. This design, depicted in Fig. 20.1(b), utilizes suitable pulse width modulating signals for optimal performance. Notably, switches S1, S2, S3, S4, and S7 operate at the switching frequency, while S5 and S6 function at the grid frequency. This unique arrangement contributes to the H6-M topology's exceptional features, enhancing efficiency and minimizing losses. However, due to space constraints in this paper, a comprehensive discussion of these features is deferred to future work beyond the current scope. By using DC decoupling principle, this circuit disengages the PV array from the grid during freewheeling modes, establishing a fresh current path in the active phase of the grid's negative half cycle. It achieves this by keeping the common mode voltage constant, effectively eliminating common mode leakage current.

2.1 Operating Modes

In Mode 1 Positive Half Cycle (PHC), S1 and S4 switches are turned ON while the remaining switches are OFF. This mode, the current starts to rise linearly and circulates via switch S1, the load resistance and switch S4, as illustrated in Fig. 20.2(a). The common mode voltage and differential mode voltage as follows.

$$V_{cm} = \frac{V_{AN} + V_{BN}}{2} = \frac{1}{2}(V_{dc} + 0) = \frac{V_{dc}}{2} \tag{1}$$

$$V_{AB} = V_{AN} - V_{BN} = V_{dc} - 0 = V_{dc} \tag{2}$$

In mode 2 during the positive freewheeling (FW) period, the DC source is completely isolated from the grid. In this period, switch S5,D1 is turned on and the remaining switches (S1,S2,

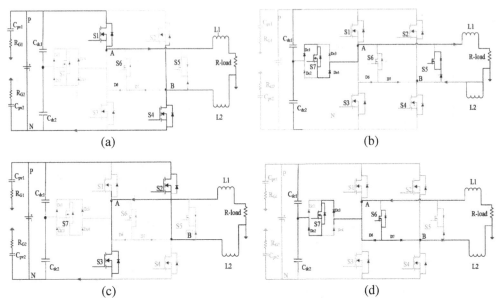

Fig. 20.2 Operating modes (a) Active-positive, (b) Positive freewheeling, (c) Active-negative, (d) Negative freewheeling

S3, S4,S6,S7) are turned off. As a result, the current in the circuit begins to diminish, and it freewheels through diode Dc1, switch S7, and diode Dc2, as shown in Fig. 20.2(b). The common mode voltage and differential mode voltage as follows

$$V_{cm} = \frac{V_{AN} + V_{BN}}{2} = \frac{1}{2}\left(\frac{V_{dc}}{2} + \frac{V_{dc}}{2}\right) = \frac{V_{dc}}{2} \tag{3}$$

$$V_{DM} = \frac{V_{dc}}{2} - \frac{V_{dc}}{2} = 0 \tag{4}$$

In mode 3 during the Negative Half period (NHP), switches S2 and S3 are ON, while the remaining power devices are OFF. Therefore, the current path is established through switch S2, the load resistance and switches S3 as shown in Fig. 20.2(c). The common mode voltage and differential mode voltage as follows

$$V_{cm} = \frac{V_{AN} + V_{BN}}{2} = \frac{1}{2}(0 + V_{dc}) = \frac{V_{dc}}{2} \tag{5}$$

$$V_{DM} = 0 - V_{dc} = -V_{dc} \tag{6}$$

In mode4 during the negative freewheeling period the dc source is completely isolated from the grid. Switch S6, D2 is ON, while the other power devices are turned off. The grid current circulates via diode D_{c2}, switch S7, and diode D_{c3}, as illustrated in Fig. 20.2(d). The common mode voltage and differential mode voltage as follows

$$V_{cm} = \frac{V_{AN} + V_{BN}}{2} = \frac{1}{2}\left(\frac{V_{dc}}{2} + \frac{V_{dc}}{2}\right) = \frac{V_{dc}}{2} \tag{7}$$

$$V_{DM} = \frac{V_{dc}}{2} - \frac{V_{dc}}{2} = 0 \tag{8}$$

From the above it is confirm that CMV is constant in mode1-mode4 and hence reduced common mode leakage current.

3. Simulation Results

This section evaluates the H5-M topology against other configurations in the H5 family through simulation studies. The system parameters are depicted in Table 20.1(Li, H.Zeng et al., 2018). Research has confirmed that H6 and H6-M use unipolar pulse width modulation, producing voltages varying between -400V, 0, and 400V. Output current waveforms show clean sinusoidal patterns without distortions, as shown in Fig. 20.3(a) – Fig. 20.3(b). The conventional H6 without clamping displays fluctuating phase leg voltages (V_{AN} and V_{BN}) between 230-400V, leading to inconsistent CMV and higher common mode ground leakage current, as depicted in Fig. 20.4(a).

Table 20.1 Simulation parameters

Parameters	Input Dc Voltage	Parasitic capacitors (C_{PV1}, C_{PV2}),	Ground resistance (R_{G1}, R_{G2}),	Filter inductors	Switching frequency	Load resistance
Value	400V	100nF	11Ω	3mH	10kHz	50Ω

Conversely, clamping-based topology like H6-M maintain stable CMV clamped at 200V, resulting in reduced common mode leakage current, shown in Fig. 20.4(a) – Fig. 20.4(b).

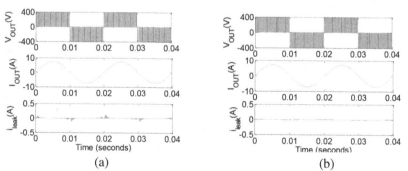

(a)   (b)

Fig. 20.3 The simulation waveforms of V_{out}, i_{out} and i_{leak} for (a) H6, (b) H6-M

This has an impact on total harmonic distortion (%THD) such as 0.99 and 1.02%. A comparative study illustrates the advantages of the H6-M topology, detailed in Table 20.2

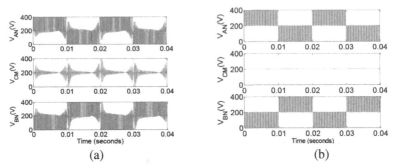

Fig. 20.4 The simulation waveforms of V_{AN}, V_{CM} and V_{BN} for (a) H6, (b) H6-M

Table 20.2 H6 family topologies summary: a comparison

Parameters	Voltage Pattern	Total switching count	No.of diodes	CMV	CMLC	%THD
H6	Unipolar	6	6	6	<300mA	0.99
H6-M	Unipolar	7	7	7	Close to zero	1.02

4. Conclusion

This paper delves into the significance of various transformerless inverters, elucidating their operational principles and structures. It emphasizes the high leakage current in decoupling topologies due to the floating Common Mode Voltage (CMV) throughout the cycle. To counter this issue, clamping based topology is proposed, either on the DC or AC side of the inverter, incorporating a new clamping circuit at the midpoint of the DC link. Here a novel modified H6 topology, integrating a rectified bridge circuit consisting of one switch and four diodes within the primary phase branches of the power circuit. This modification significantly enhances CMV performance and effectively reduces leakage current. Simulation results validate the performance of the discussed topologies in terms of CMV, common mode leakage current, switching count, and Total Harmonic Distortion (%THD), confirming that the proposed topology is highly suitable for grid-connected PV applications.

REFERENCES

1. Ahmad, S.; Mekhilef, S.; Mokhlis, H. A.(2016). Grid-tied photovoltaic transformer-less inverter with reduced leakage current. IOP Conf. Ser. Earth Environ. Sci. 673, 01.
2. G.N.(2005). Automatic Disconnection Device between a Generator and the Public Low-Voltage Grid, DIN Electro technical Standard DIN VDE 0126–1–1.
3. Khan, M.N.; Forouzesh, M.; Siwakoti, Y.P.; Li, L.; Kerekes, T.; Blaabjerg, F. (2019). Transformerless Inverter Topologies for Single-Phase Photovoltaic Systems: A Comparative Review. IEEE J. Emerg. Sel. Top. Power Electron. 8: 805–835.

4. Kumar, K.S.; Kirubakaran, A.; Subrahmanyam, N.(2020). Bidirectional Clamping-Based H5, HERIC, and H6 Transformerless Inverter Topologies With Reactive Power Capability. IEEE Trans. Ind. Appl. 56: 5119–5128.

5. Li, H.; Zeng, Y.; Zhang, B.; Zheng, Q.; Hao, R.; Yang, Z.(2018). An Improved H5 Topology with Low Common Mode Current for Transformerless PV Grid Connected Inverter. IEEE Trans. Power Electron. 34:1254–1256.

6. Mathi, A.A.D.(2019). Ramaprabha, R. Comparative Analysis of Grid Connected Transformerless Photovoltaic Inverters for Leakage Current Minimization. Indian J. Sci. Technol. 11:1–8.

7. D. Schmidt, D. Siedle, and J. Ketterer.(2003). "Inverter for transforming a DC voltage into an AC current or an AC voltage," EP Patent 1: 369 985.

8. M. Victor, F. Greizer, S. Bremicker, and U. Hubler.(2008). "Method of converting a direct current voltage from a source of direct current voltage, more specifically from a photovoltaic source of direct current voltage, into an alternating current voltage," U.S. Patent: 7 411 802.

9. Zhang, L.; Sun, K.; Feng, L.; Wu, H.; Xing, Y.(2013). A Family of Neutral Point Clamped Full-Bridge Topologies for Transformerless Photovoltaic Grid-Tied Inverters. IEEE Trans. Power Electron. 28: 730–739.

Emerging Technologies and Applications in Electrical Engineering –
Prof. Dr. Anamika Yadav et al. (eds)
© 2024 Taylor & Francis Group, London, ISBN 978-1-032-82568-7

Fuzzy Inference System-based DC Fault Detection Method for Modular Multilevel Converter HVDC System

21

Mahima Kumari[1], Anamika Yadav[2]

Department of Electrical Engineering, National Institute of Technology Raipur, CG, India

ABSTRACT: High Voltage Direct Current (HVDC) systems, particularly in complex setups such as Back-to-Back and DC line HVDC with Modular Multilevel Converters (MMC), play a pivotal role in modern power transmission networks. Ensuring the reliability and stability of these systems necessitates the development of efficient fault detection methods. In this paper, a novel fault detection approach utilizing a Fuzzy Inference System (FIS) is proposed. Simulation studies for DC faults, such as pole-to-ground and pole-to-pole faults are carried out for both the Back-to-Back connection and DC line configuration MMC-HVDC system. DC line current and voltage are taken as the inputs for the Fuzzy Logic controller with rule viewer. DC pole-to-ground and pole-to-pole short circuit faults are studied at different locations in the DC line configuration. Simulation results confirm the suitability of the proposed FIS for fault detection in the MMC-HVDC system.

KEYWORDS: MMC-HVDC, DC pole to ground fault, DC pole to pole fault, Fault Analysis, Fuzzy inference system

1. Introduction

For more than 50 years, the industry has relied on High Voltage DC Current technology (Sahu, Yadav and Pazoki 2023). It began as a specialized product but has acquired significant pace in recent decades as the demand for reliable and efficient power has increased. HVDC is frequently employed in the transmission of offshore wind energy and other renewable sources, which is especially helpful in achieving the sustainable developments goals. HVDC is based

[1]mkumari.mtech2023.ee@nitrr.ac.in, [2]ayadav.ele@nitrr.ac.in

DOI: 10.1201/9781003505181-21

on two bidirectional electric power converters, where one is acting as a rectifier and the other is acting as an inverter. It is further separated into two groups: In first group, HVDC system is Back-to-Back (BTB) connection of rectifier and inverter at two ends specifically used to connect the two asynchronous sources, whereas in second group the long distance HVDC line is utilized to transfer bulk amount of power from one end wherein converter 1 acts as rectifier to another end wherein converter 2 acts as inverter (Slettbakk 2018). The bulk amount of power is delivered through HVDC overhead lines. The Back-to-Back connection is when two converter stations are connected at the same site and often same building using direct bus bars. The converter topology of choice is the HVDC modular multilevel converter, which has added benefits of being able to connect to weak AC networks (Kurtoğlu and Vural 2022). In addition, it possesses reactive power support, low harmonic distortion as well as high modularity and reliability. The HVDC system is vulnerable to many types of faults. These faults are located in grid side, load side or in DC line or DC bus. There are two possibilities of fault in DC bus/line, either a pole is short circuited to ground or a pole is short circuited to another pole. Also, the malfunction of submodule may be considered as submodule fault. There are different types of methods for fault detection, but most evident method is FIS based method for reliable and rapid fault detection (Nadeem, et al. 2018).

In this paper, Fuzzy Inference System (FIS) is proposed to detect the fault in DC link of Back-to-Back connection and DC line of MMC-HVDC system (Dessouky, et al. 2018) (Paily, et al. 2015). The change in magnitude of DC current in positive pole and DC voltage shows the effect of fault in the system (Bhatnagar and Yadav 2020) (Fuentes-Burruel and Moreno-Goytia 2022). For fault detection the DC current and DC voltage is taken as the inputs of Fuzzy Logic Controller. FIS based method is reliable to detect most of the faults and reduces the risk of false alarm.

2. HVDC MMC Converter Topology and Modulation Strategy

The Modular Multilevel Converter (MMC) typically employs a full-bridge configuration. The full-bridge configuration in MMC consists of multiple modular subunits, often referred as arms or cells connected in series to form a multilevel structure. The MMC is a type of power converter is a relatively new and advanced HVDC converter topology known for its ability to provide high voltage quality, scalability, and improved efficiency (Kurtoğlu and Vural 2022). Each arm of the MMC consists of several submodules, and each submodule is a combination of switching devices such as insulated gate bipolar transistors (IGBTs) or other semiconductor devices. The use of multiple levels allows MMC to approximate a near-sinusoidal output voltage waveform, resulting in reduced harmonics and improved voltage quality. The modularity of MMC simplifies maintenance and repair. If a sub-module fails, it can be isolated and replaced without affecting the entire converter's operation. MMC has built-in fault tolerance capabilities. This fault tolerance enhances the reliability of the converter. MMC converters can control power flow bidirectionally, allowing them to support both power transmission from the sending end to the receiving end and power flow in the opposite direction, which can be important for grid stability and renewable energy integration MMC offers precise and rapid control of voltage and current. This control flexibility is essential for maintaining grid stability and managing power flow in real-time. Fig. 21.1 shows the basic dynamics of current and voltage between MMC converter stations.

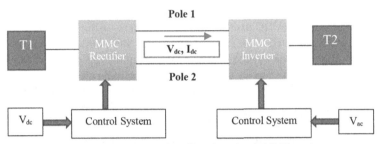

Fig. 21.1 Basic principle of back-to-back MMC system

The MMC consists of several sub-modules (SM) acting as separate units, connected together in series to form a converter arm. One submodule consists of two switches, two diodes and one capacitor. The name "modular multilevel" comes from the ability to connect these submodules in a modular way, as a piece of a larger converter, and the multilevel is defined by the amount of submodules connected in series. By supplying the capacitor's voltage to the submodule's output terminal and turning it on and off, one may regulate the submodule's operation. This method of managing several submodules allows one to maintain a steady DC voltage with the rectifier and shape the output AC voltage of the inverter.. The subscripts to be used are given as: 'j' Defines the phase (a, b, c) while 'u' means upper arm and 'l' is lower arm. 'N' Defines the amount of submodules in the converter, while 'i' is reserved for each individual module starting from 1 to 'N'.

$$Vc_{sm:i} = Ton.V \tag{1}$$

$$Varm = \sum_{i=1}^{N} Vc_{sm:i} \tag{2}$$

$$Levels = N + 1 \tag{3}$$

Looking into the definition of the control signal of the module, which essentially tells how many of the total modules of one arm is inserted at maximum AC amplitude. In an 'N+1' level converter there needs to always be 'N' modules inserted in total in the upper and lower arms of one phase. This does not mean that the upper arm must insert all modules at one point, but can keep some in the lower arm inserted as well. This is the converters ability to control. For example, with a modulation index of 0.8, the upper arm will have inserted 80% of its modules, while the lower will have inserted 20% of its modules. Next equation (4) presents the definition of modulating index with the variable T, being 1 for on-state of the submodule, and 0 for the off-state.

$$m_j = \frac{\sum_{i=1}^{N} (T_{u,ji})}{N} \tag{4}$$

For an MMC working at several levels, nearest level modulation (NLM) is the most widely used modulation approach (Jatin, Agarwal and Kumawat 2020). This effectively lowers the converter's related switching losses by enabling operation at switching frequencies that are equal to the fundamental frequency. The NLM functions by converting the continuous and sinusoidal modulating signal into a discrete stair waveform, directly informing the pulse generator how many modules to insert NLM has the advantage of being compatible with sophisticated voltage balancing algorithms based on voltage sensor measurements for each submodule. Fig. 21.2 shows the strategy of modulation. NLM is the most common strategy for higher power MMC applications in practical operation.

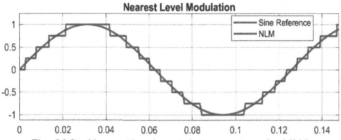

Fig. 21.2 Nearest level modulation strategy for MMC

From the network parameters, the individual rectifier and inverter should be rated for 220 MVA for this study. From theory and selection of steady state operation condition, a modulation-index of 0.8 is commonly adopted from literature as it is a good trade-off between regulation-range and efficient design of the power electronics. With this modulating index, the operating voltage of the HVDC-MMC converter is

$$V_{DC} = \frac{V_{LL}}{0.612 * m} = \frac{66\,kV}{0.612 * 0.8} = 134.8kV \tag{5}$$

The dimensioning DC voltage for the system will be 135 kV, which will be imposed on every individual arm of the converter. The total blocking voltage of all submodules should thus be rated for this as maximum DC voltage. As the calculation becomes more complex and the accuracy decreases, internal losses are not taken into consideration. Given that the system's phase currents are constant in steady state as well as during a dip and that circulating currents are suppressed, the currents at the inverter and rectifier sides are computed as follow:

$$I_{ph_{rms}} = \frac{S}{\sqrt{3} \cdot V_{LL}} = \frac{220MVA}{\sqrt{3} \cdot 66kV} = 1.92kA \tag{6}$$

The maximum nominal DC power is 220 MW, thus the maximum DC current is:

$$I_{DC_{rms}} = \frac{P_{DC}}{V_{DC_{min}}} = \frac{220MW}{107.8kV} = 2.0kA \tag{7}$$

When modulating index is at maximum (m=1.0) the DC voltage will regulate to a minimum. The number of levels must be determined using equation (8) before continuing with this calculation.

$$N = ceil\left(\frac{V_{DC}}{V_{sm}}\right) = ceil\left(\frac{135kV}{2.7kV}\right) = 50 \tag{8}$$

From the IGBT design, the 4.5kV module is chosen for conceptual design. With a derating of 60%, the submodule will be rated at 2.7 kV. The converter will have 50 modules and function as a 51 level MMC for both a rectifier and an inverter.

3. DC Faults

3.1 DC Pole to Ground Fault

A DC pole to ground fault also known as a line-to-ground faults, which occurs when one of the conductors/pole in a DC circuit comes into contact with ground or earth. This fault results in a direct current path from one pole to the ground.

3.2 DC Pole to Pole Fault

A DC pole to pole fault also referred as a line-to-line fault, arises when there is an unintended direct connection (short circuit) between two conductors or poles of a DC circuit. This fault leads to a direct current path between the two poles, by passing the normal load or circuit component. This paper discusses the pole to ground and pole to pole fault of HVDC MMC system in which the faults are taken in the DC link which is directly connected to a bus bar. As we know due to fault, current increases to its normal operating value and voltage decreases to its normal operating value (Kochar, Ekka and Yadav 2023).

4. FIS Development for Fault Detection

For the fault detection in DC link bus bar of Back-to-Back HVDC MMC system, DC current only is used as the input for pole to ground fault and DC voltage as the input for pole to pole fault (Ekka and Yadav 2022). Further for the fault detection in DC line of HVDC MMC system of the length 200km, DC current is used as an input of Fuzzy logic controller with Mamdani FIS for both pole to ground and pole to pole faults.

Table 21.1 Fault algorithm for DC current

DC Current	FAULT Detection
ILOW	0
IMID	1
IHIGH	1

Table 21.2 Fault algorithm for DC voltage

DC Voltage	FAULT Detection
VLOW	1
VMID	1
VHIGH	0

During faulty condition, DC current (I_{dc}) varies in the range of 0 to 3500A. The foremost step of Fuzzy based fault detection is to fuzzify the input variable. The membership value assignment is to be done thus the current is divided into three ranges low, medium and high as shown in Fig. 21.3. Similarly, the range of voltage is from 0 to 15×10^4 V for V_{dc} as shown in Fig. 21.4, the voltage is also divided into low, medium and high. I_{dc} is used for pole to ground

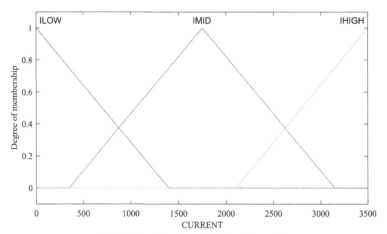

Fig. 21.3 Membership function of I_{dc}

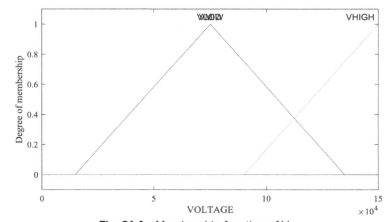

Fig. 21.4 Membership function of V_{dc}

fault detection and V_{dc} is used for pole-to-pole fault detection of the Back-to-Back HVDC MMC with bus bar connection. Table 21.1 and Table 21.2 show the membership function of input current and voltage with their respective state of fault. During the normal operating condition, the DC current magnitude will be the nominal current represented as ILOW thus fault detection output should be zero (0), whereas during a fault condition, the DC current magnitude will rise and represented as IMID or IHIGH and thus fault detection output should be high (1).

The range of current is from 0 to 30000A for I_{dcNEW1} for MMC-HVDC line Connection shown in Fig. 21.5. Here I_{dcNEW1} is used for fault detection of the Back-to-Back HVDC MMC with the DC line of 200km.

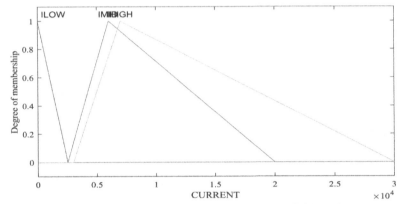

Fig. 21.5 Membership Function of I_{dcNEW1} for MMC-HVDC line Connection

5. MATLAB Simulation and Results

The Back-to-Back HVDC MMC system with bus bar and DC line has been simulated in MATLAB-Simulink including DC fault which is given through a breaker as shown in Fig. 21.6 & Fig. 21.7 and the parameters of the system are listed in Table 21.3.

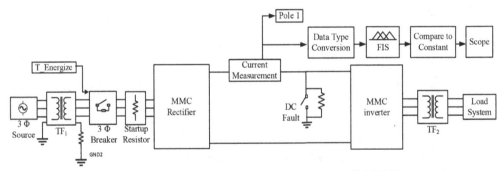

Fig. 21.6 Schematic block diagram of back-to-back HVDC MMC system

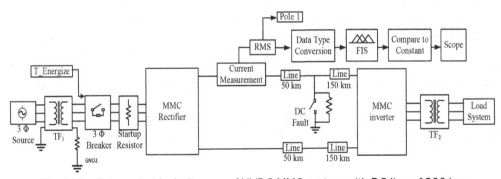

Fig. 21.7 Schematic block diagram of HVDC MMC system with DC line of 200 km

Table 21.3   Network parameters

Primary voltage	400kV
Short-circuit power	12000MVA
X/R ratio	14.28
Z0/Z1 ratio	0.8
Transformer 1 rating	110MVA
Transformer 1 voltage	400kV /66kV
Transformer 2 rating	38MVA
Transformer 2 voltage	66kV /18kV

Table 21.4   Transient state description

Time	Event
t = 0.0 sec	Energizing of transformer T1
t = 0.1 sec	Start energizing of converter with start-up resistor
t = 0.4 sec	Bypass start-up resistors, enable DC regulator
t = 0.5 sec	Enable AC regulator
t = 0.6 sec	Switch on the linear load
t = 0.7 sec	Switch on non-linear load
t = 1.0 sec	Steady state reached

As per Table 21.4, different transient states to which the HVDC MMC System is subjected are simulated. Total simulation time is 6.0 sec. from t = 0.0 sec to 1.0 sec, the system is in transient state and reaches its steady state at 1 sec as illustrated from DC current and voltage waveforms during healthy condition in Fig. 21.8 and 9 respectively. Further the fault analysis in HVDC system is carried out in steady state. A DC fault is simulated using a breaker at the instant of

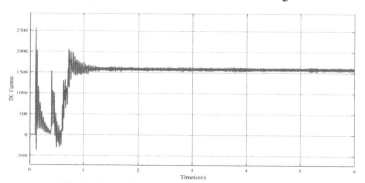

Fig. 21.8   DC current during healthy condition

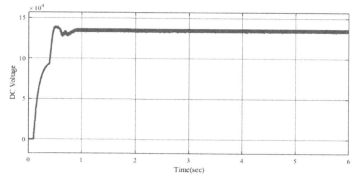

Fig. 21.9   DC voltage during healthy condition

4 sec. In first case fault is taken in between the converter station which are directly connected to a bus bar and in second case fault is taken in between the converter station of the line length of 200 km at a distance of 50 km from the MMC rectifier. The DC current and DC voltage along with the fuzzy logic controller output for different types of DC faults are exemplified in Fig. 21.10 and Fig. 21.11 respectively for the back-to back HVDC MMC System, whereas for HVDC MMC System with 200 km DC line are depicted in Fig. 21.12 and Fig. 21.13 respectively.

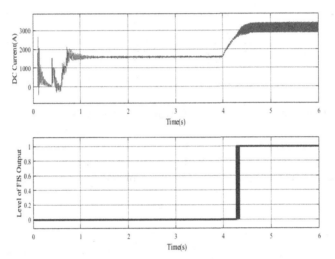

Fig. 21.10 DC current of pole to ground fault in DC bus and Fuzzy controller output

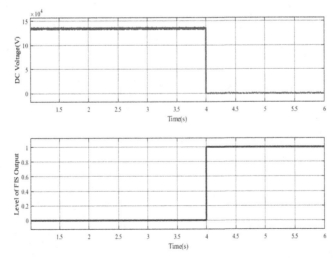

Fig. 21.11 DC voltage of pole-to-pole fault in DC bus and fuzzy controller output

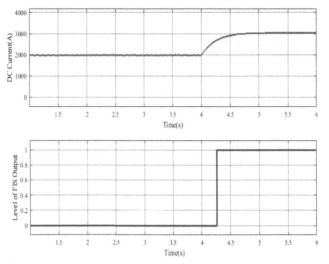

Fig. 21.12   DC current of pole to ground fault in DC line and fuzzy controller output

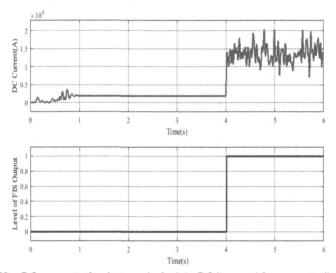

Fig. 21.13   DC current of pole-to-pole fault in DC line and fuzzy controller output

6. Conclusion

The FIS based method for detection of pole to ground and pole to pole fault for the different configuration of HVDC MMC system has been presented. Also, fuzzy logic controller can differentiate between the normal system and the system under fault. The change in DC current and DC voltage during the occurrence of a fault are the main parameters used for detection of the fault using fuzzy logic controller. The range of the membership function successfully detects the fault condition after some time from the instant of the fault occurrence. Main concern of using this method is choosing membership function range very carefully. Hence, it is concluded that the FIS based method is a suitable method for the detection of both pole to ground and pole to pole DC fault in MMC-HVDC system. This method is very efficient and accurate and provides quick response in fault condition.

REFERENCES

1. Bhatnagar, M., and A. Yadav. 2020. "Fault detection and classification in transmission line using fuzzy inference system." In 2020 5th IEEE International Conference on Recent Advances and Innovations in Engineering (ICRAIE). IEEE. 1-6.
2. Dessouky, S. S., M. Fawzi, A. Ibrahim, and N. F. Ibrahim. 2018. "DC pole to pole short circuit fault analysis in VSC-HVDC transmission system." In 2018 Twentieth International Middle East Power Systems Conference (MEPCON). IEEE. 900-904.
3. Ekka, A., and A. Yadav. 2022. "Fault identification using fuzzy in renewable energy interfaced IEEE 13 bus system." In 2022 International Conference on Intelligent Controller and Computing for Smart Power (ICICCSP). IEEE. 1-6.
4. Fuentes-Burruel, J. M. S., and E. L. Moreno-Goytia. 2022. "DC line current in a compact MTDC system for DC short-circuit fault analysis." In 2022 IEEE International Autumn Meeting on Power, Electronics and Computing (ROPEC). IEEE. 1-5.
5. Jatin, A. Agarwal, and M. Kumawat. 2020. "Mathematical Modelling and Comparative Analysis of Fault Monitoring of Modular Multilevel Converter." In 2020 First IEEE International Conference on Measurement, Instrumentation, Control and Automation (ICMICA). IEEE. 1-6.
6. Kochar, M., A. Ekka, and A. Yadav. 2023. "FIS Based Fault Identification and Classification in IEEE RTS96 System." In 2023 IEEE International Students' Conference on Electrical, Electronics and Computer Science (SCEECS). IEEE. 1-8.
7. Kurtoğlu, M., and Ahmet Mete Vural. 2022. "A novel nearest level modulation method with increased output voltage quality for modular multilevel converter topology." International Transactions on Electrical Energy Systems, 2022.
8. Nadeem, M. H., X. Zheng, N. Tai, M. Gul, and M. Yu. 2018. "Detection and Classification of faults in MTDC networks." In 2018 IEEE PES Asia-Pacific Power and Energy Engineering Conference (APPEEC). IEEE. 311-316.
9. Paily, B., S. Kumaravel, M. Basu, and M. Conlon. 2015. "Fault analysis of VSC HVDC systems using fuzzy logic." In 2015 IEEE International Conference on Signal Processing, Informatics, Communication and Energy Systems (SPICES). IEEE. 1-5.
10. Sahu, U., A. Yadav , and M. Pazoki. 2023. "A protection method for multi-terminal HVDC system based on fuzzy approach." MethodX (MethodsX).
11. Slettbakk, T. E. 2018. "Development of a power quality conditioning system for particle accelerators." Norwegian U. Sci. Tech

Emerging Technologies and Applications in Electrical Engineering –
Prof. Dr. Anamika Yadav et al. (eds)
© 2024 Taylor & Francis Group, London, ISBN 978-1-032-82568-7

Speed Control of BLDC Motor using High Gain Converter Fed Six-Step Inverter

22

Konda Naresh[1], D. Suresh[2], and Swapnajit Pattnaik[3]

National Institute of Technology, Raipur, Chhattisgarh, India

ABSTRACT: High-Gain Converter (HGC) with a Switched Inductor (SI) and a 6-step Inverter are presented to control a brushless DC (BLDC) motor speed precisely. To regulate the speed of the BLDC motor, a PI controller effectively regulates the DC link voltage. For current control, a hysteresis controller is used. Two loops control the motor: the primary control loop for controlling current and the secondary control loop for controlling speed. The main loop controls the converter's DC link voltage in this setup. Hall sensors drive the six-step inverter's electronic commutation system, which generates switching pulses. MATLAB/ Simulink system blocks were used to simulate the proposed system to validate it and HGC Converter performance can be determined by comparing it to MVD Converter performance. HGC Converters are more advantageous when used for BLDC motor speed control.

KEYWORDS: BLDC, VSI, Voltage and current control, HGC, Hysteresis control

1. Introduction

Electric motors with brushless DC motors (BLDC), also known as trapezoidal permanent magnet motors, use electronic switching rather than mechanical switching. Moreover, they offer higher speeds, greater efficiency, and superior heat dissipation. Permanent magnets are used in the rotor of BLDC motors, while coils are used in the stator. An electronic switch controls the rotor's armature coils, which are controlled by transistors or rectifiers. A rotor's position is usually sensed by hall sensors or rotary encoders. Modern drive technology relies heavily on BLDC motors.

[1]knaresh.phd2022.ee@nitrr.ac.in, [2]dsuresh.ee@nitrr.ac.in, [3]spattnaik.ele@nitrr.ac.in

DOI: 10.1201/9781003505181-22

BLDC motors are powered by DC or renewable energy sources through a power-electronic inverter and a DC-DC converter with high voltage gain [A. B. Kancherla, 2020]. These converters are valuable for renewable energy applications as they facilitate voltage boosting, making them suitable for distribution systems [K. Naresh, 2023]. They are used in various fields, including aerospace, medical, robotics, servo appliances, electric vehicles, and electric motors [D. G. Jegha, 2020]. DC-DC boost converters are used for high gain and higher voltage applications [Konda Naresh, 2023]. DC motors use brushes to switch, while electronic controllers switch motor windings. Step-up voltage uses traditional or classical converters, while transformer-free Double Boost Converter (DBC) [Sunil Kumar Sahare, 2023] is a potential solution for unconventional power sources. This article discusses the use of proportional-integral (PI) controllers [M. R. Haque, 2021] and hysteresis control in DC-to-DC power converters, with PI controllers being popular due to high sampling frequency and hysteresis current control [A. Usman, 2020] is admired for its quick response to divergence from control references. This article introduces a modified boost converter topology for high gain at lower switching stress, demonstrating a promoting response to input and suitable for double voltage applications.

The rest of the paper is structured as follows. Section 2 describes the conventional boost converter, and modified boost converter, that is voltage double converter (VDC), and High Gain Converter (HGC) converter. Section 3 describes the BLDC motors fed by six-step inverters and fed by HGC converters. Section 5 discusses the article containing the simulation results. Section 6 summarises the paper of the conclusion.

2. DC-DC Converters

2.1 Modified Boost Converter

The Modified Voltage Doubler (MVD) converter is a boost converter circuit that adds three capacitors and three uncontrolled diodes to increase voltage gain. It operates in two modes, depending on the power-controlled switch's ON or OFF conditions. The inductor is charged through the SW switch, while capacitors C1 and C3 are drained to the load, resulting in a C1 + C3 output voltage. The SW switch charges capacitor C1 by connecting the input supply voltage to D1 while it is off.

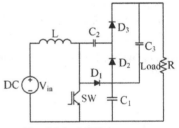

Fig. 22.1 MVD converter

2.2. High Gain Converter (HGC) Converter

We proposed only one circuit here: nothing more than an inductor switched for a switched inductor (SI) [P. K. Maroti, 2017]. Because the switched inductor (SI) (Fig. 22.5) structure adds the extra gain of the (1+D) times to the converter. The proposed HGC converter configurations

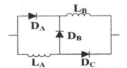

Fig. 22.2 Switched inductor structure

have the same workings, but the HGC converter depends on the power-controlled switch's state. The switched inductor structure operates in parallel when the circuit is in charging mode, and in series when it is in discharging mode. When the power-controlled switch is on, the SI structure is in charging mode, and when it is off, the remaining two modes of operation are the same as those of the MVD converter.

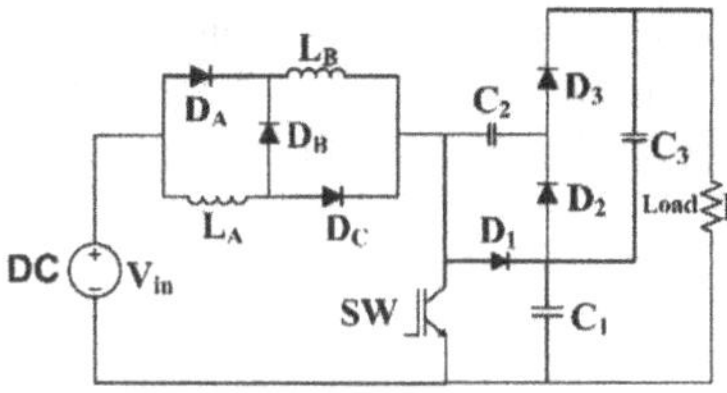

Fig. 22.3 High HGC converter

2.3 Mathematical Analysis of Voltage Conversion Ratio of MVD Converter and HGC Converter

Modified Voltage Doubler (MVD) converter: Where V_{in} is the input supply voltage, V_L is the Inductor voltage, V_{C1} is the Capacitor C_1 voltage, V_{C2} is the capacitor C_2 voltage, V_{C3} is the Capacitor C_3 voltage, and V_o is the Load Voltage.

The power-controlled switch S_W On Condition:

KVL, for mesh 1:

$$\Delta_{iL_{on}} = \frac{V_{in}}{L} \cdot DT \tag{1}$$

KVL, for mesh 2:

$$V_{C1} = V_{C2} \tag{2}$$

KVL, for mesh 3:

$$V_{C0} = V_{C1} + V_{C3} \tag{3}$$

The power-controlled switch S_W Off Condition:

KVL, for mesh 1:

$$\Delta i_{Loff} = \frac{V_{in} - V_{C1}}{L} \cdot (1 - D) \cdot T \tag{4}$$

KVL, for mesh 2:

$$\Delta i_{Loff} = \frac{V_{in} - V_{C1} + V_{C2} - V_{C3}}{L} \cdot (1 - D) \cdot T \tag{5}$$

KVL, for mesh 3:

$$V_0 = V_{C1} + V_{C3} \tag{6}$$

With respect to the Volt-Second balance equation, from equation 1 and 4

$$\frac{V_{C1}}{V_{in}} = \frac{1}{1 - D} \tag{7}$$

From equation 7

$$V_{in} + (1 - D) \cdot \left(V_O - \frac{V_{in}}{1 - D} \right)$$

$$\frac{V_O}{V_{in}} = \frac{2}{1 - D} \tag{8}$$

High Gain Voltage Doubler converter: Where V_{LA} is the Across Inductor L_A voltage of the SI structure, V_{LB} is the Across Inductor L_B voltage of the SI structure, The power-controlled switch S_W ON Condition: In SI structure, Inductors are in parallel So, $V_{LA} = V_{LB} = V_L$

KVL, for mesh 1:

$$\Delta_{iL_{on}} = \frac{V_{in}}{L} \cdot DT \tag{9}$$

KVL, for mesh 2:

$$V_{C1} = V_{C2} \tag{10}$$

KVL, for mesh 3:

$$V_{C0} = V_{C1} + V_{C3} \tag{11}$$

The power-controlled switch S_W Off Condition: In SI structure, inductances are in series so, $V_{LA} + V_{LB} = 2V_L$

KVL, for mesh 1:

$$\Delta i_{Loff} = \frac{V_{in} - V_{C1}}{2L} \cdot (1 - D) \cdot T \tag{12}$$

KVL, for mesh 2:

$$2V_L = V_{in} + V_{C2} - (V_{C1} - V_{C3}) \tag{13}$$

KVL, for mesh 2:

$$V_0 = V_{C1} + V_{C3} \tag{14}$$

Equations 14 and 13 can be written as

$$\Delta i_{Loff} = \frac{V_{in} + V_{C2} - V_o}{2L} \cdot (1 - D) \cdot T \tag{15}$$

With respect to the Volt-Second balance equation, from equations 9 and 12

$$\frac{V_{C1}}{V_o} = \frac{1 + D}{1 - D} \tag{16}$$

With respect to the Volt-Sec balance equation

$$\frac{V_O}{V_{in}} = 2\frac{1 + D}{1 - D} \tag{17}$$

3. BLDC motor Drive fed with Six–Step Inverter and HGC Converter

The BLDC motor fed with an HGC Converter circuit is shown in Fig. 22.4. and the BLDC motor ideal back-EMF, stator current and hall signal waveforms in Fig. 22.8. Mathematical modelling equations of 3-Θ BLDC Motor expressed in equations 18,19, and 20 are

$$V_a = i_a R_a + L\frac{d}{dt}i_a + e_a \tag{18}$$

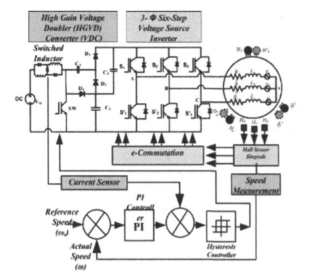

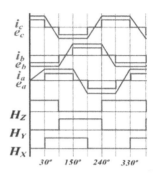

Fig. 22.5 BLDCM's waveforms of stator current, back-EMF and hall signals

Fig. 22.4 HGC-fed BLDC motor fed with six-step inverter

$$V_b = i_b R_b + L \frac{d}{dt} i_b + e_b \tag{19}$$

$$V_c = i_c R_c + L \frac{d}{dt} i_c + e_c \tag{20}$$

Were R_a, R_b and R_c Stator resistance, i_a, i_b and i_c, e_a, e_b and e_c Stator Back-EMF, of BLDC Motors phase A, B and C; L is the Inductance of the Stator winding.

The electromagnetic torque equation of the BLDC Motor is given by

$$\tau = \frac{e_a \cdot i_a + e_b \cdot i_b + e_c \cdot i_c}{\omega} \tag{21}$$

Where ω is the rotor speed of the BLDC Motor. The mechanical torque is

$$T_m = \mathcal{I} \frac{d\omega}{dt} + B \cdot \omega + T_L \tag{22}$$

Where B is the Damping Constant, is $\mathcal{I}$ the Rotor Inertia of the BLDC Motor, and T_L is the Load Torque.

The article presents a brushless DC motor drive based on voltage double converters fed with a six-step inverter. An inverter's DC link voltage is regulated by a high-gain converter. A Hall effect sensor senses the position of the rotor in the six-step inverter, which generates the switching pulses. High-gain converters regulate the DC link voltage of BLDCs using a six-step inverter with an output speed loop and an inner current loop. A comparison is made between the BLDC motor fed by a boost converter and the BLDC motor fed by a HGC converter. The switching pulses are generated as per the ideal waveforms of the BLDC motor are shown in Fig. 22.5.

4. Simulation Results and Discussion

The BLDC motor speed control is simulated using Simpower system blocks in MATLAB, with an outer speed control loop and inner current control loop developed using the MATLAB Simulink environment. The simulation parameters of the BLDC motor are compared with those of the MVD and HGC converters, and DC-DC converter simulation parameters are used to perform the simulations and responses are shown in Table 22.1.

Table 22.1 Simulation results and gains of the different converters

S. No.	Name of the Converter	Voltage Gain	DC-DC Converter	
			Inductor Current (A)	Converter Output Voltage (V)
1	Conventional boost converter	$\dfrac{1}{1-D}$	8.67A	199 V
2	VDC Converter	$\dfrac{2}{1-D}$	5.8A	394.9 V
3	HGC Converter	$2\dfrac{1+D}{1-D}$	3.58A	450.6

Source: Author's compilation

Table 22.2 Simulation results of different converters fed BLDC motor

BLDC Motor					
S. No.	Converter-fed BLDC	Current	Back-EMF	Speed	Torque
1	VDC fed BLDC	2.2 A	183.2 V	2499 rpm	2.789 N-m
2	HGC fed BLDC	0.8 A	183.1 V	2499 rpm	3.334 N-m

Source: Author's compilation

From the Fig. 22.6 to 10 are the simulation characteristics of VDC fed BLDC Motor. Figure 22.6 represents the VDC-fed BLDC motor stator current and back-EMF is shown in Fig. 22.7. These waveform simulation values are shown in Table 22.1, the BLDC motor fed with the VDC, the motor rotor speed is shown in Fig. 22.8., and the torque of the motor is represented in Fig. 22.9 and these values are represented in Table 22.1.

From the Figs. 22.10 to 22.13 are the simulation characteristics of the HGC-fed BLDC Motor. From the Figs. 22.10 to 22.11 are the simulation characteristics of the HGC-fed BLDC Motor. Figure 22.10 represents the HGC-fed BLDC motor stator current and back-EMF is shown in Fig. 22.11. These waveform simulation values are shown in Table 22.2, the BLDC motor fed with the HGC, the motor rotor speed is shown in Fig. 22.12., and the torque of the motor is represented in Fig. 22.13 and these values are represented in Table 22.2.

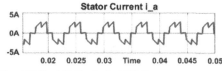

Fig. 22.6 VDC fed BLDC motor stator current (Amp)

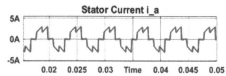

Fig. 22.10 HGC fed BLDC motor stator current (Amp)

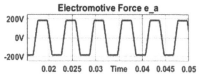

Fig. 22.7 VDC fed BLDC motor back – EMF (Volts)

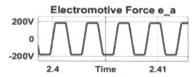

Fig. 22.11 HGC fed BLDC motor back – EMF (Volts)

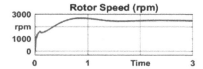

Fig. 22.8 VDC fed BLDC motor rotor speed (rpm)

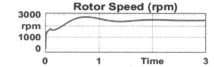

Fig. 22.12 HGC fed BLDC motor rotor speed (rpm)

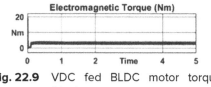

Fig. 22.9 VDC fed BLDC motor torque (N-m)

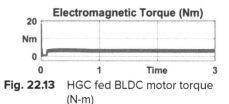

Fig. 22.13 HGC fed BLDC motor torque (N-m)

5. Conclusion

Brushless DC motor speeds are regulated with HGCs (High-Gain Converters) and six-step inverters. This setup enhances BLDC motor speed control, reducing current and torque ripples. BLDC motors perform better than HGCs and Modified Voltage Converters (MVCs) compared to HGCs and MVCs, according to an analysis of performance parameters. With HGC, not only is the motor speed more efficiently controlled but the torque ripple is also reduced, outperforming the MVC converter to improve the overall efficiency of BLDC motor systems.

REFERENCES

1. A. B. Kancherla and D. R. Kishore, "Design of Solar-PV Operated Formal DC-DC Converter Fed PMBLDC Motor Drive for Real-time Applications," 2020 IEEE International Symposium on

Sustainable Energy, Signal Processing and Cyber Security (iSSSC), Gunupur Odisha, India, 2020, pp. 1-6, doi: 10.1109/iSSSC50941.2020.9358813.

2. A. Usman and B. S. Rajpurohit, "Design and Control of a BLDC Motor drive using Hybrid Modeling Technique and FPGA based Hysteresis Current Controller," 2020 IEEE 9th Power India International Conference (PIICON), Sonepat, India, 2020, pp. 1-5, doi: 10.1109/PIICON49524.2020.9112895.

3. Darcy Gnana Jegha, A., M. S. P. Subathra, Nallapaneni Manoj Kumar, Umashankar Subramaniam, and Sanjeevikumar Padmanaban. 2020. "A High Gain DC-DC Converter with Grey Wolf Optimizer Based MPPT Algorithm for PV Fed BLDC Motor Drive" *Applied Sciences* 10, no. 8: 2797. https://doi.org/10.3390/app10082797.

4. K. Naresh, D. Suresh and S. Pattnaik, "Comparative Analysis of High Gain Boost Converter (HGBC) fed BLDC Motor," 2023 IEEE Renewable Energy and Sustainable E-Mobility Conference (RESEM), Bhopal, India, 2023, pp. 1-6, doi: 10.1109/RESEM57584.2023.10236055.

5. Konda Naresh, D. Suresh and S. Pattnaik, 2023, High Gain DC-DC Converter Fed Six-Step Inverter Based BLDC Motor, Proceedings of the Second International Conference on Emerging Trends in Engineering (ICETE 2023),662-672, doi=10.2991/978-94-6463-252-167, Atlantis Press.

6. M. R. Haque and S. Khan, "The Modified Proportional Integral Controller for the BLDC Motor and Electric Vehicle," 2021 IEEE International IOT, Electronics and Mechatronics Conference (IEMTRONICS), Toronto, ON, Canada, 2021, pp. 1-5, doi: 10.1109/IEMTRONICS52119.2021.9422548.

7. P. K. Maroti, S. Padmanaban, P. Wheeler, F. Blaabjerg and M. Rivera, "Modified boost with switched inductor different configurational structures for DC-DC converter for renewable application," 2017 IEEE Southern Power Electronics Conference (SPEC), Puerto Varas, Chile, 2017, pp. 1-6, doi: 10.1109/SPEC.2017.8333674.

8. S. K. Sahare and D. Suresh, "Modified Boost Converter-Based Speed Control of BLDC Motor," 2023 IEEE International Students' Conference on Electrical, Electronics and Computer Science (SCEECS), Bhopal, India, 2023, pp. 1-5, doi: 10.1109/SCEECS57921.2023.10062973.1.

Emerging Technologies and Applications in Electrical Engineering –
Prof. Dr. Anamika Yadav et al. (eds)
© 2024 Taylor & Francis Group, London, ISBN 978-1-032-82568-7

Investigation of Topologies and PWM Schemes of RSC-MLI for Asymmetrical Configurations

23

Supriya Choubey[1], Hari Priya Vemuganti[2]
NIT Raipur

ABSTRACT: Reduced switch count multilevel inverter (RSC-MLI) has developed as a fascinating field in power electronic converters. Numerous topologies such as T-type MLDCL, SSPS, and Cascaded T-type for different symmetrical and asymmetrical RSC MLI configurations are described with various modulation strategies such as Reduced Carrier and Multireference for controlling RSC-MLI has been reported. This paper aims to investigate the performance and features of various asymmetrical RSC MLI with different novel PWM techniques in MATLAB/Simulink environment

KEYWORDS: Reduced switch count (RSC), Multilevel inverter (MLI), Pulse width modulation (PWM), Multilevel DC link (MLDCL), SSPS, T-type, LSPWM, PSPWM

1. Introduction

MLIs apply the principle of integrating small several levels of DC voltage and have several advantages that include reduced dv/dt, EMI, and THD. Multilevel inverters are widely utilized in many different applications, including electric motor drives, UPS systems, renewable energy systems, and high-voltage power transmission. Among all of the reported MLI configurations the majority of attention has been focused toward the conventional MLI topologies of cascading H-bridge (CHB), flying capacitor (FCMLI), and diode clamped (DCMLI)(Krishna R and L. P. Suresh.2016). But for the higher number of levels, many switches are required, thus the circuit becomes complex. To overcome these constraints (McGrath, Holmes, 2002). A variety of symmetrical and asymmetrical RSC MLI arrangements have been published, including T-type,

[1]supriyachoubey@gmail.com, [2]hpvemugantiee@nitrr.ac.in

DOI: 10.1201/9781003505181-23

packed unit cell (PUC), switched capacitor (SC) topologies, stacked arrangement, multilevel DC link (MLDCL), switched series parallel sources (SSPS), series connected switched sources (SCSS), and several three-phase configurations. Among them Type and MLDCL are most widely used RSC MLI configurations(Hari Priya, Sreenivasarao, Siva Kumar.2021)

Controlling the inverter's output voltage is frequently required in industrial applications to handle changes in the DC input voltage, regulate the inverter's voltage and meet the drives' demand for consistent voltage/frequency control. Numerous modulation methods have been created to regulate the output voltage of MLI. Level-shifted (LSPWM) and phase-shifted (PSPWM), carrier-based modulation algorithms, constitute PWM control methods because of their simple switching logic (Supriya, and Hari Priya 2021). However sym./asym. RSC MLI configuration have large switch count reductions, the modified functioning of the inverter cannot be realized using the classic carrier-based PWM techniques, such as LSPWM and PSPWM(Mohammed, Mustafa F., and Mohammed A..2022). Therefore, several carrier-based PWM schemes, including MRC, switching-function and RCPWM, are used to control any RSC-MLIs. In order to get the inverter to switch in the desired manner, unified logic expressions are combined with modulating signal design, unipolar level shifted carrier, and reduced carrier PWM.

RCPWM involves a unipolar modulating and $(k-1)$/two unipolar level shifted carrier signals to obtain k-levels in phase-voltage. The MRC PWM technique is a generalised scheme easily expandable to various groups. By comparing several unipolar sinusoidal DC-shifted ref. with a unipolar carrier signal, this technique generates switching pulses (Reddy, Hari Priya, and Kasireddy. 2022). To generate k phases in phase-voltage, this method employs $(k-1)/2$ DC shifted ref. The unipolar sinusoidal ref. is shifted with the carrier signal peak to produce the DC-shifted ref. (McGrath, Holmes, 2002). The hybrid switching technique is one of the novel modulation schemes. PWM has been reported for several types of sym. and asym. RSC-MLIs, including hybrid T-type topologies, switching DC sources, and PUC.

The objective of the present investigation is to investigate the attributes and constraints of the proposed structure of the widely used Asymmetrical RSC MLI configurations constructed for seven levels. To implement the suggested configuration, the widely used modified RCPWM technique with unified switching logic is employed. The structure of the paper is as follows: The features, limits, and switching function of Asym. configuration of RSC MLI unit topologies are covered in Section II. The proposed RSC-MLI topologies for seven levels and modified reduced carrier PWM is described in Section-III, along with how it is applied to the suggested architecture. In the end, Section IV Simulation outcomes for the reported state of Asym. RSC MLI.RSC MLI Topologies.

2. RSC MLI Topologies

T-type RSC MLI- Because of the large drop in switch count this structure is one of the most popular RSC-MLI topologies shown in Fig. 23.1(c). This MLI employs both unidirectional and bidirectional switches to connect the load terminals to various dc-link nodes. T-type uses m+1 switches, (m-1)/2 dc sources, and m-levels on output voltage.

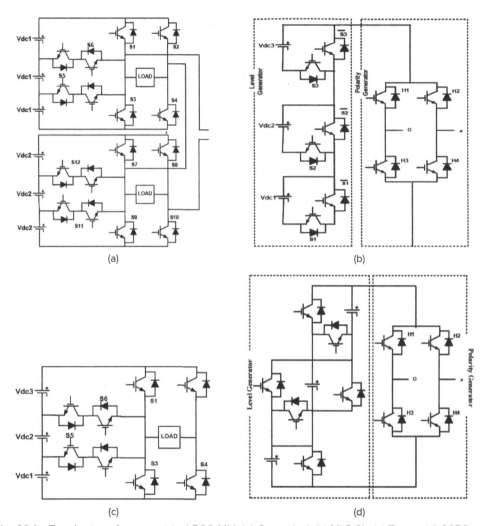

Fig. 23.1 Topologies of asymmetrical RSC MLI: (a) Cascaded, (b) MLDCL, (c) T-type, (d) SSPS

2.1 Multilevel dc-link RSC-MLI

By controlling the switching states of the semiconductor switches and the voltage levels in the DC link circuit, the MLDCL can generate output voltage waveforms with multiple voltage levels. The level generator generates a unipolar voltage having m+1 levels and range of zero to mV_{dc} with m basic units. The polarity generator converts this unipolar voltage into bipolar voltage with 2m+1 phases in phase voltage by using an H-bridge with switches operating at fundamental frequency. Figure 23.2(b) represents the asymmetrical configuration of MLDCL RSC MLI.Switching state of MLDCL

2.2 Switched Series Parallel Sources (SSPS) RSC-MLI

Switched Series Parallel Sources (SSPS) RSC-MLI structure increases the number of levels by implementing series/parallel switching of DC source of voltage shown in Fig. 23.1(d). SSPS have a modular design with individual level and polarity generators. Polarity generator generates zero voltage while level generators are unable to generate it. SSPS generates 2m+1 levels in phase-voltage with m identical dc voltages by using 3(m-1) switches in the level generator and 4 switches in the polarity generator. The seven-level SSPS configuration is shown in Fig. 23.1(d)

Table 23.1 Switching state of asymmetrical MLDCL RSC MLI

s. no.	Voltage Combination	Switches in Conduction		Output Voltage $3V_{dc1}=2V_{dc2}=V_{dc3}==V_{dc}$
		Level Gen.	Polarity Gen.	
1	$V_{dc1}+V_{dc3}+V_{dc2}$	S1-S2-S3	H_1-H_4	$+6V_{dc}$
2	$V_{dc3}+V_{dc2}$	S1'-S2-S3	H_1-H_4	$+5V_{dc}$
3	$V_{dc1}+V_{dc3}$	S1-S2'-S3	H_1-H_4	$+4V_{dc}$
4	$V_{dc1}+V_{dc2}$	S1-S2-S3'	H_1-H_4	$+3 V_{dc}$
5	V_{dc2}	S1'-S2-S3'	H_1-H_4	$+2 V_{dc}$
6	V_{dc1}	S1-S2'-S3'	H_1-H_4	$+ V_{dc}$
7	0	S1'-S2'-S3'	H_2-H_3	0
8	V_{dc1}	S1-S2'-S3'	H_2-H_3	$- V_{dc}$
9	V_{dc2}	S1'-S2-S3'	H_2-H_3	$-2 V_{dc}$
10	$V_{dc1}+V_{dc2}$	S1-S2-S3'	H_2-H_3	$-3 V_{dc}$
11	$V_{dc1}+V_{dc3}$	S1-S2'-S3	H_2-H_3	$-4 V_{dc}$
12	$V_{dc3}+V_{dc2}$	S1'-S2-S3	H_2-H_3	$-5 V_{dc}$
13	$V_{dc1}+V_{dc3}+V_{dc2}$	S1-S2-S3	H_2-H_3	$-6 V_{dc}$

2.3 Cascading of T-Type RSC MLI

The most common way to achieve mn−1 levels in phase voltage is to cascade an identical number of 'n'-level inverters. In order to create a 13-level T-type structure with a seven-level cascaded T-type configuration, or n=7, m must be 2. The 13-level cascaded T-type configuration produced by cascading two seven-level T-type units in each phase is shown schematically in Fig. 23.1(a). This 13-level structure, which is being evaluated, consists of m(n+1) switches, or 16 switches in each phase, and 3*m(n+1), or 48 switches total across all phases. Table 1 shows the switching state of Thirteen-level T-type asymmetrical RSC MLI.

3. PWM and Control

The topology must be implemented using an appropriate PWM control techniques to confirm its effectiveness of achieving the necessary voltage levels. Carrier-based PWM schemes are the most fundamental and extensively used of the reported PWM techniques. This section discusses the drawbacks associated with applying conventional PWM techniques on RSC-MLI and implementing new unified modulation addresses that are specifically designed for these specialized inverters. Figure 23.2 represents MRC modulation scheme for seven and Thirteen-level RSC-MLI. It uses three dc shifted ref. and one unipolar carrier signal to obtain seven level in phase voltage. Figure 23.3 shows the RCPWM' carrier and modulating signal configuration controls sym. and asym. RSC MLI. The switching pulse is obtained by comparing ref. and carrier. The pulse actively varying from 0 to 1, when the ref. is in carrier limits similar manner others pulses can be produced. Further, these pulses are operated with user defined logical expressions such that they meet the switching action of the inverter to be controlled.

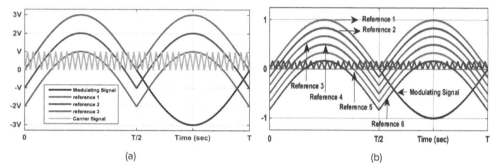

(a)   (b)

Fig. 23.2   Multi-reference PWM (a) Mutireference PWM for Seven-level, (b) Mutireference for Thirteen level

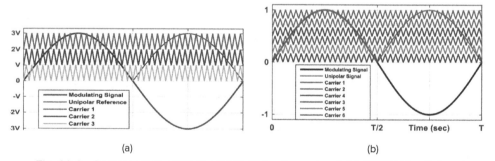

(a)   (b)

Fig. 23.3   Reduce carrier PWM (a) RCPWM for Seven-level, (b) RCPWM for 13-Level

4. Simulation Results

This section shows the simulation results of RSC MLI configurations for their state of art reported in order to analyse the operation of Asymmetrical RSC-MLIs and their PWM techniques. Various topological, including novel RSC-MLIs like T-type and MLDCL, SSPC, and SSPS are imposed. Additionally, the effects of a voltage increase levels, distinguishing modulation index, amplitude and switching frequency, the produced output voltage, and its harmonic performance are investigated using several kinds of topologies that implement popular carrier-based PWM techniques. Asymmetrical RSC MLI with MRC and RCPWM scheme is evaluated in MATLAB/Simulink environment. Each DC source voltage of 100V, f_{cr} and f_{sw} are selected as 2kHz and 50Hz res. Simulation is carried out at m_a of 1.

The obtained output voltage for equal voltage source and dissimilar voltage source are shown in Fig. 23.4, which depict the waveform and harmonic spectra of phase-voltage, line-voltage. The following observations are made, phase-voltage THD is obtained as 17.95 % Line-voltage THD is 16.78% for Seven-level multi ref. PWM and phase & line voltage THD is 17.91% and 16.73% for seven-level RCPWM. The harmonic spectra of RC and MRC PWM Thirteen-level RSC MLI is obtained phase and line-voltage. From these results, it is observed that MRC and RCPWM scheme results THD of 8.56% and 8.53%% in phase-voltage respectively.

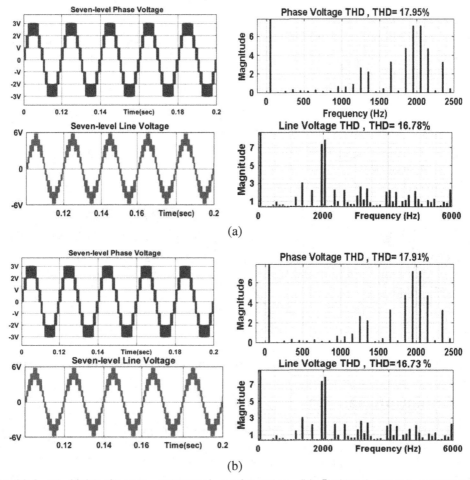

Fig. 23.4 (a) Multi-reference symmtrical configuration, (b) Reduced carrier symmetrical configuration

5. Conclusion

This paper investigates the performance of Phase & line voltage and harmonic performance for T-type, MLDCL. Cascaded, SSPS of asymmetrical RSC MLI by multireference and Reduce Carrier modulation methods are utilized for controlling RSC-MLIs. Asymmetrical MLI achieve the desired output voltage levels with fewer components than their symmetrical RSC MLI. This configuration can lead to lower costs, reduced complexity, and improved reliability, offering greater modularity and allowing for easier expansion or modification of the system.

REFERENCES

1. Vemuganti, Hari Priya, Dharmavarapu Sreenivasarao, Siva Kumar.2021. "A survey on reduced switch count multilevel inverters." *IEEE Open Journal of the Industrial Electronics Society* 2 (2021): 80-111.
2. Mohammed, Mustafa F., and Mohammed A..2022 Qasim. "Single phase T-type multilevel inverters for renewable energy systems, topology, modulation, and control techniques: A review." *Energies* 15, no. 22 (2022): 8720
3. Choubey, Supriya, and Hari Priya Vemuganti.2021. "Performance Investigation of Three-phase Inverters with Conventional and Novel Topologies of MLI.". *2nd International Conference for Innovation in Technology (INOCON)*, pp. 1-8. IEEE, 2023.
4. Krishna R and L. P. Suresh.2016. "A brief review on multi-level inverter topologies," International Conference on Circuit, Power and Computing Technologies (ICCPCT), Nagercoil, India, pp. 1-6, doi: 10.1109/ICCPCT.2016.7530373.
5. O. Arslan, M. Kurtoglu, F. Eroglu and A. M. Vural, 2018. "Comparison of phase and level shifted switching methods for a three-phase modular multilevel converter," 5th International Conference on Electrical and Electronic Engineering (ICEEE), Istanbul, Turkey, pp. 91-96, doi: 10.1109/ICEEE2.2018.8391307.
6. Reddy, R., V. Hari Priya, and I. Kasireddy. 2022."Simulink implementation of fifteen-level CHB MLC for active front end application." In *AIP Conference Proceedings*, vol. 2418, no. 1. AIP Publishing.
7. P. McGrath, D. G. Holmes, 2002. "Multicarrier PWM Strategies for Multilevel Inverters", IEEE Trans. Ind. Electron., Vol. 49, N o. 4, pp. 858-867.

Emerging Technologies and Applications in Electrical Engineering –
Prof. Dr. Anamika Yadav et al. (eds)
© 2024 Taylor & Francis Group, London, ISBN 978-1-032-82568-7

Isolation between Internal and External Fault in DC Microgrid using Artificial Neural Network

24

Aditi Barge[1], Ebha Koley[2]

Dept. of Electrical Engineering,
National Institute of Technology, Raipur, Raipur (C.G.), India

ABSTRACT: DC microgrid have gained significant attention in recent years because of its benefits related to high efficiency, low losses, improved power transfer capability and its integration with DERs. However, the widespread use of Microgrid has been hampered by the complexities involved in developing a resilient protection technique which results in high fault current value without zero-crossing. In this regards an ANN technique (Neural Net Fitting) is applied for protection of DC microgrid. The proposed ANN based method isolates between internal and external faults in the microgrid to ensure protection against internal fault within stipulated time. The test results confirm the suitability of the proposed scheme in differentiating between internal fault and external faults in DC microgrid.

KEYWORDS: DC microgrid, Distributed energy resources, Neural network fitting, Artificial neural network

1. Introduction

The advent of technological advancement and increase in environmental concerns has increased the demand for sustainable and renewable energy to an exceptional degree, as a result there is a steady rise in the utilization of microgrid systems. A microgrid is a small power network integrated with distributed energy resources which includes photovoltaic array, wind turbine, synchronous generator, battery system and AC, DC loads. The microgrid system have certain capabilities i.e., can operate in grid connected and in islanded mode (disconnected from main

[1]bargeaditi12@gmail.com, [2]ekoley.ele@nitrr.ac.in

DOI: 10.1201/9781003505181-24

grid) and provide smooth transition between the grid connected and islanded mode. Microgrids system are basically of three types AC microgrid, DC microgrid and Hybrid microgrid (both AC and DC) (Hooshyar and Iravani, 2017). Among all three, DC microgrids have emerged as highly efficient system, it provides improved power quality, low losses, is compatible with electrochemical storage system and highly robust which makes this system adequate for its use (Tiwari et al., 2021).

As DC microgrids incorporates different DERs, therefore, ensuring its safety during both normal and faulty condition becomes necessary. In DC lines faults are usually due to lightning strikes, mechanical stress and pollution. The most frequent and dominant faults in DC system are Pole to Pole (PP) and Pole to Ground (PG) fault, where PG fault is frequently encountered issue in comparison to PP fault (Khairnar and Shah, 2016), (Ali et al., 2021). These are permanent faults and it is required to remove the faulty section from the system soon the fault is detected. DC PP fault are the rare faults occurring in the DC system which is generally caused due to insulation failure between two DC cables while the DC PG fault occurs in DC system due to insulation failure between DC cable and ground.

To provide protection against these faults and to ensure the reliable and efficient operation, a reliable and accurate protection scheme is needed. One of the important issues of the DC microgrids protection is that as DC microgrids support bidirectional power flow. Therefore, it is essential to differentiate between the internal faults and external faults so that only the respective faulty section must be disconnected from the system while the microgrid can supply the remaining sections in service. Since, the microgrid can work as grid connected as well as in islanded mode. Some of the of the protection schemes employed for protection of DC microgrid reported earlier are (Telukunta et al., 2017), (Chen and Mei, 2015), (Habib et al., 2019).

This paper presents a protection technique for DC microgrid which is able to differentiate between internal faults and external fault. This protection technique has been developed using artificial neural network (ANN). The instantaneous DC voltage and DC current signals are given as input to the ANN based protection scheme. Various types of internal and external fault have been created in DC microgrid. The proposed scheme has been tested for the internal and external fault scenarios and some of the results are depicted in result section.

2. DC Microgrid Model Description

For fault analysis, a DC microgrid with ring main topology is considered as shown in Fig. 24.1. The microgrid system is integrated with different distributed energy resources (DERs) i.e., Photovoltaic array of 250KW, PMSG wind turbine of 500KW and Diesel generator of 186KVA along with energy storage system (Li-ion battery) with AC and DC load of 500KW are connected. The ring voltage of grid is 350 V. All the DERs are connected to suitable converters, for their synchronization with DC line. The Photovoltaic array and the Energy storage system (Li-ion battery) is connected to bidirectional DC-DC converter and the Main/ Utility grid, PMSG wind turbine and Diesel generator is connected to AC-DC converters of suitable rating. The DC Microgrid can operate in both grid-connected and in islanded mode

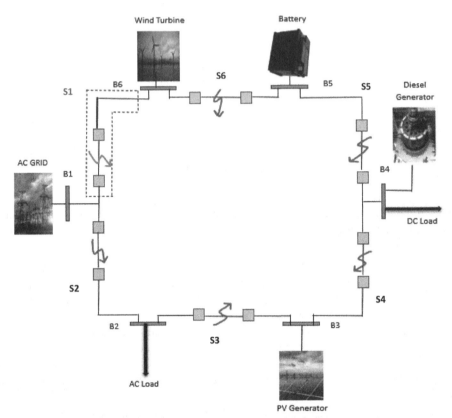

Fig. 24.1 DC microgrid ring main configuration

using the PCC (point of common coupling) connected to the Utility grid. In grid connected mode, the utility grid supplies power to the load and in Islanded mode all the DERs and battery acts as sources. The DC microgrid consists of six sections (S1-S6), each section is of 2km and six buses (B1-B6) each bus is connected to DER, Bus1 (B1) is connected to Main grid, Bus2 (B2) is connected to AC load, Bus (B3) is connected to Photovoltaic array, Bus4 (B4) is connected to DC load, Bus5 (B5) to Battery Storage system, Bus (B6) is connected to PMSG wind turbine, A Pie model transmission line considered for cable. Suitable parameters are considered for simulation of DERs, Battery and convertors (Mohanty and Pradhan, 2018), (Muhtadi and Mortuza). For study purpose, Fault is created in each section of microgrid and its effects are studied using instantaneous current and voltage samples at B1. Here, section 1 (S1) is considered as Internal section and other five (S2-S6) as external section for relay.

3. Proposed Fault Detection Scheme

To overcome the limitations of conventional relays, more effective and robust protection strategies are now been used. Machine learning possesses the ability to integrate domain knowledge with real-world data, leading to versatile computing solutions. In the present scheme, ANN a class of soft computing technique is applied. This scheme can flexibly adapt operating condition of system at high speed (Nagam et al., 2017), (Koley et al., 2016). A layout of proposed technique is shown in Fig. 24.2.

The procedure followed for the development of this protection technique is as follows:

A. Fault Analysis

B. ANN architecture selection

C. Training of Data

A. Fault Analysis

To study the characteristics of microgrid under no fault and fault conditions, fault analysis is carried out. Whenever there is a PP or PG fault in the microgrid, its voltage and current characteristics changes, these characteristics varies with change in fault locations and fault resistances. The voltage and current samples of Bus 1 in no fault condition as well as in fault conditions are obtained. For training, current and voltage samples at Bus 1 are accurately recorded with variation in fault location, distance from the bus and fault resistance for both type of faults.

B. ANN architecture selection

ANN are mathematical models incepted by human brain. ANN have found wide application in medical field, image recognition, speech recognition for classification, regression etc. For training weights are adjusted to form a relationship between input and target matrix. In Neural Network fitting, three training algorithms are there, Levenberg-Marquardt, Scaled Conjugate Gradient and Bayesian Regularization, among these three Levenberg-Marquardt is the found to be most efficient and faster algorithm.

C. Training

For training of network voltage and current samples of bus1 for different fault types, locations and resistance are gathered and a matrix is generated, along with a target matrix to train the ANN for fault classification. A three-layer feed forward network with two input, one output and one hidden layer is formed. The count of hidden layer neuron is 10 as shown in Fig. 24.3. Table 24.1 depicts different fault parameters. Fault parameters include variation in fault locations and fault resistances.

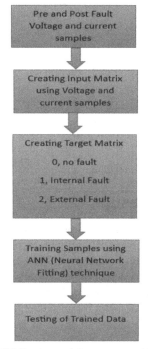

Fig. 24.2 Layout for proposed fault detection scheme using neural network fitting technique

Table 24.1 Fault parameters used for matrix formation

Fault location	200 m, 400 m, 600 m, 800 m
Fault Resistance	PP fault: 0.5Ω
	PG fault: 5Ω, 10Ω, 15Ω, 20Ω

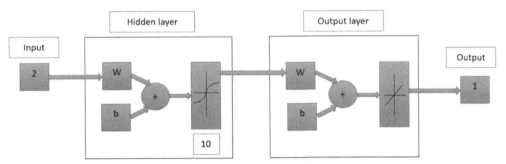

Fig. 24.3 Architecture of used network

4. Results

The trained neural network is tested using test matrix consisting of random fault cases (voltage and current data samples) which are not considered during training. The competency of this method is established through testing of data samples. The tested data comprises of DC line faults (internal and external). The faults are simulated by varying fault resistances and fault locations in each section of DC microgrid. No fault cases are also included in test matrix. The regression plot for proposed ANN based method for test matrix is shown in Fig. 24.4. Some of the test results are given in Table 24.2.

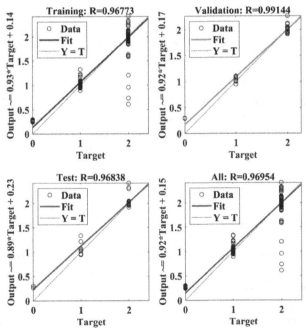

Fig. 24.4 Regression plot of neural network Fitting technique applied on voltage and current samples

Table 24.2 Response of ANN classifier

Fault Type	Section	Location	Output of ANN classifier
PG fault	S1	125m	1
PG fault	S1	225m	1
PP fault	S1	355m	1
PP fault	S1	125m	1
PG fault	S2	225m	2
PG fault	S3	355m	2
PG fault	S4	250 m	2
PP fault	S5	550 m	2
PP fault	S6	600 m	2
No fault	-	-	0

5. Conclusion

In this paper, a reliable ANN based protection scheme for a DC microgrid is proposed. The proposed method is able to detect the fault in internal section (S1) and external sections (S2-S6) of the DC Microgrid grid. The test results show the proposed protection technique is reliable and able to identify DC line faults (internal or external faults). Further, the proposed method is immune to the variation in fault parameters such as fault locations and fault resistances.

REFERENCES

1. Ali, Zulfiqar., Terriche, Yacine., Hoang, Le Quang Nhat., Abbas, Syed Zagam., Hassan, Mustafa Alrayah., Sadiq, Muhammad., Su, Chun-Lien. and Guerrero, Josep M. (2021). "Fault Management in DC Microgrids: A Review of Challenges, Countermeasures, and Future Research Trends." IEEE Power & Energy Society Section 9.
2. Chen, Laijun. and Mei, Shengwei. (2015). "An Integrated Control and Protection System for Photovoltaic Microgrids." CSEE Journal of Power and Energy Systems, Vol. I, No. I.
3. Habib, Hany Fawzy., Esfahani, Mohammad Mahmoudian. and Mohammed, Osama A. (2019). "Investigation of Protection Strategy for Microgrid System Using Lithium-Ion Battery During Islanding." IEEE Transactions on Industry Applications, Vol. 55, No. 4.
4. Hooshyar, Ali. and Iravani, Reza. (2017). "Microgrid Protection." Proceedings of the IEEE, vol. 105, no. 7: pp 105.
5. K. Khairnar, Ashwini. and Shah, Dr. P. J. (2016). "Study of Various Types of Faults in HVDC Transmission System". International Conference on Global Trends in Signal Processing, Information Computing and Communication.
6. Koley, Ebha., Verma, Khushaboo. and Ghosh, Subhojit. (2016). "A modular neuro-wavelet based non-unit protection scheme for zone identification and fault location in six-phase transmission line." The Natural Computing Applications Forum.

7. Mohanty, Rabindra. and Pradhan, Ashok Kumar. (2018). "Protection of Smart DC Microgrid With Ring Configuration Using Parameter Estimation Approach." IEEE Transactions on Smart Grid, Vol. 9, No.6.

8. Muhtadi, Abir. and Saleque, Ahmed Mortuza. "Modeling and Simulation of a Microgrid consisting Solar PV & DFIG based Wind Energy Conversion System for St. Martin's Island"

9. Nagam, Sai Sowmya., Koley, Ebha. and Ghosh, Subhojit. (2017). "Artificial Neural Network Based Fault Locator for Three Phase Transmission Line with STATCOM." IEEE International Conference on Computational Intelligence and Computing Research.

10. Telukunta, Vishnuvardhan., Pradhan, Janmejaya., Agrawal, Anubha., Singh, Manohar. and Srivani, Sankighatta Garudachar. (2017)."Protection Challenges Under Bulk Penetration of Renewable Energy Resources in Power Systems: A Review." CSEE Journal of Power and Energy Systems, Vol. 3, No. 4: pp 15.

11. Tiwari, Shankarshan Prasad., Koley, Ebha. and Ghosh, Subhojit. (2021). "Communication-less ensemble classifier-based protection scheme for DC microgrid with adaptiveness to network reconfiguration and weather intermittency." Sustainable Energy, Grids and Networks 26: pp 100460.

Emerging Technologies and Applications in Electrical Engineering –
Prof. Dr. Anamika Yadav et al. (eds)
© 2024 Taylor & Francis Group, London, ISBN 978-1-032-82568-7

Modelling and Examination of Symmetric and Asymmetric CHB Inverter for Multiple Output Levels

25

B. Sathyavani[1]

Dept. of Electrical and Electronics Engineering SR University Warangal, India

Tara Kalyani S[2]

Dept. of Electrical and Electronics Engineering JNTUH, India

K Sreedevi[3]

Dept. of Electrical and Electronics Engineering SR University Warangal, India

ABSTRACT: For high-power applications, multi-stepped inverters have often been used. Multilayer inverters operate with reduced harmonic production, little electromagnetic interference, and an improved output voltage waveform. Most of these inverters are used in industries, including traction, mills, cranes, and hoists. Using asymmetric voltage ratios, the research study demonstrates how a five-level cascaded H bridge (CHB) operates for output voltage levels of seven and nine. Three single-phase, five-level inverter topologies with equal and different voltage ratios have been developed using MATLAB simulation to show the inverter's performance. The results from employing the generalized sine PWM in each configuration are evaluated in terms of total harmonic distortion (THD) and the quantity of switching devices.

KEYWORDS: PWM, LDN, THD, CHB

1. Introduction

Inverters are now used extensively in industrial utilities, variable frequency drives, and transmission lines, where conventional inverters have been overruled with multilevel inverters

[1]bvaniddn@gmail.com, [2]tarakalyani@gmail.com, [3]sreedevikunumalla@gmail.com

DOI: 10.1201/9781003505181-25

[1-2]. Compared to two-level inverters, MULTILEVEL inverters (MLI) have many benefits, such as decreased dv/dt, total harmonic distortion (THD), and electromagnetic interference (EMI). The three often utilized typical topologies are diode clamped (DCMLI), flying capacitor (FCMLI), and cascaded H-bridge (CHB) [3-4]. Though the CHB inverter is efficient compared to other conventional topologies, this topology is also not preferred for higher levels. However, in comparison to two-level inverters, the increased number of switches in MLIs has complicated their circuit configuration and raised a number of size, cost, and dependability issues. Consequently, the need to decrease the scale of circuits has given rise to a novel domain of MLIs known as reduced switch count. In recent days researchers are developing novel inverter topologies with a smaller number of power switching devices [5], few topologies named, T type, Envelop type, Multi DC link, and Reverse Voltage (RV)type and Level doubling network are well-known RSC-MLI topologies [6,11].

2. Cascaded H bridge Inverter

A frequently implemented topology for multilevel inverters is the cascaded H-bridge (CHB). Three voltage levels can be generated by a single CHB cell, which facilitates single-phase to three-phase conversions [6-7]. An H-bridge with four power relays and one DC voltage source is utilized in this configuration. It is required that a multilayer inverter feature three voltage levels. One H-bridge unit in the cascaded H-bridge multilevel inverter satisfies this criterion. The DC-link voltage is utilized by each inverter in this configuration to modulate the voltage at its output terminals. The sum of every voltage level constitutes the output voltage. The voltage steps generated by a multi-stepped H-bridge inverter are +Vdc, -Vdc, and 0; the inverter has a total of 2k+1 voltage steps; and its maximum output is (k*Vdc). Vdc represents the voltage of a solitary CHB cell, while k denotes the overall quantity of CHB cells arranged in the system.

Known for its adaptability, this topology offers a practicable and varied solution for applications that require multiple inverters.

2.1 Simulation of Five-level Symmetrical CHB Inverter

A five-level CHB inverter is a type of multilayer inverter that makes a five-level output voltage by stacking two H-bridge inverters cascading each other as shown in Fig. 25.1. In order to acquire all five possible voltage levels, the functioning of the inverter is detailed in Table 25.1. Considering DC link voltage magnitude V_{dc1} = Vdc2 = Vdc = 10 V.

The simulation diagram of five level CHB inverter given in Fig. 25.1 is developed in MATLAB environment, using 8 IGBT switching devices, and two DC link voltage sources. The switching pulses

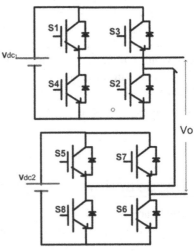

Fig. 25.1 Five level CHB inverter

Table 25.1 Five-level CHB inverter

Voltage level	Conduction switches from CHB1 module	Conduction switches from CHB 2 module
$2 V_{dc}$	S1, S2	S5, S6
V_{dc}	S1, S2	S6, S8
0 V	S2, S4	S6, S8
$- V_{dc}$	S3, S4	S6, S8
$-2 V_{dc}$	S3, S4	S7, S8

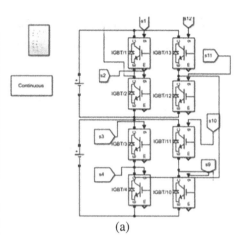

(a)

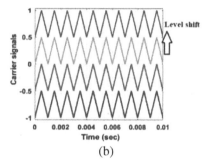

(b)

Fig. 25.2 Simulation diagram of five-level CHB inverter (a) Simulation circuit, (b) Carrier signals

are applied to the inverter using Pulse width modulation method [8-9]. With this technique, the output voltage level and frequency are controlled by altering the pulse width that is applied to the switches. PWM can lessen harmonic distortion and enhance the output waveform's power quality. PWM techniques come in various forms, including carrier-based PWM, space vector PWM, and sinusoidal PWM [10-14]. Figure 25.1a and 25.1b shows the simulation circuit diagram of 5 level CHB inverter and required carrier waveform to obtain five level output. For a CHB inverter of five-level two of the carrier signals are placed up to the zero reference, and another two signals are below the zero reference as shown in Fig. 25.1b. The results obtained from the simulation is presented in Fig. 25.3. The THD is observed as 26.83%.

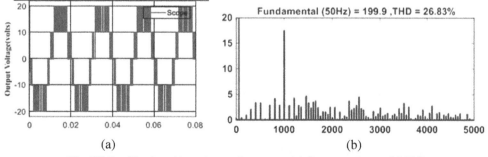

(a)   (b)

Fig. 25.3 Five level inverter performance (a) Output voltage, (b) THD

2.2 Simulation of Seven Level Asymmetrical CHB Inverter

2.2a Binary asymmetry DC voltage ratio

Developing the five-level inverter shown in Fig. 25.1 in MATLAB simulation and operating it with a DC voltage source in binary ratio makes the five-level inverter work like a seven-level inverter. The DC voltage for the H bridge inverter, Vdc1 and Vdc2 are 10 V and 20 V respectively. The operation under binary DC voltage rations is different from symmetric voltage ratio [9, 14-17]. The switching states to obtain seven level with the asymmetric and binary ratio is presented in Table 25.2.

Table 25.2 Seven level asymmetrical CHB inverter switching states

Voltage level	Conduction switches from CHB1 module	Conduction switches from CHB 2 module
30 V	S1, S2	S5, S6
20 V	S2, S4	S5, S6
10 V	S3, S4	S5, S6
0	S2, S4	S6, S8
-10 V	S3, S4	S6, S8
-20 V	S2, S4	S7, S8
-30 V	S3, S4	S7, S8

The inverter is controlled using sinusoidal PWM method for better performance, the generalized level shifted pulse width modulation is implemented. The modulation index (ma) is set to unity. The obtained results after simulation are presented in Fig. 25.4. From Fig. 25.4a, the seven various voltage levels as well as the maximum, minimum voltage levels are 30 V and -30 V in the waveform can be observed. The output voltage THD in percentage is given in Fig. 25.4b. The THD% is 19.14 which is less than five level inverter.

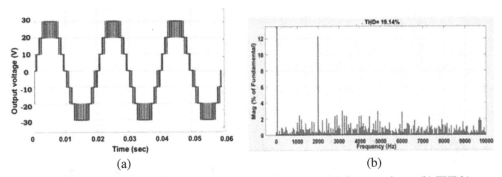

(a)   (b)

Fig. 25.4 Performance of the seven-level CHB inverter (a) Output voltage, (b) THD%

2.2b Trinary asymmetry DC voltage ratio

The more number of voltage levels can be obtained with trinary voltage ratio as compared with binary and symmetrical dc voltage formation. Therefore for the network shown in Fig. 25.1 for nine various voltage levels, the DC link voltage is set to Vdc1=10 V, Vdc2 = 30 V. Table 25.3 shows the switching states used to create nine levels with asymmetric and trinary voltage ratios.

Table 25.3   Nine level asymmetrical CHB inverter switching states

Voltage level	Conduction switches from CHB1 module	Conduction switches from CHB 2 module
40 V	S1, S2	S5, S6
30 V	S4, S2	S5, S6
20 V	S3, S4	S5, S6
10 V	S1, S2	S8, S6
0	S2, S4	S6, S8
-10 V	S3, S4	S6, S8
-20 V	S1, S2	S7, S8
-30 V	S2, S4	S7, S8
-40 V	S3, S4	S7, S8

The sinusoidal pulse width modulation (PWM) approach is used to operate the inverter. In order to get higher performance, the generalized level shifted pulse width modulation (GLSPM) technique is applied. Eight carrier waveforms and a single reference sine waveform are necessary for a nine-level inverter to function properly. With unity modulation index (ma).

Figure 25.5 shows the results that were gathered after the simulation run. The output voltage pattern has nine different voltage levels, and the highest and lowest voltage levels are 40 V and -40 V, respectively, each step voltage variation is 10 volts, this data can be seen in Fig. 25.5a. The resulting voltage and total harmonic distortion (THD) is shown as a percentage in Fig. 5b. The THD% is 13.45, which is less than what an inverter with five and seven levels would produce.

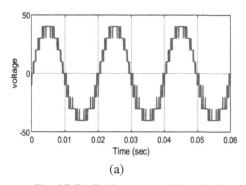

(a)

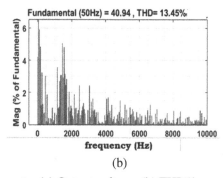

(b)

Fig. 25.5   Performance of the nine-level CHB inverter (a) Output voltage, (b) THD%

Results comparison:

At higher voltage levels, the operation of a five-level CHB inverter with a symmetrical voltage ratio is contrasted with that of one with an asymmetric voltage ratio. The evaluation of the inverter's performance is conducted by calculating the percentage of total harmonic distortion in the output voltage. The outcomes of a comparison of the inverter's overall efficacy at various voltage levels are presented in Table 25.4.

Table 25.4 Comparison of inverter

DC voltage ratios	Required DC voltage source magnitudes.	Number of output voltage levels	THD%	Maximum voltage stress on the switch
Symmetrical	$V_{dc1} = V_{dc2} = 10$ V	5	26.83	10 V
Asymmetrical (binary ratio)	$V_{dc1} = 10$ V, $V_{dc2} = 20$ V	7	19.14	20 V
Asymmetrical (Trinary ratio)	$V_{dc1} = 10$ V, $V_{dc2} = 30$ V	9	13.45	30 V

In Fig. 25.6, the voltage level and THD% curve are depicted. When employing an equal number of switching devices and DC voltage sources, the nine-level inverter with a trinary voltage ratio exhibits the lowest THD at 13.45%, while the five-level inverter with an equal voltage source demonstrates the highest THD at 26.83%. Simulation parameters for the operation of the inverter is given in Table 25.5.

Table 25.5 Simulation parameters

Switching frequency	1500 Hz
Modulation index(m_a)	1
Operating frequency	50 Hz
Load resistance	10 ohms
Number of DC voltage sources	2
Number of IGBT switching devices	8

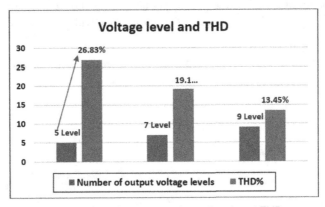

Fig. 25.6 CHB inverter voltage level and THD

3. Conclusion

This research article describes how a five-level CHB inverter becomes a seven- and nine-level inverter utilizing different voltage ratios. MATLAB simulation was used to create and operate an inverter employing generalized level-shifted PWM for all voltage level formations. Results are examined using THD, voltage levels, and voltage stress. Five, seven, and nine-level inverters share switching devices and DC sources. The nine-level inverter has higher maximum voltage stress but lower THD than the five and seven-level inverters. The inverter's voltage and THD were improved by adding an asymmetrical DC source ratio while keeping the same number of switching and source devices.

REFERENCES

1. J. Rodriguez, L. Jih-Sheng, and P. Fang Zheng, "Multilevel inverters: a survey of topologies, controls, and applications," IEEE Transactions on Industrial Electronics, vol. 49, no. 4, pp. 724-738, 2002.

2. R. Agrawal, and S. Jain, "Comparison of Reduced part count multilevel inverters (RPC-MLIS) for integration to the grid," electrical Power and Energy System, vol. 84, pp. 214-224, 2017.

3. J. Rodriguez, J.-S. Lai, and F. Z. Peng, "Multi-level inverter: a survey of topologies, controls, and applications," IEEE Trans. Ind. Electron, Vol. 49, no. 4, pp. 724-738, Aug. 2002.

4. S. K. Chattopadhyay and C. Chakraborty, "A New MultilevelInverterTopology With Self- Balancing Level Doubling Network," IEEE Trans. On Ind. Electron., vol.61, no.9, pp.4622-4631, Sept. 2014.

5. Gerardo Ceglia, Víctor Guzmán, Member, IEEE, Carlos Sánchez, Fernando Ibáñez, Julio Walter, and María I. Giménez, Member, IEEE, "A New Simplified Multilevel Inverter Topology for DC-AC Conversion," IEEE Transactions on Power Electronics, Vol. 21, no. 5, Sep. 2006.

6. P.Thongprasri," A 5-Level Three-Phase Cascaded Hybrid Multilevel Inverter," International Journal of Computer and Electrical Engineering, Vol. 3, No. 6, December 2011.

7. Sathyavani B, and Tara Kalyani S. (2020) Implementation of LDN to MLI and RSC-MLI configurations with a simple carrier-based modulation. IOP Conf. Series: Materials Science and Engineering 981 (2020) 042071 IOP Publishing doi:10.1088/1757-899X/981/4/042071

8. D. G. Holmes and T. A. Lipo, Pulse With Modulation for Power Converters. New York: Wiley, 2003.

9. Dhanamjayulu, C., & Meikandasivam, S. (2017). Implementation and comparison of symmetric and asymmetric multilevel inverters for dynamic loads. IEEE Access, 6, 738-746.

10. Vadhiraj, S., Swamy, K. N., & Divakar, B. P. (2013, December). Generic SPWM technique for multilevel inverter. In 2013 IEEE PES Asia-Pacific Power and Energy Engineering Conference (APPEEC) (pp. 1-5). IEEE.

11. Vemuganti H P,Sreenivasarao D and Kumar G S 2017 Improved pulse-width modulation scheme for T-type multilevel inverter IET Power electronics 10(8) 968-976

12. Kumar, K. V., Michael, P. A., John, J. P., & Kumar, S. S. (2010). Simulation and comparison of SPWM and SVPWM control for three phase inverter. ARPN journal of engineering and applied sciences, 5(7), 61-74.

13. Ibrahim, Z. B., Hossain, M. L., Bugis, I. B., Mahadi, N. M. N., & Hasim, A. S. A. (2014). Simulation investigation of SPWM, THIPWM and SVPWM techniques for three phase voltage source inverter. International Journal of Power Electronics and Drive System, 4(2), 223-232.

14. R. A. Ahmed, S. Mekhilef, and H. W. Ping, " New multilevel inverter topology with minimum number of switches," in Proceedings of the 14th International Middle East Power Systems Conference (MEPCON'10), pp. 1862–1867, Cairo University, Cairo, Egypt, December 2010.

15. Sathyavani B, and Tara Kalyani S. (2020).Single and double LDN configuration analysis with cascaded H bridge multilevel inverter. IOP Conf. Series: Materials Science and Engineering 981 (2020) 042066 IOP Publishing doi:10.1088/1757-899X/981/4/042066

16. Bandela, S., Sandipamu, T. K., Vemuganti, H. P., Rangarajan, S. S., Collins, E. R., & Senjyu, T. (2023). An Efficacious Modulation Gambit Using Fewer Switches in a Multilevel Inverter. Sustainability, 15(4), 3326.

17. Rajababu D, Sudhakar AVV and Sathyavani B 2019 Development of technology for highpower industry converters Int. J.Innov. Technol.Explor. Eng.8(10)3130-3132.

Emerging Technologies and Applications in Electrical Engineering –
Prof. Dr. Anamika Yadav et al. (eds)
© 2024 Taylor & Francis Group, London, ISBN 978-1-032-82568-7

A Hybrid Feature Selection based Ensemble Model to Detect IoT Security Attacks

26

Ritik Ranjan[1], Tirath Prasad Sahu[2]
Department of Information Technology
National Institute of Information Technology, Raipur, CG-492010, India

ABSTRACT: As there is advancement in IoT and it has become integrated to every business operation, the challenges faced due to security threats to IoT networks has also increased. This has heightened the need of processing data securely at the edge of networks using fog nodes. To address it, this research paper proposes an Intrusion Detection System (IDS) using ensemble learning approach optimized using hybrid feature selection (FS) strategy. This paper uses a correlation-based filter FS approach combined with a modified version of Binary Crow Search Algorithm, an evolutionary algorithm and then a multi-layered ensemble model for final prediction. The model is evaluated using modern dataset UNSW-NB15 showing promising results with high detection rates for different attack types.

KEYWORDS: Hybrid feature selection, Ensemble learning, IoT, Intrusion detection system, Binary crow search algorithm

1. Introduction

As the Internet of Things (IoT) continues to grow, there is a need for IDS solutions that can protect IoT devices and networks. An IDS is a security mechanism that monitors a network or systems for malicious activity or policy violations. However, as the threat landscape continues to evolve, so too must IDS technology. Traditional security measures are ineffective in detecting and preventing cyberattacks, as cybercriminals are intelligent enough to bypass them. So various ML based solutions have been widely employed in cybersecurity applications

[1]1312.ritik@gmail.com, [2]tpsahu.it@nitrr.ac.in

DOI: 10.1201/9781003505181-26

to detect IoT attacks. AI and ML-powered IDS solutions can learn from historical data to identify patterns of behavior that are associated with known malware or attack vectors. In Anomaly detection, they can be used to identify anomalous network traffic that may indicate an attack by being able to detect change in the traffic patterns of a particular user.

Key Contributions

- This paper proposed an ensemble model for IDS which combines Gaussian Naïve Bayes (GNB), Random Forest(RF) and k-Nearest Neighbours(KNN) using eXtensive Gradient Boosting(XGB) to produce better results. All the three classifiers being of very different nature from one another.
- The paper uses wrapper FS technique Binary Crow Search Algorithm (BCSA) with varying flight length on top of correlation-based filter technique to get optimal subset of features.
- 10 Fold cross validation technique has been used as a sampling method to avoid overfitting of training data.
- The model is tested on UNSWNB15 dataset having modern IOT attacks and produced better results than most of the existing IDS.

2. Literature Review

(Nimbalkar and Kshirsagar, 2021) used a FS method using Information Gain(IG) and Gain Ratio(GR) and evaluated on IoT-BoT and KDD Cup 1999 datasets using a JRip classifier. It achieved higher performance than the original feature set and traditional IDSs, using 16 features for IoT-BoT and 19 features for KDD Cup 1999. (V. Kumar, Das, and Sinha, 2021) proposed

Table 26.1 Summarization of related ML techniques on IDS on IoT systems

Author	Method	Dataset	Pros.	Cons.
(Pajouh, Dastghaibyfard, and Hashemi, 2017)	Used PCA and LDA with NB and kNN for two-tier classification	NSL-KDD	Good detection rates	Old dataset, rare attacks detection only
(V. Kumar, Das, and Sinha, 2021)	Rule-based classification model	UNSW-NB15	High detection rates for some attacks	Low average accuracy
(Nimbalkar and Kshirsagar, 2021)	IG and GR based FS with JRip classifier	IoT-Bot and KDD-Cup 1999	High accuracy and detection rates	Outdated datasets
(Ren et al., 2019)	Hybrid data optimization based IDS	UNSW-NB15	Better detection rates	For rare attacks only
(P. Kumar, Gupta, and Tripathi, 2021)	Ensemble model using correntropy based FS	DS2OS and UNSW-NB15	High accuracy and better detection rates for 30 features	Low performance for less features and less precision

a rule-based IDS classification model, which used information gain based filter FS method on UNSWNB15 dataset. Various Decision tree models were used for generating integration rules. It improved the detection rates for some attacks to 90.32% but have low average accuracy of 65.21%. (P. Kumar, Gupta, and Tripathi, 2021) proposed a distributed ensemble design based intrusion detection system (IDS) using Fog computing to detect IoT attacks. The model used only filter FS and the overall detection rate and precision achieved was 76.57% and 48.81% respectively with an accuracy of 93.21% for 30 features.

3. Proposed Methodology

3.1 Pre-processing

In the dataset, any missing numerical value is replaced with the mean of that column and any missing categorical value is replaced with '*unknown*' class. The categorical features are

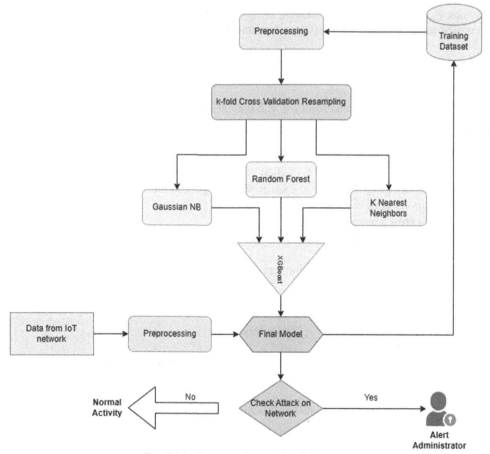

Fig. 26.1 Proposed model architecture

converted into numerical features using Label Encoder and One Hot Encoding technique. The data is then scaled down using Standard scaler technique.

3.2 Correlation-based Filter Method

Correlation coefficient method, being a filter technique, is faster FS method as it is independent of the training method. Correlation is used to determine similarity between every feature variable and target variable. Then the features are ranked on the basis of this value, with higher correlation value being ranked higher and the top ranked features are selected (Nimbalkar and Kshirsagar 2021).

3.3 BCSA-TVFL Wrapper Method

Crow Search Algorithm(CSA) has been used as a wrapper FS technique to further enhance the performance of model (Meraihi et al. 2021). In CSA, there are two parameters – flight length(fl) and awareness probability(AP). They have fixed values. To balance the trade-off between global search and local search, BCSA with time varying flight length (TVFL) has been used in the proposed model (Chaudhuri and Sahu, 2021) as shown in equation (3).

$$tvfl = \left(1 - \frac{t}{T}\right) tvfl_{max} + \frac{t}{T} tvfl_{min} \qquad (1)$$

Assumptions:

Number of features = Z. Then crows are in Z-dimension space.

Number of crows in the population space = N and AP = 0.1.

Position of a crow is given by 'm'-dimensional vector X ($x_1, x_2, x_{3,...}, x_Z$).

Algorithm 1: Steps in BCSA-TVFL method
Initialise a random position vector to each crow in the population
Evaluate positions using the fitness function
Initialize each crow's memory
while iterations < max_no_of_iterations
    for n ← 1 to N do
        crow$_n$ follows a random crow$_p$
        if $r_p \geq AP$
            $x_{new} = x_{curr} + r_n * tvfl * (m_{p,curr} - x_{n,curr})$
            Transform x_{new} to binary vector
        else
            x_{new} = a random binary position vector in z-dimension
        end
    end
    Check if the new positions are feasible
    Evaluate new positions
    Then each crows memory is updated with the best solutions so far
end

The fitness function used is described in Eq. (2) where Υ_R is classification error rate, $|X|$ is selected number of features and N is the total number of features. α and β are constants with $\beta = 1 - \alpha$. The final position vector is converted into an n-dimensional binary solution using equations (3) and (4).

$$Fitness(X) = \alpha\Upsilon_R(X) + \frac{\beta|X|}{N}, where\ \alpha, \beta \in [0,1] \tag{2}$$

$$T(x) = \frac{1}{1 + e^{-x}} \tag{3}$$

$$X_{new} = \begin{cases} 1, if\ r_i < T \\ 0, if\ r_i > T \end{cases} \tag{4}$$

3.4 Proposed Ensemble Model

Algorithm 2: Steps in the proposed ensemble model

Input data: $S = \{X_i, Y_i\}_{i=1}^n$, where $X_i \in R^n$ and $Y_i \in L$
Step-1: Split S into K equal-sized subsets to use 10 fold cross validation.
$$S = \{S_1, S_2, S_3, ..., S_K\}$$

for $k = 1$ to K **do**

      for $\quad m = 1\ to\ M$ do
            Train a classifier for first level C_{km} on $S - S_k$
            Make predictions $C_{km}(C_k)$
      end
      Create new training set by combining predictions
      for $\quad i = 1\ to\ n$ do
            $X_i' = \{C_{k1}(X_i), C_{k2}(X_i), C_{k3}(X_i), ..., C_{kM}(X_i)\}$
      end

            Training set for next level: $S' = \{X_i', Y_i\}_{i=1}^n$
end
Step-2: Train classifier C' on S' for the next level
Step-3: Train base classifiers on entire dataset S
for $\quad\quad m = 1\ to\ M$ do
      Train classifier C_m on S
end

Output: $\quad E(S) = \{C_1(S), C_2(S), C_3(S), ..., C_M(S)\}$

Notations used are S: Training Dataset, R^n: Feature space, L: Labels, K: Number of folds, C: Classifier, n: Training samples count and E: Ensemble classifier.

4. Results and Analysis

The UNSWNB15 dataset (Moustafa and Slay, 2015) has been used for prediction which has 43 features.

Table 26.2 Model performance and comparison

Author	Method	Selected Feature Count	Accuracy	Detection Rate
(V. Kumar, Das, and Sinha, 2021)	Rule based IDS (RIDS)	13	65.21	90.32
(V. Kumar, Das, and Sinha, 2021)	C5 model (C5)	13	75.8	83.47
(Ren et al., 2019)	Data-optimization based IDS (DOIDS)	43	92.8	63.09
(P. Kumar, Gupta, and Tripathi, 2021)	Ensemble model with filter FS (EFFS-30)	30	93.21	76.57
Proposed	Ensemble model with hybrid FS (EHFS)	10	94.28	94.93

5. Conclusion

This paper proposed an ensemble model with hybridized FS to detect IoT attacks on modern UNSW-NB15 dataset. It has been reduced greatly from 43 to 10 features using filter-wrapper FS. The model achieved high accuracies and detection rates for each class. On comparing with the existing state-of-the-art IDS techniques, our model out performed others.

REFERENCES

1. Chaudhuri, Abhilasha, and Tirath Prasad Sahu. 2021. "Feature Selection Using Binary Crow Search Algorithm with Time Varying Flight Length." Expert Systems with Applications 168: 114288.
2. Kumar, Prabhat, Govind P Gupta, and Rakesh Tripathi. 2021. "A Distributed Ensemble Design Based Intrusion Detection System Using Fog Computing to Protect the Internet of Things Networks." Journal of Ambient Intelligence and Humanized Computing 12: 9555–72.
3. Kumar, Vikash, Ayan Kumar Das, and Ditipriya Sinha. 2021. "UIDS: A Unified Intrusion Detection System for IoT Environment." Evolutionary Intelligence 14: 47–59.
4. Meraihi, Yassine, Asma Benmessaoud Gabis, Amar Ramdane-Cherif, and Dalila Acheli. 2021. "A Comprehensive Survey of Crow Search Algorithm and Its Applications." Artificial Intelligence Review 54 (4): 2669–2716.
5. Moustafa, Nour, and Jill Slay. 2015. "UNSW-NB15: A Comprehensive Data Set for Network Intrusion Detection Systems (UNSW-NB15 Network Data Set)." In 2015 Military Communications and Information Systems Conference (MilCIS), 1–6. IEEE.
6. Nimbalkar, Pushparaj, and Deepak Kshirsagar. 2021. "Feature Selection for Intrusion Detection System in Internet-of-Things (IoT)." ICT Express 7 (2): 177–81.

7. Pajouh, Hamed Haddad, GholamHossein Dastghaibyfard, and Sattar Hashemi. 2017. "Two-Tier Network Anomaly Detection Model: A Machine Learning Approach." Journal of Intelligent Information Systems 48: 61–74.
8. Ren, Jiadong, Jiawei Guo, Wang Qian, Huang Yuan, Xiaobing Hao, and Hu Jingjing. 2019. "Building an Effective Intrusion Detection System by Using Hybrid Data Optimization Based on Machine Learning Algorithms." Security and Communication Networks 2019.

Emerging Technologies and Applications in Electrical Engineering –
Prof. Dr. Anamika Yadav et al. (eds)
© 2024 Taylor & Francis Group, London, ISBN 978-1-032-82568-7

Enhancement of Grid Integrated Hybrid Renewable Energy Sources with Plug-In Hybrid Electric Vehicles using Novel Control Techniques

27

Hira Singh Sachdev[1]

Department of Electrical Engineering, NIT Raipur

Miska Prasad[2]

Department of Electrical & Electronics Engg., ACE Engineering College, Hyderabad

Parvinder Dung[3]

Department of Hospital Adminstration, Devi AhilyaVishwavidyalya, Indore

ABSTRACT: Voltage fluctuations have grown to be major problems in industrial and commercial applications. In this paper, a booster converter (BC)-based Dynamic Voltage Restorer (DVR) is discussed for mitigating voltage swell problem. The Maximum Power Point Tracking (MPPT) technique based on booster converter is used to maximize the output power from the solar panel under fluctuating sun irradiation and ambient temperature. The highest output power from the solar panel is obtained using the booster converter based Maximum Power Point Tracking (MPPT) technique under varying solar irradiance and ambient temperature. Hence, it proved the finest performance of the proposed control technique in grid integration system.

KEYWORDS: Boost converter, Voltage swell, DVR, Photovoltaic

1. Introduction

Solar energy is one of the most dependable sources for producing electricity from renewable sources. Solar power generation is a very promising source of electricity, especially in nations with abundant daytime sunlight. Ali et al. (2018). However, two-stage converters are necessary for the common PV power converter circuit Umarani et al. (2016). Prior to connecting to the electrical grid, the DC input is first converted to AC after which the PV voltage is increased

[1]singhhira10@gmail.com, [2]pmiska26@gmail.com, [3]parvinderdung@gmail.com

DOI: 10.1201/9781003505181-27

to the necessary amount. Farhat et al. (2015). ZSI is a new topology that is now being used in power converters, particularly DC-AC converters. A single-stage power converter with buck-boost characteristics, low ripple input current, and high voltage gain is employed Wang et al. (2019). Since ZSIs are thought to be a new kind of inverter, numerous studies have been done in this field. None of these methods will be effective without changing MPPT because we won't be able to draw more power from the PV system for grid supply or maintain a constant DC-link voltage at the inverter input Carrasco et al. (2016). The literature reports many MPPT techniques and an emphasis has been placed on improving MPPT in Z-source inverters. The classic and the soft computing methods are the two different kinds of MPPT methodologies. Currently, typical MPPT algorithms such as Perturb and Observe (P&O) Shrikant et al. (2015), Incremental Conductance (InC) Gheibi et al. (2016), and others are heavily utilised to track the electricity generated by solar panels, Hill Climbing (HC), and others. Although the perturbation and observation approach is extremely straightforward, the oscillations and tracking speed play a significant role in how well it performs. In addition, the quick variation in sun irradiation causes the P&O algorithm to lose its direction. Although the Incremental Condense (InC) algorithm does not lose tracking speed, its performance is similarly influenced by oscillations and tracking speed. In this paper, a booster converter (BC)-based Dynamic Voltage Restorer (DVR) is discussed for mitigating voltage swell problem. The highest output power from the solar panel is obtained using the booster converter based Maximum Power Point Tracking (MPPT) technique under varying solar irradiance and ambient temperature.

2. Dynamic Voltage Restorer with Z-Source Inverter

Figure 27.1 depicts the 3-leg ZSI-based DVR design. As an interface transformer, the isolated ZSI-based DVR topology has the benefit that the voltage rating of the ZSI can be optimally constructed. 6 switches are needed for the 1-Φ 3-Leg ZSI based DVR.

3. Results of the Simulation and Discussion

Using the MATLAB/SIMULINK platform, the proposed PSO-fed dynamic voltage restorer system shown in Fig. 27.1 was constructed to show how well it functioned when a balanced three-phase nonlinear load was abruptly switched. The various features of the solar cell under various sun exposures are depicted in Fig. 27.2(a-b). At 25 °C, the temperature is maintained constant and irradiation modifications are rendered (0.1kW/m2, 0.5kW/m2, 0.7kW/m2, 1W/m2).

Figure 27.2(a–b) the results of the MATLAB program display the P-V and I-V charac-teristics under these circumstances, respectively. All that determines radiation is the current that the incident light produces; the more radiation, the more current.

Conversely, the voltage remains relatively constant throughout time. It offers a comprehensive breakdown of the efficiency and potential of its solar energy conversion. A solar cell or solar panel's Pmax is one of the electrical I-V factors that significantly affects the output performance and solar efficiency of the device. The I-V and P-V curves for the module are

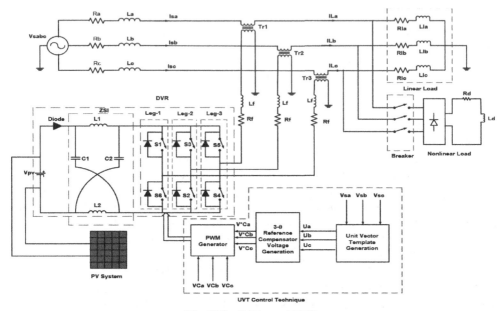

Fig. 27.1 BC based DVR

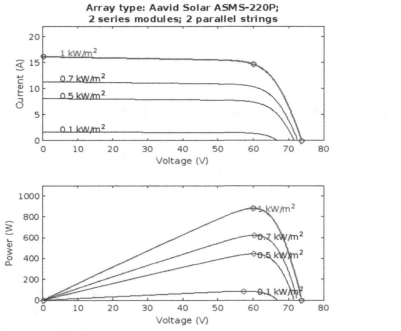

Fig. 27.2 Solar Photovoltaic Characteristics under Different Irradiations V-I, and P-V characteristics are (a), and (b).

shown in Figs. 27.4a and b. The panel offers its maximum output at a specific voltage and current level. The output power of the panel increases as the module voltage rises, peaks (sometimes referred to as the maximum power point (MPP) in the module), and then decreases as the voltage rises approaching the open circuit voltage. A photovoltaic (PV) system's output depends on environmental parameters such as solar irradiance. It is evident how radiation affects the highest point of power: the greater the overall point of power, the higher the radiation. Figure 27.5 illustrates how solar power varies as sun irradiance changes. As can be seen from Fig. 27.3, electrical power output is directly correlated with solar irradiation and rises as it does. A maximum solar power of 80 kW is noted when irradiance is 1 kW/m2 from 0.0 seconds to 0.4 seconds. At 0.5 seconds, solar power decreases from 80 kW to 60 kW because solar irradiance decreases from 1kW/m2 to 0.8 kW/m2.

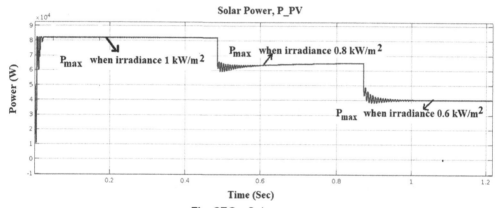

Fig. 27.3   Solar power

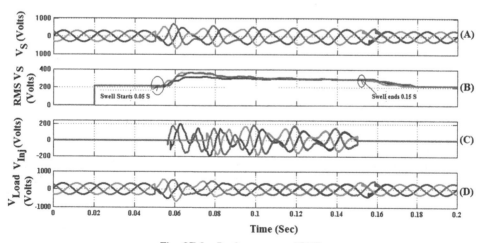

Fig. 27.4   Performance of DVR

4. Performance of DVR

The 30% voltage increase that happens when a nonlinear load is shifted quickly in a distribution network is shown in Figs. 27.4a and b. which begins at time t=50 milliseconds and lasts 150 milliseconds. Figure 27.4c shows how the BC-DVR can increase voltage while decreasing voltage swell. The sinusoidal load voltage is shown in Figure 27.4d after the voltage swell effect has been reduced.

5. Conclusion

This work developed a method for reducing voltage surge caused by the abrupt addition of a balanced three-phase nonlinear load that combines a solar system with a dynamic voltage restorer with boost converter. Results indicated that by injecting a precise magnitude of compensated voltage, the suggested restorer may lower potential voltage changes in the swell situation. To have a further investigation in this area, new hybrid control strategies can develop to attain an optimal harmonic power flow and finest accuracy in coordinated system. The suggested method exhibits greater ability to provide maximum power output.

REFERENCES

1. Ali, U.S., (2018). Impedance source converter for photovoltaic stand-alone system with vanadium redox ow battery storage. Materials Today Proceedings. 5:24-247.
2. Umarani, D. and Seyezhai, R., (2016). Modeling and control of quasi Z-source cascaded H-bridge multilevel inverter for grid connected photovoltaic systems. Energy Procedia. 90:250-259.
3. Farhat, M., Barambone S.O and Sbita, L., (2015). Efficiency optimization of a DSP-based standalone PV system system using a stable single input fuzzy logic controller. Renewable and Sustainable Energy Reviews. 49:907-920.
4. Wang, Y., Luo, H. and Xiao, X.Y., (2019). Voltage sag frequency kernel density estimation method considering protection characteristics and fault distribution. International Journal of Electric Power Systems Research. 170:128-137.
5. Carrasco, M. and Mancilla-David, F., (2016). Maximum power point tracking algorithms for single-stage photovoltaic power plants under time-varying reactive power injection. Solar Energy. 132:321-331.
6. Shrikant, M. and Nagaraj Rao, S., (2015). Comparative Analysis of a 3-Phase, 3-Level Diode Clamped ZSI Based on Modified Shoot Through PWM Techniques. IEEE International Conference on Signal Processing, Informatics, Communication and Energy Systems (SPICES), 6:19-21.
7. Gheibi, A., Mohammadi, S.M.A. and Farsangi, M., (2016). A proposed maximum power point tracking by using adaptive fuzzy logic controller for photovoltaic systems. Scientia Iranica, Transactions D: Computer Science & Engineering and Electrical Engineering, 23:1272-1281.

Emerging Technologies and Applications in Electrical Engineering –
Prof. Dr. Anamika Yadav et al. (eds)
© 2024 Taylor & Francis Group, London, ISBN 978-1-032-82568-7

Comparative Analysis of Grid Forming and Grid following Solar Farm Inverter in Islanded Mode using SRF-PLL Control Method

28

Sumit Srivastava[1]
M. Tech, IET, Lucknow, India

Pushkar Tripathi[2]
Assistant Professor, IET, Lucknow, India

Guguloth Ravi[3]
E.O.2, C.P.R.I., Bhopal, India

ABSTRACT: To improve the stability of power grids that have a substantial number of inverter-based resources (IBRs), recent global research has led to the development of advanced control methods incorporated into both grid-following (GFL) and grid-forming (GFM) inverters, addressing a range of operational scenarios. The paper utilizes electromagnetic transient (EMT) time domain simulations and conducts small signal analysis based on linearized state space modelling to validate the stability of the Synchronous Reference Frame - Phase Locked Loop (SRF-PLL) control strategy and establish a method for defining interconnection requirements that are compatible with various GFL and GFM IBR control methods. Ultimately, the simulation results demonstrate that inverters exhibit greater stability in response to load disturbances in islanded conditions, highlighting differences in power system stability when comparing GFL and GFM inverters in diverse load conditions within an isolated system.

KEYWORDS: Grid Forming Inverter, Grid Following Inverter, Synchronous Reference Frame - Phase Locked Loop, Small signal model

[1]srisumit.11@gmail.com, [2]pushkar.tripathi@ietlucknow.ac.in, [3]gravi@cpri.in

DOI: 10.1201/9781003505181-28

1. Introduction

The growing presence of inverter-based resources (IBR) in global transmission systems is altering the dynamic attributes of the power grid. As synchronous machines are phased out or retired to accommodate the expansion of IBRs, the consequence is a decrease in the available short-circuit current within the grid. (Sanford, 2018).

The traditional GFL inverters, due to their lack of inertia, have raised issues related to system stability, reliability, and control. They contribute to a reduction in the total inertia of power systems, rendering them susceptible to grid variations. Incorporating droop control mechanisms (Golsorkhi & Lu, 2015) into grid-following (GFL) inverters has played a significant role in configuring the power grid for diverse situations. Nevertheless, these functions are essentially supplementary features to GFL inverters and are constrained by their own control parameters, leading to notable operational limitations. Recently, there has been a surge of interest in grid-forming (GFM) technology (Unruh et al., 2020), (Ramasubramanian, 2022).

Because its main purpose is to enhance power system functionality, grid-forming (GFM) inverters are frequently built as voltage sources capable of adapting their voltage and frequency in synchronization with power grids, offering a range of GFM features. Phase locked loop-based control function is one of them; it helps power grids with voltage, frequency, and inertia and makes it possible for parallel inverters to run smoothly. By putting these features into practice, GFM inverters can regulate the grid, improving its stability and dependability under a range of operational scenarios.

The rest of this paper is structured as outlined below: In Section 2, a system configuration is established. Section 3 constructs a linearized state space model for an IBR control system with PLL control techniques at the inverter level. Moving on to Section 4, it offers simulation results that assist in defining the interconnection prerequisites related to voltage control for future grid-forming inverters, along with presenting a scenario to illustrate the advantages of connecting grid-forming inverters to the power system and distinguishing the stability conditions between GFM and GFL inverters. Finally, Section 5 comprises concluding remarks.

2. Model Overview

The model depicts a 100 MVA solar PV farm connected to a grid system that can be GFL or GFM, as well as a very simple power system between them. The power system modelled in Fig. 28.1 is a relatively simple one that consists of loads and a PI section line.

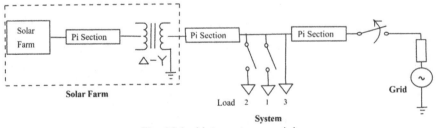

Fig. 28.1 Main system model

For grid following mode, Q_{flag} =1, V_{flag} =1 and Q_{flag} = 0, V_{flag} = N/A. The model will attempt to regulate P and Q injection in this mode, with some inverter-level frequency support. For grid forming mode, Q_{flag} =1, V_{flag} = 0.

3. SRF-PLL Control Method

To achieve grid synchronization in grid-connected inverters, encompassing both GFL and GFM inverters, it is essential to obtain an estimate of grid parameters.

Consequently, synchronization methods (Venkatramanan & John, 2019) are applied in these inverters to gauge grid attributes like voltage magnitude, phase angle, and frequency. For a GFM inverter, the precise assessment of these grid characteristics is of utmost importance, not only for its regular operation of supplying power to the grid but also for determining the optimal functioning of GFM features in instances of abnormal grid conditions and islanding scenarios. Fig. 28.2, illustrates the schematic representation of an SRF-PLL, where V_{abc} represents the three-phase voltage measurements. θ_{pll} signifies the calculated phase position of the grid voltage, while ω_{nom}, $\Delta\omega$ and ω_{pll} denote the grid's initial angular frequency, the angular frequency deviation, and the estimated angular frequency, respectively. Additionally, v_q refers to the q-axis voltage that is taken out using Park's transformation and subsequently forwarded to a proportional-integral (PI) controller for further treatment.

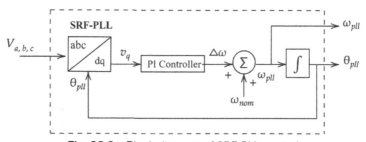

Fig. 28.2 Block diagram of SRF-PLL control

4. Small Signal Model

The small signal model can be created by utilizing P_{inv}^{ref}, Q_{inv}^{ref}, and V_{inv}^{ref} as reference (and constant) values, while the inverter-injected current is considered the output variable (y). In the study conducted, the plant-level controller is presumed to be non- operational, representing its slower operational time scale. Therefore, the reference values from the plant controller are treated as constants. The inverter's generated current can be designated as i_d and i_d when represented in the inverter's rotating dq reference frame for control, and in a shared network's rotating xy reference frame when developing the small signal model. Along with the seven state variables ($s_1 - s_7$) depicted in Fig. 28.3, the elements of the network's injected current within the xy frame are also identified as state variables, specifically s_8 and s_9. This simplifies the evaluation of small signal stability in a network that encompasses interconnected inverters. (Ramasubramanian et al., 2023)

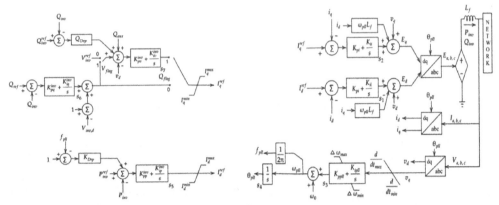

Fig. 28.3 Detailed inverter-level control arrangement for small-signal analysis controls

5. Results

The simulations of the GFL-GFM model are conducted in the electromagnetic transient (EMT) domain at a time step of 5 microseconds. In order to demonstrate the model's performance, the initial scenario involves the examination of a single inverter linked to a grid. The network configuration is illustrated in Fig. 28.1. The inverter has P_{ref} = 80MW, Q_{ref} = 20MVAR, V_{ref} = 1.05pu, P_{load} = 30MW per phase, and Q_{load} = 7MVAR per phase at the beginning of the simulation. The grid supplies the excess power since the IBR's generation is less than the entire load. At t=3.0s, the circuit breaker that links to the grid is disconnected, resulting in the formation of an islanded mode inverter network. Subsequently, at t=5.0s and t=7.0s, a load of P_{load} = 3MW per phase and Q_{load} = 0.7MVAR per phase is introduced into the network. Fig. 28.4 provide a comparative analysis of the responses in the electromagnetic transient (EMT) domain between the GFL and GFM inverter modes of the model. These figures enable a direct side-by-side examination of their behaviours. Initially, when the grid is disconnected, a power generation gap is evident in the system since the IBR resource was deployed at 80 MW, 20MVAR while the load connected was 30MW per phase, 7MVAR per phase. In the Grid Forming mode, this gap in generation (encompassing both active and reactive power) is effectively managed and sustained through the use of reactive droop in V-control, and P-control architecture. In contrast, the Grid Following mode experiences deficits in voltage and active power. When additional loads are introduced into the network at t=5.0s and t=7.0s, there is a minimal variation in active power and voltage observed, with negligible variations in reactive power within the Grid Forming mode shown in Fig. 28.4 (right side). Conversely, in the Grid Following mode with Q_{flag} = 0, as Fig. 28.4 (left side) illustrates, there are some changes with distortion observed in active power, reactive power, and voltage. Upon an overall examination of voltage and current waveforms, a waveform distortion emerges from t=3.0s onwards in the GFL mode (Fig. 28.4, left side), while a stable waveform is consistently observed in the Grid Forming inverter (Fig. 28.4, right side) throughout the entire simulation duration during islanded conditions. These findings suggest that the system exhibits greater stability in the Grid Forming mode compared to the Grid Following mode in islanded conditions.

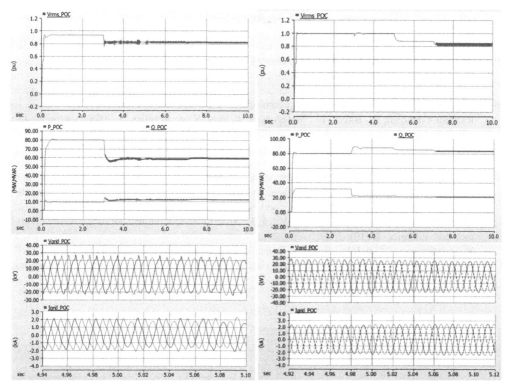

Fig. 28.4 Grid following (left side) and Grid forming (Right side) solar farm inverter output

Here, the gains are $K_{pq}^{inv} = 0.5$, $K_{iq}^{inv} = 0.5$, $K_{pv}^{inv} = 0.5$, $K_{iv}^{inv} = 0.01$, $K_{pp}^{inv} = 0.5$, $K_{ip}^{inv} = 0.5$, $K_{ppll} = 20$, $K_{ipll} = 0.005$, $K_{Drp} = 30$ and $Q_{Drp} = 0.05$, $L_{pu} = 0.2$pu and f = 60Hz.

6. Conclusion

The paper formulates three-phase electromagnetic models for both GFL and GFM inverters, incorporating them into a grid. This enables the small signal stability analysis of main system model. These inverter models accurately depict the control techniques employed by standard GFL and GFM inverters. The validity of the suggested inverter models is confirmed through EMT simulations carried out with PSCAD software, and their performance is assessed in an islanded mode operation. The paper's time-domain outcomes offer valuable insights into the intricacies of voltage control and the waveforms of active and reactive power associated with the behaviours of GFM and GFL inverters. The study's findings demonstrate that GFM inverters exhibit greater stability when dealing with load disturbances in an islanded condition and significantly enhance the system's voltage stability compared to GFL inverters. This insight can contribute to the development of more cost-effective systems and help establish standardized dynamic responses for inverter-based resources in future power networks.

REFERENCES

1. Golsorkhi, M. S., & Lu, D. D. 2014. A control method for inverter-based islanded microgrids based on VI droop characteristics. IEEE Transactions on power delivery. 30(3): 1196-1204.

2. Ramasubramanian, D. 2022. Differentiating between plant level and inverter level voltage control to bring about operation of 100% inverter-based resource grids. Electric Power Systems Research. 205: 107739. https://www.sciencedirect.com/science/article/pii/S0378779621007203.

3. Ramasubramanian, D., Baker, W., Matevosyan, J., Pant, S., & Achilles, S. 2023. Asking for fast terminal voltage control in grid following plants could provide benefits of grid forming behavior. IET Generation, Transmission & Distribution. 17(2): 411-426. https://doi.org/10.1049/gtd2.12421.

4. Sanford, D. 2018. System strength requirements methodology system strength requirements & fault level shortfalls. https://www.aemo.com.au/-/media/Files/Electricity/ NEM/Security_and_ Reliability/ System-Security-Market-Frameworks-Review/2018/System_Strength_Requirements_ Methodology _PUBLISHED .pdf. (Accessed July 15, 2021).

5. Unruh, P., Nuschke, M., Straub, P., & Welck, F. 2020. Overview on grid-forming inverter control methods. Energies. 13(10): 2589.

6. Venkatramanan, D., & John, V. 2019. Dynamic phasor modeling and stability analysis of SRF-PLL-based grid-tie inverter under islanded conditions. IEEE Transactions on Industry Applications. 56(2): 1953-1965.

Emerging Technologies and Applications in Electrical Engineering –
Prof. Dr. Anamika Yadav et al. (eds)
© 2024 Taylor & Francis Group, London, ISBN 978-1-032-82568-7

DC-DC SMPS with Full Bridge Rectifier at the Secondary

29

Himanshu Yadu[1], D. Suresh[2]
Department of Electrical and Electronics Engineering,
National Institute of Technology, Raipur, India

ABSTRACT: In this paper, implementation of a full bridge converter based switched mode power supplies (SMPS) is presented. The SMPS is integrated with DC supply and photovoltaic source. The mathematical model of the system is developed. The root locus and pole zero diagrams are implemented for the SMPS study the performance characteristics using PV source and DC supply. The simulation is carried out using MATLAB/Simulink software to validate transfer function model.

KEYWORDS: SMPS, Uncontrolled rectifier, IGBT

1. Introduction

A switched-mode power supply (SMPS), also known as a switch-mode power supply or simply a switching power supply, is a type of electronic power supply that efficiently converts electrical power from one form to another. Unlike traditional linear power supplies, which regulate voltage by dissipating excess energy as heat, SMPS devices use high-frequency switching techniques to control the output voltage (A Azis H*, 2019). This results in greater efficiency and smaller, lighter power supplies, making them a popular choice in a wide range of electronic devices, from laptops and smartphones to industrial equipment and LED lighting. SMPSs work by rapidly switching the input voltage on and off using semiconductor switches, typically transistors, IGBTs or MOSFETs. This high-frequency switching allows them to control the output voltage with minimal energy loss, making them significantly more efficient than linear power supplies, which operate by regulating voltage through variable resistance.

[1]yaduh69@gmail.com, [2]dsuresh.ee@nitrr.ac.in

DOI: 10.1201/9781003505181-29

The primary components of an SMPS include the input rectifier, filter, power switch, transformer, output rectifier, and feedback control circuit. The feedback circuit is crucial for maintaining stable output voltage and current, even when the input voltage or load varies. Switched-mode power supplies offer several advantages, such as higher efficiency, reduced size and weight, and the ability to handle a wide range of input voltages. These features make SMPSs suitable for various applications, from consumer electronics to industrial systems, where space and energy efficiency are essential. However, they can be more complex to design and may produce electromagnetic interference (EMI), which may require additional filtering or shielding. Nonetheless, SMPS technology continues to be a cornerstone of modern power supply design due to its numerous benefits.

2. High Switching Converter Topology

The proposed topology is a modified version of the full bridge SMPS. The only difference is that, in full bridge, centre tapped rectification method is used while in proposed topology full bridge rectifier is used at the secondary side of transformer as shown in Fig. 29.1. Here, we have considered the leakage reactance and winding resistance to be zero. The switches used here are IGBTs. The IGBT has advantages over the power MOSFET and BJT (Chitra Natesan, 2015) (Tang Yong, 2020). It has a very low 'ON'-state voltage drop and better current density in the 'ON' state. Driving IGBTs is simple and requires low power. At the secondary side there is full bridge rectifier instead of centre tapped rectifier bridge that is present in conventional full bridge SMPS. The reason behind is that the peak inverse voltage of diode in the centre tapped rectifier is double of the maximum voltage across secondary winding of the transformer while in full bridge it is equal to maximum voltage across secondary winding of the transformer. Table 29.1 shows the parameters of the model to be studied.

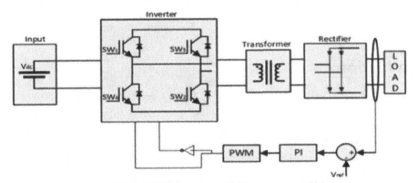

Fig. 29.1 DC-DC SMPS fed with DC supply and PV source

3. Calculation of Transfer Function

While calculating the transfer function of the system the equivalent circuit of the transformer is considered as shown in the Fig. 29.2. Here all the parameters are referred to the primary side of the transformer by using the transformation ratio.

$$\left.\begin{array}{l} R_1 = R/K^2 \\ X_1 = X_c/K^2 \\ C_1 = CK^2 \end{array}\right\} \tag{1}$$

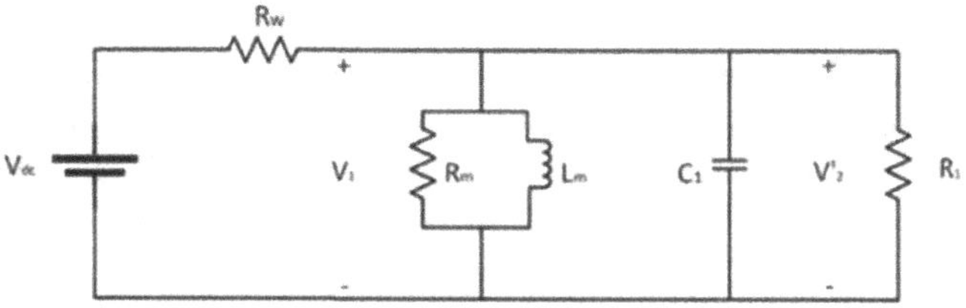

Fig. 29.2 Equivalent circuit of system referred to primary side

Here Z'_L is the parallel combination of R_1 and X_1. In the calculation the secondary winding resistance and leakage inductance of both the primary and secondary winding of the transformer are neglected. During the switching when switches SW_1 & SW_2 are on, the equivalent circuit whole system can be considered as circuit given below. In this state the input voltage V_{dc} is directly applied across transformer and is equal to the summation of voltage drop across R_w and the primary winding voltage V_1. R_w is the winding resistance of primary side and is equal to 1 mΩ. At the same time diode D_1 & D_2 are also on and therefore V_2 is equal to V_o. If we refer all the secondary parameter to the primary then we can calculate transfer function very easily and since we are considering approximately ideal condition same transfer function for the next switching state. Here,

$$\left.\begin{array}{l} K = N_2/N_1 \\ V'_2 = V_1 = V_2/K \end{array}\right\} \tag{2}$$

Since L_m, R_m, C_1 and R_1 all the elements are in parallel, for simplification the equivalent resistance of parallel combination of R_m & R_1 can be calculated and the circuit is further simplified as given below. Here,

$$R' = \frac{R_1 R_m}{R_1 + R_m} \tag{3}$$

Now the calculation of Z which is the parallel combination of R', L_m and C_1 can be done. After applying Laplace transformation equation

$$Z(s) = \frac{sL_m R'}{s^2 R' L_m C_1 + sL_m + R'} \tag{4}$$

The relation between V_2' and V_{dc} can be obtained by applying the voltage division rule.

$$V'_2 = \frac{V_{dc}(L_m R')}{s^2(R' R_w L_m C_1) + sL_m(R' + R_w) + R' R_w} \tag{5}$$

The transfer function of above system is found to be:

$$\frac{V'_2}{V_{dc}} = \frac{1.883}{9.24*10^{-12} s^2 + 1.883s + 0.077} \tag{6}$$

4. Waveform

Here the forward resistance of the switches is considered as $1m\Omega$. From Fig. 29.3 it can be seen that harmonics are present in the system, but as the order of the harmonics is increases the magnitude decreases. Here, up to third harmonics the magnitude is considerable and beyond the third harmonic the impact is very less therefore it can be neglected. Even in the ideal consideration the problem of harmonics is there and it will increase with the implementation of practical parameter so the design of the filter circuit needs to made in such a manner that it will increase system robustness and tuning capability along with reduction in the system harmonics.

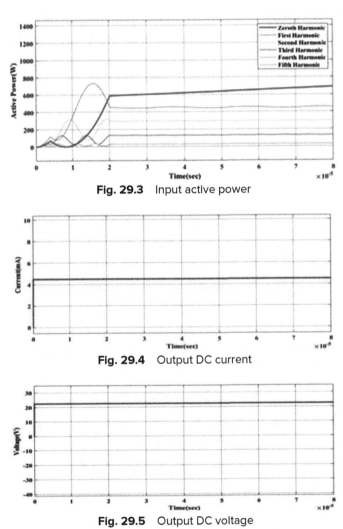

Fig. 29.3 Input active power

Fig. 29.4 Output DC current

Fig. 29.5 Output DC voltage

Table 29.1 Values of parameter

Parameters	Values
Input voltage (V_{dc})	340 V
Transformation Ratio ($N_2:N_1$)	1:2
Magnetization Resistance(R_m)	77 Ω
Magnetization Inductance(L_m)	0.245 mH
Capacitance(C)	100 * 10⁻⁶ F
Resistance(R)	5 Ω
Switching Frequency(f)	50000 Hz

5. Result

The root locus of the transfer function as stated in equation (6) is shown in the Fig. 29.7 to focus the stability zone of the topology under the ideal condition. The Pole zero map of the transfer function as stated in equation (6) is shown in the Fig. 29.6 to find out the stability zone of the converter, from both the plots we can conclude that system is stable under ideal considerations.

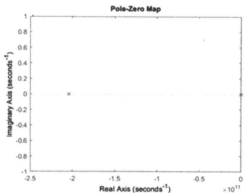

Fig. 29.6 Pole zero map of the transfer function of the converter

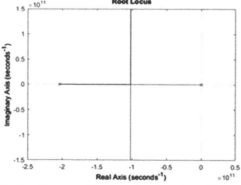

Fig. 29.7 Root Locus of the transfer function of the converter

6. Conclusion

The simulation study of the SMPS is carried out using MATLAB/Simulink software. The high current SMPS is designed and simulated to study its characteristics. From Fig. 29.5 we can see that 24 V output is generated with 340 V as input. Further in above system switching function for the IGBTs can be generate using Sliding Mode Control so that the system stability will increase and these techniques will also help in increasing the accuracy and robustness of the system which again help in easy tuning of the system so that the implementation becomes achievable under practical condition (Guipeng chen, 2019).

REFERENCES

1. A Azis H*, C. R. (2019). Comparative analysis between the switch mode power supply (SMPS) using IC Tl494cn transformer based on power supply linear. IOP Conference Series: Materials Science and Engineering.
2. Bose, B. (2020). Modelling of Microinverter and PushPull Flyback Converter for SPV Application. 8th International Conference on Reliability, Infocom Technologies and Optimization (Trends and Future Directions) (ICRITO).
3. Chitra Natesan, A. D. (2015). IGBT and MOSFET: A comparitive study of power electronics inverter topology in distributed generation. International conference on circuit, Power and Computing Technologies ICCPCT.
4. Guipeng chen, j. d. (2019). ingerated dual-output synchronous DC-DC buch converter. IEEE.
5. Halder, T. (2020). State Space Modelling and Stability analysis of the flyback converter. IEEE.
6. Madisa V G Varaprasad, B. A. (2023). Comparision of Fly-back Converter and Reverse Fly-back Converter for PV Application. IEEE Renewable Energy and Sustainable E-Mobility conference (RESEM).
7. Suresh Kumar Tummala, L. D. (2022). Switched Mode Power Supply: A High efficient low noise forward converter design topology. IEEE 2nd international conference on sustainable energy and future transpotation (SeFeT).
8. Tang Yong, W. B. (2020). Mechanism and Model Analysis of IGBT DIsplacement Current. 3rd International Conference on Advanced Electrronics Materials, Computer and Software Engineering (AEMCSE).

Emerging Technologies and Applications in Electrical Engineering –
Prof. Dr. Anamika Yadav et al. (eds)
© 2024 Taylor & Francis Group, London, ISBN 978-1-032-82568-7

Topological Investigation of HSC Based RSC-MLI Topologies

30

K. Sheshu Kumar[1]

Dept of Electrical Engineering GIET University,
Gunupur, Odisha, India

Hari Priya Vemuganti[2]

Department of Electrical Engineering,
National Institute of Technology Raipur

N. Bhanu Prasad[3]

Dept of Electrical Engineering GIET University,
Gunupur, Odisha, India

ABSTRACT: Objective to reduce the size, cost and complexity of classical Multilevel inverter (MLI) topologies has led to the development of various reduced switch count (RSC) MLI configurations. In brood, most of the RSC-MLIs reported are H-bridge based, with inherent level and polarity generator or with separate level and polarity and level generator. However, as alternative to the basic four-switch H-bridge structure, a novel six- switch Hexagonal switched cell (HSC) structure is reported. The basic HSC unit operates as simple five-level inverter with symmetrical dc voltage ratios and, seven level inverter with binary voltage ratios. Emphasizing the prominence of HSC structure, various RSC-MLI configurations are reported. This paper aims to present an investigative report on the popular HSC based RSC-MLI topologies and further evaluate their performance with a simple carrier based PWM in obtaining satisfactory phase and line voltage performance

KEYWORDS: Hexagonal switched cell (HSC), H-bridge, Reduced carrier PWM, Switching function PWM, T-type, Reduced switch count (RSC), Multilevel inverter (MLI)

[1]vvksheshu17@gmail.com, [2]hpvemuganti.ee@nitrr.ac.in, [3]vvksheshu17@gmail.com

DOI: 10.1201/9781003505181-30

1. Introduction

The feature of Multilevel inverter (MLI) to synthesize high AC output voltage using matured medium power electronic devices has enhanced MLI significance in medium-voltage high-power applications (Franquelo, L.G et.al., 2008). However, enormous raise in the switch count of the popular classical MLI configurations such as CHB, DCMLI and FCMLI for higher levels has imposed size cost and complexity limitations and restricted their deep penetration into industrial markets Thus, various alterative trials to overcome the challenges of classical MLI configurations preserving their merits are explored. Among them reduced switch count configurations (RSC) and asymmetrical configurations are popular. RSC MLI configurations aim to operate with reduced device count to achieve higher voltage levels, with respect to the classical MLI configurations. In contrast asymmetrical configurations, involve the concept of imposing non-identical dc voltage ratios and produce a greater number of levels for additive and subtractive combinations of dc sources (Rech, C 2007). This method enhancing the number of voltage level reduces the size and device count limitations of MLIs, however creates non-uniform device ratings and increases voltage/current ratings of incorporated switches.

It is to be noted that most of the reported RSC-MLI configurations incorporates either the concept of H-bridge or Half-bridge structure at the AC side. However, the purpose and operation of H-bridge may vary with topological orientation. For example, RSC-MLI such as MLDCL, SSPS, SCSS, RV and Unit based MLI are few H-bridge based RSC-MLI configurations with separate level and polarity generator (Omer, P., et.al., 2020) Thus, the H-bridge incorporated in these configurations is responsible for polarity generation converting the unipolar stair case output of level generator to bi-polar output voltage (Vijayaraja, L., et.al., 2016). In contrast, topologies such as T-type incorporate H-bridge which is responsible for both polarity and level generator (Park, S.-J., et al., (2003). In detail, one leg of H-bride operates at carrier frequency and is responsible for ±peak voltage level generation, whereas the other leg of H-bridge operates a fundamental frequency and operates as per the required output voltage polarity (Gupta, K.K., et al., 2015)

It is be noted that, though the various reported RSC-MLI configurations offers appreciable reduction in switch count over the classical configurations, the basic CHB MLI configurations, still claims to be superior in terms of fault tolerant ability, switching redundancies, and uniform device ratings (Salem, A., et al., 2022). Thus, alterative to the basic three-level four switch H-bridge structure, a novel six switch Hexagonal Switched Cell (HSC) is reported (Odeh, C.I.,2014). The basic HSC unit is framed by antiparallel integration of two-half bridge cell via a pair of uni-directional switches. This hexagonal six-switch structure of HSC can support additive and substrative combinations of dc sources and operates as dedicated five-level inverter for symmetrical voltage ratios and seven-level inverter for binary voltage ratios (Odeh, C.I. et.al., 2016)

Imposing the novel concept of HSC, various RSC-MLI configurations with HSC structure on AC side (Vemuganti, H.P., et al., (2021). Among the various HSC based configurations reported, Hybrid T-type (HT-type) is the framed by imposing the concept of HSC on the popular T-type configuration, such that the basic HSC unit can dedicatedly act as nine-level inverter for symmetrical voltage ratios and, 13-level, 17-level for binary and trinary voltage ratios respectively. Thus, motivated by the merits of basic HSC unit, various HSC and Extended HSC based RSC-MLI configurations reported, which offers appreciable reduction in switch count preserving switching redundancies. This proposed work aims to present an investigative discussion on the features and topological operation on popular HSC based RSC-MLI configuration and, further evaluate their performance in achieving satisfactory output performance with simplified carrier based PWM. However, the switching logic of the popular carrier based PWM schemes of LSPWM and PSPWM of classical MLIs are not suitable for controlling the either symmetrical or asymmetrical topologies of novel HSC configuration. Thus, the proposed work uses a simplified reduced carrier based PWM followed unified logical expressions to control the considered HSC configurations and obtain desired performance.

Organization of the paper as follows: Section-II presents significance of HSC over the various H-bridge based RSC-MLI configurations. Section-III presents the features and topological operation of popular HSC topologies. Section-IV discus the implementation of popular Reduced carrier PWM to realize HSC topologies. Finally, the simulation performance is presented in Section-V.

2. RSC-MLIs Topologies: HSC Significance

The most popular H-bridge based RSC-MLIs such as MLDCL, SSPS and MLM are shown in Fig. 30.1(a)-(c) and, T-type is shown in Fig. 30.1(d). The topologies depicted in Fig. 30.1(a)-(c) possess separate level and polarity generator, with H-bridge acting as polarity generator. However, the purpose of H-bridge in T-type shown in Fig. 30.1(d) is to support both level and polarity generation.

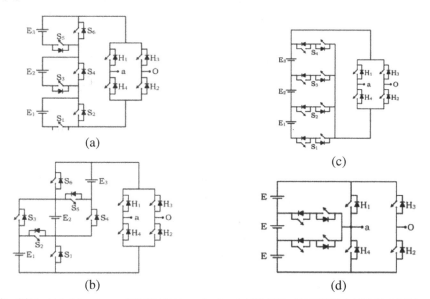

(a)

(b)

(c)

(d)

Fig. 30.1 H-bridge based RSC-MLI topologies (a) MLDCL, (b) SSPS, (c) MLM, (d) T-type

The primary unit of HSC shown in Fig. 30.2(a), is actually evolved by anti-parallel connection of two half-legs (H_1-H_6 & H_3-H_4) through a pair of unidirectional switches (H_2 and H_5). It is to be noted that with E_S=E_R, HSC operate as simple five-level inverter, with voltage magnitude, $\pm E_S/E_R$ and $\pm(E_S+E_R)$. But with E_S= 2 E_R, this HSC module operates as seven-level inverter. It is to be noted that the device voltage rating of H_1 & H_6 is E_S; H_3 & H_4 is E_R; and H2, H5 is E_S+E_R. To emphasize the superiority of six-switch HSC module over the 4 switch H-bridge, The basic three-level H-bridge unit is shown in Fig. 30.2(b).

3. HSC based RSC-MLI Topologies

These configurations employ unidirectional switches to form HSC structure and, utilize bi-directional switches to link HSC and DC link. HSC arrangement enables the topology to operate with switching redundancies. Depending on the integration of bi-directional switches to DC link, numerous HSC-based configurations shown in Fig. 30.3 are reported. Fig. 30.3(a) depicts Topology-I of the HSC, featuring a bi-directional switch positioned on one side of the HSC. As this arrangement represents a hybrid connection of the T-type with the HSC, this is referred as Hybrid T-type (HT-type) or Enhanced T-type RSC-MLIs. The topological layout of this RSC-MLI includes two rigid DC sources, E_S and E_R, situated on either side of the HSC, as illustrated in Fig. 30.3(a). This topology can be expanded either by adding more bi-directional switches or by cascading multiple modules.

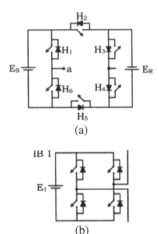

(a)

(b)

Fig. 30.2 The Basic HSC & H-bridge unit (a) The Basic HSC unit, (b) The Basic H-Bridge unit

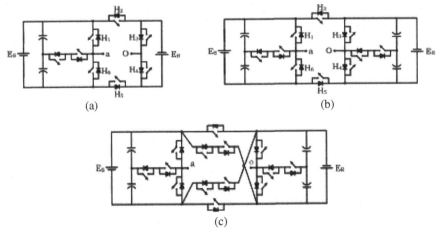

(a)   (b)

(c)

Fig. 30.3 HSC based RSC-MLI topologies (a) HT-type: Topology-I, (b) HT-type: Topology-II, (c) HT-type: Topology-III

The HSC Topology-II, depicted in Fig. 30.3(b), connects both sides of the HSC with bi-directional switches. This configuration resembles a back-to-back connection of two half-leg T-type modules through a pair of unidirectional devices. This setup enhances the asymmetrical capabilities of the inverter and allows for the generation of voltage levels with a notable reduction in switch count. However, the remaining characteristics and operation of this topology remain similar to Topology-I. The extension of the HSC with bi-directional switches i.e., two cross-connected switches inside the HSC is illustrated in Fig. 30.3(c). This configuration generates a maximum output voltage of (Es + ER). However, the inclusion of cross-connected switches increases the number of redundant switching states, allowing the configuration to produce outputs with up to 25 switching combinations.

4. PWM Implementation

To analyze the ability of the popular HSC and HT-type module in producing satisfactory performance and level generation in the out voltage, this section presents the implementation of simplified carrier based PWM i.e., Reduced carrier PWM to realize the considered HSC based Topologies. However, the sample implementation of the PWM is demonstrated for the 9-level HT-type topology-II, imposed with symmetrical voltage ratios. The carrier arrangement of the considered modified reduced carrier PWM to realize 9-level inverter is shown in Fig. 30.4(a) and the corresponding switching logic implementation is shown in Fig. 30.4(b). The logical operations involved to generate the desired switching pulses is given below and the look table is shown in Table 30.1

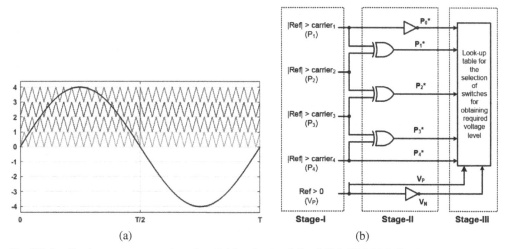

(a)   (b)

Fig. 30.4 Carrier arrangement and switching logic of the MRC PWM (a) Carrier arrangement 9-level Inverter, (b) Switching logic implementation

Table 30.1   Look-up table for 9-level HT-type Fig. 30.3(b)

Switch	Corresponding Pulse
S_1	$(V_P \cdot P_4{}^*) + (V_N \cdot (P_0{}^* + P_1{}^* + P_2{}^*))$
S_2	V_N
S_3	$(V_N \cdot P_4{}^*) + (V_P \cdot (P_0{}^* + P_1{}^* + P_2{}^*))$
S_4	V_P
S_5	$(V_P \cdot P_3{}^*)$
S_6	$(V_N \cdot P_3{}^*)$
S_7	$(V_P \cdot (P_2{}^* + P_3{}^* + P_4{}^*)) + (V_N \cdot P_0{}^*)$
S_8	V_N
S_9	$(V_N \cdot (P_2{}^* + P_3{}^* + P_4{}^*)) + (V_P \cdot P_0{}^*)$
S_{10}	V_P

5. Simulink Implementation

The simulation performance of basic HSC unit and extended HSC top logy i.e., HT-type with the considered PWM is evaluated here. The basic module of HSC depicted in Fig. 30.2 produce 5-level with symmetrical and can produce 7 and

Table 30.2   Simulation parameters

Voltage of each dc-source	V_{dc}	100 V
Amplitude-modulation index	m_a	0.95
Frequency - carrier	F_{sw}	100 Hz ; 2000 Hz
Sampling time	T_s	1 μS

binary voltage ratios (asymmetrical). However, the HT-type of Fig. 30.3 can produce 9-level with symmetrical and can produce 13 and 17-level with asymmetrical. Thus, with four dc sources with 10 switches, Fig. 30.3 can synthesize 9-level output voltage with symmetrical and up to 31-level with asymmetrical for higher voltage ratios.

Table 30.3   Output Harmonic performance comparison of topology-1 and topology-II

			Phase Voltage THD	
			Carrier frequency	
	Level	Nature	50Hz	2000 Hz
Topology-I	7-level	Symm	13.33%	20.62%
Topology-I	13-level	Asymm	7.45%	10.39%
Topology-II	9-level	Symm	10.51%	15.60%
Topology-II	15-level	Asymm	6.52%	8.58%

Operating Fig. 30.3 with symmetrical and asymmetrical of 1: 2: 3 produces 9-level and, 15-level output voltages respectively. Further, opting higher carrier frequency to 2000 Hz, the attained output voltages for the considered cases of Fig. 30.6(a) and (b) are shown in Fig. 30.5 (a) and (b) respectively. The corresponding harmonic spectrum of output voltages of Fig. 30.5 are depicted in Fig. 30.6 respectively. Though Table 30.3 claims to have improved

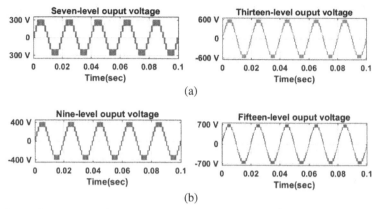

Fig. 30.5 Out voltage wave forms of topology-I and II (a) Output voltage of Topology-I for symmetrical and asymmetrical, (b) Output voltage of topology-II for symmetrical and asymmetrical

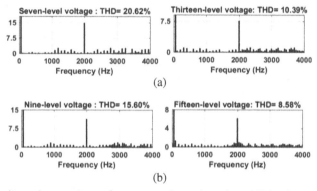

Fig. 30.6 Out voltage harmonic performance of topology-I and II (a) Output voltage THD of Topology-I for symmetrical and asymmetrical, (b) Output voltage THD of topology-II for symmetrical and asymmetrical

harmonic performance with low frequency carrier, it is to be noted that the increasing the carrier frequency suppress the low order harmonics, by shifting them to carrier frequency. Hence all the carriers around fundamental frequency are shifted to carrier frequency, which increases the cut of frequency and thus reduces the filter size.

6. Conclusion

This paper demonstrated an investigative report significance of HSC based RSC-MLI configurations over the classical H-bridge structure and conventional H-bridge based RSC-MLIs. The key merit of HSC to promote symmetrical and asymmetrical voltage ratios, where the basic 6 switch unit can act as dedicate five-level inverter with symmetrical voltages and 7-level inverter with binary voltage ratios turned the HSC superior over the H-bridge. Further, the ability of the basic HSC unit and popular HSC -based RSC-MLI configurations in attaining desired out voltage performance is demonstrated using the most simplified

REFERENCES

1. Franquelo, L.G. (2008). The age of multilevel converters arrives. IEEE industrial electronics magazine, **2**(2): 28-39.
2. Gupta, K.K., et al., (2015). Multilevel inverter topologies with reduced device count: A review. IEEE transactions on Power Electronics. **31**(1): 135-151.
3. Odeh, C.I.,(2014). A cascaded multi-level inverter topology with improved modulation scheme. Electric Power Components and Systems. **42**(7): . 768–777.
4. Odeh, C.I. et.al., (2016): Topology for cascaded multilevel inverter. IET Power Electronics. **9**(5): 921–929.
5. Omer, P., J. Kumar, and B.S. Surjan, (2020). A Review on Reduced Switch Count Multilevel Inverter Topologies. IEEE Access **8**: 22281-22302.
6. Park, S.-J., et al., (2003). A new single-phase five-level PWM inverter employing a deadbeat control scheme. IEEE Transactions on power electronics. **18**(3): 831–843.
7. Rech, C. and J.R. Pinheiro, (2007). Hybrid multilevel converters: Unified analysis and design considerations. IEEE Transactions on Industrial Electronics **54**(2): 1092-1104.
8. Salem, A., et al., (2022). Hybrid Three-Phase Transformer-Based Multilevel Inverter with Reduced Component Count. **10**: p. 47754-47763
9. Vijayaraja, L., S.G. Kumar, and M. Rivera (2016). A review on multilevel inverter with reduced switch count. in IEEE International Conference on Automatica (ICA-ACCA) IEEE.
10. Vemuganti, H.P., et al., (2021). A survey on reduced switch count multilevel inverters. **2**: 80-111.

Emerging Technologies and Applications in Electrical Engineering –
Prof. Dr. Anamika Yadav et al. (eds)
© *2024 Taylor & Francis Group, London, ISBN 978-1-032-82568-7*

Novel Hybrid PWM for Non-Cascaded RSC-MLI Converters

31

Disha Tiwari[1] and V. Hari Priya[2]

Department of Electrical Engineering, National Institute of Technology, Raipur

ABSTRACT: A novel PWM scheme to realize asymmetrical non-cascaded RSC-MLI with separate level and polarity generator is proposed in this paper. Most often reduced switch count MLI are preferred to obtain desired output voltage levels with reduction in switch count w.r.t the classical MLI configurations. In addition, imposing asymmetrical dc voltage ratios on RSC-MLI further reduces the switch count and turns the configuration more compact and simpler. Most often hybrid scheme is the simplest PWM scheme reported to realize asymmetrical MLI configurations such as CHB. However, implementation of this hybrid PWM scheme to realize asymmetrical configurations is limited. This is mixed frequency scheme which imposes the concept of forcing the lower voltage module to obtain the levels not achieved by higher voltage module. To ensure this, the modulating signal of the lower voltage module is obtained by comparing the output of higher voltage module with actual modulating signal. Thus, this hybrid PWM scheme is suitable for an MLI/ RSC-MLI configuration with cascaded structure and not suitable for non-cascaded RSC-MLI such as MLDCL. To elevate and address this limitation a novel hybrid PWM scheme to implement non cascaded RSC-MLI such as MLDCL is proposed in this paper further the performance of proposed scheme in achieving desired output voltages is validated on asymmetrical seven level MLDCL (operated with binary voltage ratios) using MATLAB.

KEYWORDS: RSC-MLI, Hybrid PWM, MLDCL

1. Introduction

The demand for efficient power converters for PV/wind/grid connected systems, industrial applications, energy storage systems, electric vehicles and active front end converters etc.

[1]dishatiwari28@gmail.com, [2]hpvemuganti.ee@nitrr.ac.in

DOI: 10.1201/9781003505181-31

has enhanced the impact of multilevel inverter in high power medium voltage applications. Significant feature of multilevel inverter (MLI) to synthesize high ac output voltage with matured medium power electronic devices has gathered great attention from both academy and Industry. In addition, MLI offers various other merits of less dv/dt, reduced common mode voltages, improve harmonic performances, less filtering requirements and poses fault tolerant ability.

Among the various topologies reported for MLI, Diode clamped (DCMLI), Flying Capacitor (FCMLI) and Cascaded H-Bridge (CHB) are the classical configurations which are widely penetrated in industrial markets and explained by J. Rodriguez, J. S. Lai, and F. Z. Peng (2002) in their paper. DCMLI achieve required voltage levels by interconnecting multiple DC sources in series and uses diodes to clamp voltages at different levels. FCMLI uses the concept of pre-charging capacitors and achieves the output voltages by switching flying capacitors. CHB uses the concept of floating capacitors and connects multiple H-Bridges in cascade to achieve desired output voltage levels. Thus, the modular fault tolerant topological structure of CHB with adequate switching redundancies, even power distribution, natural voltage balancing ability has it superior over DCMLI and FCMLI. Among the various modulation schemes reported to realize these MLIs, carrier based PWM schemes are the simplest. Further, among the carrier-based sine PWM techniques, Phase shifted (PSPWM) and Level shifted (LSPWM) are the most popular involving (n-1) carrier to realize 'n' level inverter (phase voltage).

Though MLIs offers various merits, their increase in device count with increase in output voltage levels, increases the switch count, requirement of other auxiliary components in associated driver and protection circuits, thus imposing size, cost and control limitations on MLI. Thus, great exploration carried to reduce the size of the MLI at higher levels has led to the development of Reduced switch count (RSC) MLI configurations explained by G.-T. Kim and T. A. Lipo (1996). In contrast, the size and device of MLI can also be reduced by imposing asymmetrical dc voltages (binary, trinary) rather than conventional symmetrical dc voltage ratios. Thus RSC-MLI with asymmetrical voltage ratios offers effective reduction in switch count and size, with respect to its corresponding symmetrical configurations.

Among the various RSC-MLI configurations reported, MLDCL is one most popular configuration with viable ability to serve as an alternative to CHB. Though various PWM carrier based PWM schemes reported to implement RSC-MLI, none of them could be as dominant as Hybrid PWM scheme. Hybrid PWM is the mixed frequency switching scheme used to realize asymmetrical MLI configurations and well reported for CHB MLI for asymmetrical configurations This mixed frequency scheme imposes the concept of forcing the lower voltage module of MLI to obtain the levels not achieved by higher voltage module. To ensure this, the modulating signal / reference of the lower voltage module is obtained by comparing the output of higher voltage module with actual modulating signal. Thus, this scheme is only suitable for cascaded MLI configurations such as CHB and not suitable for non-cascaded MLI/RSC-MLI configurations such as MLDCL. So, in this paper to eliminate this limitation of Conventional Hybrid PWM, a novel hybrid PWM is proposed. The proposed scheme is suitable for Non-cascaded RSC-MLI configurations with separate level and polarity generator.

The paper organization is as follows: Section-2 Topological description of MLDCL based RSC-MLI and elevates the limitation of conventional Hybrid PWM scheme. Sections-3 Proposed hybrid PWM scheme for low frequency switching. Section-4 Simulink implementation of proposed PWM scheme MLDCL Section-5 Conclusions

2. Topological Description of MLI

2.1 Topological Description

Cascaded H-Bridge multilevel inverters utilize multiple H-bridge inverter cells connected in series, each producing a specific voltage level. The choice of topology depends on factors such as desired output voltage levels, efficiency, complexity, and cost considerations. Each topology has its advantages and disadvantages, and the selection depends on specific application requirements and constraints. Fig. 31.1 shows the per phase topological structure of CHB with H-bridges cascaded and its corresponding switching operation to generate seven-level output voltage for binary voltage ratios is shown in Table 31.1.

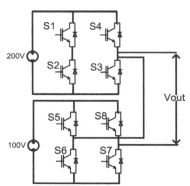

Fig. 31.1 Cascaded H-Bridge with two modules

Table 31.1 Switching state for cascaded H-bridge hybrid PWM

S. No.	V0	V1	V2	Switch path for upper H-Bridge	Switch path for lower H-Bridge
1.	3V	V	2V	S_1&S_3	S_5&S_7
2.	2V	0	2V	S_1&S_3	S_5 &S8 or S_6&S_7
3.	V	V	0	S_1&S_4 or S_2&S_3	S_5&S_7
4.	0	0	0	S_1&S_4 or S_2&S_3	S_5 &S_8 or S_6&S_7
5.	-V	-V	0	S_1&S_4 or S_2&S_3	S_8&S_6
		V	-2V	S_4&S_2	S_5&S_7
6.	-2V	0	-2V	S_4&S_2	S_5 &S_8 or S_6&S_7
7.	-3V	-V	-2V	S_4&S_2	S_8&S_6

Asymmetrical topologies are preferred where high number of levels of output voltages are required with a smaller number of switches. In asymmetrical voltage configuration dc input voltages with magnitude in ratios has been used. If the source voltage ratio is in 2:1 then it is binary dc input voltages. If the source voltage ratio is in 3:1 then it is trinary voltage source arrangement

2.2 Multilevel DC Link (MLDCl)

Multilevel Dc link inverter is used to generate desired levels in output voltage using RSC-MLI topology. MLDCL consists of level generator and polarity generator. In level generator two half bridges are being used with binary voltage ratios in source voltage. Polarity generator consists of H-bridge to convert the unipolar output voltage generated from level generator to bipolar output voltage of desired level. Figure 31.2 shows the per phase topological structure of MLDCL with two modules and its corresponding switching operation to generate five-level as seven-level output voltage for binary voltage ratios is shown in Table 31.2.

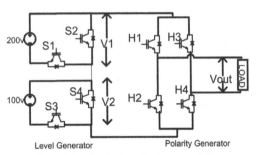

Fig. 31.2 MLDCL with two modules

Table 31.2 Switching state for MLDCL

S. No.	V0	V1	V2	Switch Path for Upper module	Switch path for lower module	Switch path for polarity generator
1.	3V	V	2V	S_1	S_3	$H_1\&H_2$
2.	2V	0	2V	S_1	S_4	$H_1\&H_2$
3.	V	V	0	S_2	S_3	$H_1\&H_2$
4.	0	0	0	S_2	S_4	$H_1\&H_2$
		V	2V	S_1	S_3	$H_1\&H_3$
		V	0	S_2	S_3	$H_1\&H_3$
		0	2V	S_1	S_4	$H_2\&H_4$
5.	-V	V	0	S_2	S_3	$H_3\&H_4$
6.	-2V	0	2V	S_1	S_4	$H_3\&H_4$
7.	-3V	V	2V	S_1	S_3	$H_3\&H_4$

2.3 Limitation of Conventional PWM – Hybrid PWM

Basically, the hybrid PWM technique is used for asymmetrical configurations (C. Rech and J. R. Pinheiro. 2007). In CHB topology to generate high level of output voltages with a smaller number of switches hybrid PWM technique is being used. Commonly to generate seven level output voltage there is a requirement of three H-Bridges with individual voltage source of same magnitude comprises with twelve switches, having six carrier signals to compare with modulating/reference signal but after applying hybrid PWM technique to generate same seven level inverter there is a requirement of two H-Bridges with binary voltage ratio of order 2:1 for dc voltage source with only eight switches

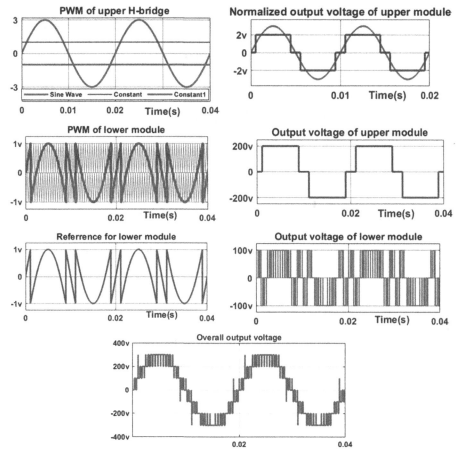

Fig. 31.3 Output voltages of conventional hybrid PWM

This is a mixed frequency scheme which imposes the concept of forcing the lower voltage module to obtain the levels not achieved by the higher voltage module. To ensure this, the modulating signal/reference of the lower voltage module is obtained by comparing the output of the higher voltage module with the actual modulating signal. Thus, this hybrid PWM scheme is suitable for an MLI/ RSC-MLI configuration with a cascaded structure and not suitable for non-cascaded RSC-MLI such as MLDCL

3. Proposed Hybrid PWM Scheme for Non-cascaded RSC-MLI

The schematic diagram for the proposed hybrid PWM scheme is shown in Fig. 31.5. For non-cascaded RSC-MLI most popular method i.e., MLDCL is being used. MLDCL is comprised of a level and polarity generator. So, in this paper, we have proposed a novel hybrid PWM.

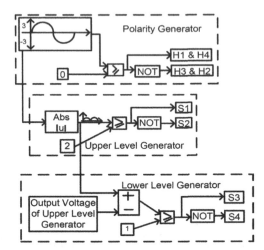

Fig. 31.4 Schematic implementation of proposed low-frequency hybrid PWM for seven-level MLDCL

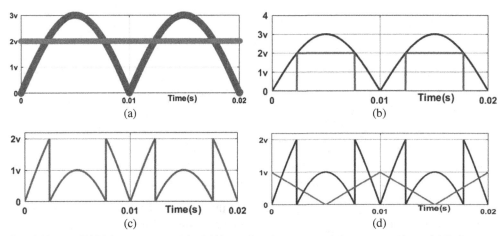

Fig. 31.5 (a) PWM for upper module (b) Normalized upper module output voltage (c) Reference for lower module (d) PWM for lower module

This paper proposes a hybrid PWM scheme with reduced switch count composed of level generator and polarity generator. For level generators there is two half bridge modules Upper-level generator is for high voltage low frequency whereas lower-level generator is for low voltage high frequency. There are two switches in upper and low-level generators. For upper-level generator the unipolar sinusoidal voltage is compared with constant two to turn on and off switches S1 and S2 respectively. Similarly for lower-level generator the output voltage of upper-level generator is subtracting from unipolar sinusoidal voltage. The overall output voltage of level generator is unipolar in nature to get proper bipolar output voltage polarity generator is being used. Polarity generator comprises of H-bridge with four switches H1, H2, H3&H4. The Simulink results of corresponding scheme should be shown in further Fig. 31.5

4. Simulink Performance of the MLDCL with the Proposed PWM

In this hybrid seven-level inverter is being developed by using two half bridges as a level generator and one full bridge as a polarity generator by using asymmetrical voltage sources of having binary ratio of 2:1. Here RSC-MLI is being implied for developing seven-level inverter. In conventional method for generating seven-level inverter there is a requirement of six carriers to compare with modulating signal. The first result shows the output voltage generated from upper-level module of level generator which possess high output voltage with low frequency. The second result shows the output voltage generated from lower-level generator. The output voltages from upper module, lower module and overall output voltage are shown in Figs. 31.6, 31.7, 31.8, 31.9 respectively.

Table 31.3 Simulation parameters

Sample time	Ts	1e-6
Power converter Dc voltage ratios	Vdc1 Vdc2	100 200
PWM – Amplitude modulation index Frequency modulation index Fundamental frequency	 ma mf fs	 1 2 50 Hz

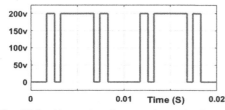

Fig. 31.6 Upper-level generator output voltage

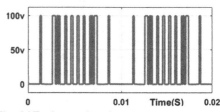

Fig. 31.7 Lower-level generator output voltage

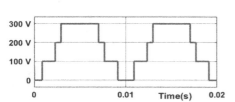

Fig. 31.8 Output voltage of level generator

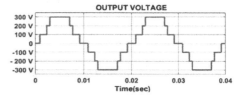

Fig. 31.9 Output voltage of polarity generator

The second result is the simulation result of overall output voltage of polarity generator. For polarity generator we are using H-bridge. To achieve the desired levels on output voltage with a smaller number of switches the combination of level generator and polarity generator is required.

5. Conclusion

The work provides an analysis of non-cascaded RSC-MLIs with proposed hybrid PWM. This paper attempts to reach the seven-level output voltage by approaching MLDCL method using RSC-MLI topology having two separate generators, level and polarity simultaneously, to overcome the limitations of cascaded RSC-MLI. In addition, RSC-MLI decreases the count of switch and the configuration becomes simple and compact. These converters offer advantages like reduced DC sources, improved efficiency, lower cost, and small size. MLDCL has merits of fault tolerant, same device rating for any level, redundancy, dc-link voltage balancing.

REFERRENCES

1. C. Rech and J. R. Pinheiro. (2007), "Hybrid Multilevel Converters: Unified Analysis and Design Considerations," in *IEEE Transactions on Industrial Electronics, vol. 54, no. 2,* pp. 1092-1104, *Doi:* 10.1109/TIE.2007.892255.
2. F. Z. Peng, J. S. Lai, J. W. McKeever, and J. Van Coevering. (1996) "A multilevel voltage-source inverter with separate dc sources for static var generation," IEEE Trans. Ind. Appl., vol. 32, no. 5, pp. 1130–1138.
3. G.-T. Kim and T. A. Lipo (1996) "VSI-PWM rectifier/inverter system with a reduced switch count," IEEE Trans. Ind. Appl., vol. 32, no. 6, pp. 1331–1337.
4. H. Abu-Rub, J. Holtz, J. Rodriguez, and G. Booming (2010) "Medium voltage multilevel converters—State of the art, challenges, and requirements in industrial applications," IEEE Trans. Ind. Electron. Vol. 57, no. 8, pp. 2581–2596.
5. H. Akagi. (2011) "Classification, terminology, and application of the modular multilevel cascade converter (MMCC)," IEEE Trans. Power Electron. Vol. 26, no. 11, pp. 3119–3130.
6. H. Akagi, M. Hagiwara, and R. Maeda (2012) "Negative-sequence reactive power control by a PWM STATCOM based on a modular multilevel cascade converter (MMCC-SDBC)," IEEE Trans. Ind. Appl., vol. 48, no. 2, pp. 720–729.
7. H. Akagi, S. Inoue, and T. Yoshii. (2007) "Control and performance of a transformer less cascade PWM STATCOM with star configuration," IEEE Trans. Ind. Appl., vol. 43, no. 4, pp. 1041–1049.
8. J. Rodriguez, J. S. Lai, and F. Z. Peng (2002) "Multilevel inverters: A survey of topologies, controls, and applications," IEEE Trans. Ind. Electron. Vol. 49, no. 4, pp. 724–738.
9. J. Rodriguez, S. Bernet, Bawku, J. O. Pont, and S. Kour (2007) "Multilevel voltage-source-converter topologies medium-voltage drives," IEEE Trans. Ind. Electron., vol. 54, no. 6, pp. 2930–2945.
10. J. S. Lai and F. Z. Peng. (1996) "Multilevel converters-a new breed of power converters," IEEE Trans. Ind. Appl., vol. 32, no. 3, pp. 509–517.
11. L. Maharjan, S. Inoue, and H. Akagi. (2008) "A transformer less energy storage system based on a cascade multilevel PWM converter with star configuration," IEEE Trans. Ind. Electron., vol. 44, no. 5, pp. 1621–1630.

12. M. I. Marei, A. B. Tantawi, and A. A. El-Sattar (2012) "An energy optimizes control scheme for a transformer less DVR," Electric Power Syst. Res., vol. 83, no. 1, pp. 110–118.
13. M. Malinowski, K. Gopa Kumar, J. Rodriguez, and M. A. Pérez (2010)"A survey on cascaded multilevel inverters," IEEE Trans. Ind. Electron., vol. 57, no. 7, pp. 2197–2206.
14. S. Bernet (2000) "Recent developments of high-power converters for industry and traction applications," IEEE Trans. Power Electron., vol. 15, no. 6, pp. 1102–1117.
15. S. Kour et al. (2010) "Recent advances and industrial applications of multilevel converters," IEEE Trans. Ind. Electron., vol. 57, no. 8, pp. 2553–2580
16. T. A. Meynard, H. Foch, P. Thomas, J. Cour Ault, R. Jakob, and M. Narrated (2002). "Multicell converters: Basic concepts and industry applications," IEEE Trans. Ind. Electron., vol. 49, no. 5, pp. 955–964.

Emerging Technologies and Applications in Electrical Engineering –
Prof. Dr. Anamika Yadav et al. (eds)
© 2024 Taylor & Francis Group, London, ISBN 978-1-032-82568-7

Design and Simulation of Electrical (EV) Vehicle Power Supply

32

Duddu Rajesh[1]
PhD Scholar, EE Department, NIT Raipur

D. Suresh[2]
Assistant Professor, EE Department, NIT Raipur

ABSTRACT: Electric vehicles (EVs) are revolutionizing transportation by replacing traditional internal combustion vehicles, promoting clean and efficient modes of transportation. This paper discusses design of Power Supplies for Electric Cars that is divided into two sections: an AC to DC rectification stage and a DC-to-DC converter stage. In this paper, the DC/DC converter is implemented using a full-bridge LLC resonant converter. Compared to standard LC series resonant converters, LLC resonant converters provide a few benefits. With the First Harmonic Approximation (FHA), the gain equation also incorporates the secondary leakage inductance as part of the gain equation. Using the gain equation, the design technique for optimizing resonant networks for specified input and output criteria is examined. The effectiveness of the 600W full-bridge LLC resonant converter design procedure is validated through simulation results.

KEYWORDS: Electric vehicle power supply, DC-DC, FHA, LLC resonant converter

1. Introduction

Gasoline-powered vehicles are becoming more popular in the market due to their cost-effective cost, and quick fuelling times. However, the natural environment is badly impacted by the greenhouse gases these vehicles emit, leading to acid rain, climate change, increased pollution in the atmosphere, etc. Manufacturers of automobiles are turning to electric vehicles to solve

[1]drajesh.phd2023.ee@nitrr.ac.in, [2]dsuresh.ee@nitrr.ac.in

DOI: 10.1201/9781003505181-32

the issue (Akash Gangwar, N. J. Merlin Mary & Shelas Sathyan 2023). However, consumers are unable to switch to electric vehicles due to their higher cost, longer charging times, and lower efficiency. Modern power electronics have allowed automobile designers to develop onboard chargers that are efficient, compact, and can handle a wide range of input-output voltages. Because switching losses in classic converters are too high to provide soft-switching, power density is limited and high-frequency switching operation is not permitted.

High-frequency power converters have reduced the size of filter capacitors and magnetic components in the converter design while also increasing efficiency. The converter's overall size and weight are reduced as a consequence [1].

Figure 32.1 shows a common Electrical vehicle Battery charger arrangement, which incorporates an EMI filter to reduce input ripple, a circuit for power factor correction (PFC), DC link capacitor and a DC/DC converter (Sumantra Bhattacharya, Caroline Willich and Josef Kallo, 2022). PFCs are typically powered by a 230V AC source, and a boost converter is used in most of them. The heart of the EV charging arrangement is the DC-DC converter. Fig. 32.2(a) and Fig. 32.2(b) depict the series and parallel resonant converters, which are the fundamental resonant topologies (Hong Huang, 2010).

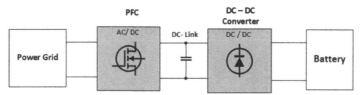

Fig. 32.1 Basic EV battery charger arrangement

The combination of series and parallel configurations is known as a series-parallel resonant converter, or SPRC. It is possible to modify it to utilize 2 inductors and 1 capacitor. and one inductor and two capacitors, forming an LLC (Fig. 32.2(c)) and LCC (Fig. 32.2(d)) resonant

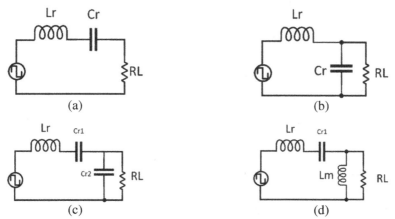

Fig. 32.2 Configurations of resonant converters (a) Resonant converter in series, (b) Resonant converter in parallel, (c) Resonant LCC converter, (d) Resonant LLC converter

converters respectively. Due to the high AC currents, this LLC structure requires large and costly separate physical capacitors, however, by adding more resonant frequencies, it gets over the limitations of basic SRC or PRC. The two physical inductors—the transformer's magnetizing inductance (L_m) and the series resonant inductance (L_r)—can be combined into a single component, which is an advantage of the LLC topology over the LCC topology.

This paper presents the practical design technique for optimizing the resonant network for specified input/output criteria is examined using the gain equation. Section 2 describes the operation and gain equation of the LLC resonant converter. Section 3 discusses the converter design technique, and Section 4 displays simulation results. Section 5 concludes this paper. The effectiveness of 600W full-bridge LLC resonant converter design procedure is validated through simulation results.

2. LLC Resonant Converter

The resonant converter used by LLCs output stage can be either a full-bridge or a center-tapped rectifier with a capacitive output filter, while the supply side stage can be either a full-bridge or half-bridge. Figure 32.3 illustrates LLC resonant converter's full-bridge implementation with a full-bridge output rectifier. Where the leakage inductances in the primary and secondary are denoted by L_{lp} and L_{ls}, respectively, and the magnetizing inductance L_m.

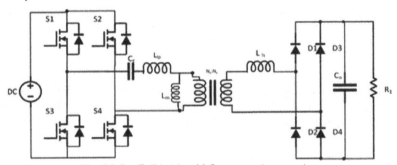

Fig. 32.3 Full-bridge LLC resonant converter

Applying a voltage in the form of a square wave (V_d) to the resonance network is done through complete bridge, which consists of S_1, S_2, S_3, and S_4. By applying the First Harmonic Approximation, which makes the assumption that only the fundamental component of the square-wave voltage input to the resonant network contributes to power transfer to the output, we can determine the voltage gain of the resonant converter because of the filtering action of the resonant network.

The calculation of the equivalent load resistance is shown in Fig. 32.4. A sinusoidal current source (I_{ac}) and a square wave of voltage (V_{RI}) at the rectifier's

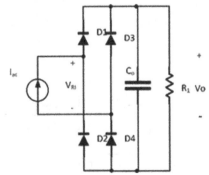

Fig. 32.4 Circuit for equivalent resistance R_{ac}

input replace the primary side circuit. Taking transformer turns ratio ($n = N_p/N_s$) into account, The primary's indicated equivalent load resistance is computed as

$$R_{ac} = \frac{8n^2}{\pi^2} \times R_1 \tag{1}$$

The fundamental components of output voltage and current or resonant converter are

$$V_{RI} = \frac{4}{\pi} \times V_0 \times \sin(wt)$$

$$I_{ac} = \frac{\pi}{2} \times I_0 \times \sin(wt)$$

When the transformer turns ratio is considered, the equivalent resistance

$$n = \frac{N_p}{N_s}; \quad R_{ac} = \frac{8n^2}{\pi^2} \times R_1$$

With the use of AC equivalent circuit, the voltage gain equation may be determined as follows.

$$G = \frac{nV_0}{V_{in}} = \left| \frac{\omega^2 L_m R_{ac} C_r}{j\omega\left(1 - \frac{\omega^2}{\omega_0^2}\right)\left(L_m + n^2 L_{ls}\right) + R_{ac}\left(1 - \frac{\omega^2}{\omega_p^2}\right)} \right| \tag{2}$$

Where,

$$L_p = L_m + L_{lp}; \quad L_r = L_{lp} + L_m \| (n^2 L_{ls}); \quad R_{ac} = \frac{8n^2}{\pi^2} \times R_1 \quad \omega_0 = \frac{1}{\sqrt{L_r C_r}} \quad \omega_p = \frac{1}{\sqrt{L_p C_r}}$$

There are two resonant frequencies, as shown in (2). L_r and C_r decide one, whereas L_p and C_r determine the other. L_p and L_r may be measured on the primary side of a transformer with the secondary side windings short and open, respectively. At resonant frequency the gain equation becomes

$$G_{@\omega=\omega_0} = \frac{L_m}{L_p - L_r} = \frac{L_m + n^2 L_{ls}}{L_m} \tag{3}$$

Excluding leakage inductance on the transformer's secondary side results in a gain of unity in (3). Further simplification of equation (2) becomes by assuming that $L_{lp} = n^2 L_{ls}$

$$G = \frac{nV_0}{V_{in}} = \left| \frac{\left(\frac{\omega^2}{\omega_p^2}\right)\frac{k}{k+1}}{j\omega\left(\frac{\omega}{\omega_0}\right)\cdot\left(1 - \frac{\omega^2}{\omega_0^2}\right)\cdot Q\frac{(k+1)^2}{2k+1} + \left(1 - \frac{\omega^2}{\omega_p^2}\right)} \right| \tag{4}$$

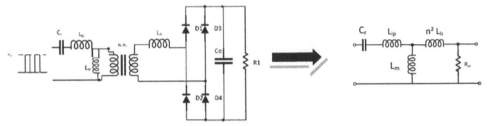

Fig. 32.5 Circuit equivalent for the LLC resonant converter in AC

where $k = \dfrac{L_m}{L_{lp}}$ (5)

$$Q = \frac{\sqrt{L_r \Big/ C_r}}{R_{ac}}$$ (6)

At resonant frequency the gain equation becomes

$$G_{@\omega=\omega_0} = \frac{L_m + n^2 L_{ls}}{L_m} = \sqrt{\frac{L_p}{L_p - L_r}} = \frac{k+1}{k}$$ (7)

As shown in Fig. 32.6, the LLC resonant converter of Fig. 32.5's AC equivalent circuit may be made simpler using L_p and L_r. $L_r = L_{lp} + L_m \parallel L_{lp}$ $L_p = L_{lp} + L_m$

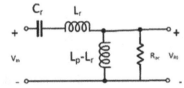

Fig. 32.6 LLC resonant converter's simplified AC equivalent circuit

3. Design Procedure

With a clearer grasp of the gain behavior, it is now possible to construct the design. So far, all discussion has focused on understanding the parameters of Q, k, and n and how they affect circuit operation (Hang-Seok Choi, 2007). The input DC voltage is 220 – 240V and output is 24V/25A. The gain in equation (4) for different k values, changes with Q. For various k values, the gain changes with Q. It appears that decreasing k or Q values might result in higher peak gain. Reducing k or Q results in a decrease in the magnetizing inductance, which raises the circulating current when the resonant frequency (ω_0) is fixed. Conduction loss and the available gain range must thus be matched. K and Q values are obtained from the graph. If the converter is intended to operate just below resonance frequencies, the high circulating current may deteriorate efficiency due to the wide fluctuations in the input voltage. It is computed that under full load, the minimum gain is 1.0. The smallest gain is used to compute the transformer turn ratio as $n = \dfrac{N_p}{N_s} = \dfrac{G_{min} \cdot V_{in}^{max}}{V_0 + V_F}$ the diode forward voltage drop is represented by V_F.

$$n = \frac{N_p}{N_s} = \frac{1 \times 240}{30 + 0.8} = 7.79$$

When 10kHz is used as the resonant frequency, the resonant parameters are calculated using the following formulae. The determination of the resonant parameters are made as $L_m = 693\mu H$, $L_{lp} = 107\mu H$ ($L_p = 800\mu H$, $L_r = 100\mu H$). From then, equivalent load resistance is calculated using the $R_{ac} = \dfrac{8n^2}{\pi^2} \times R_1$ where $R_1 = \dfrac{V_0}{I_0}$

$$R_{ac} = \frac{8 \times 7.79^2}{\pi^2} \times \frac{30}{20} = 73.78\Omega$$

The resonant circuit parameters $C_r = \dfrac{1}{Q\omega_0 R_{ac}}$ (8)

$$C_r = \frac{1}{0.4 \times 2\pi \times 10k \times 73.78} = 5.39 \times 10^{-7} F$$

4. Simulation Results

The suggested system was simulated using MATLAB/Simulink system blocks to verify it. The output voltage measured at the resonance condition in the proposed system is 30V and the output current measured at the same is 20A. as shown in Fig. 32.7. The calculated power in this model is 600W.

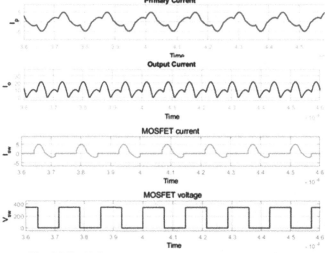

Fig. 32.7 LLC resonant converter output waveforms

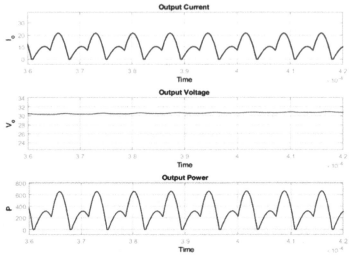

Fig. 32.8　Switch waveforms

The voltage observed across MOSFET is identified in Fig. 32.8 and current through the is shown in the same figure.

5. Conclusion

The leakage and magnetizing inductances of the transformer were employed as resonant parameters in the design process for the LLC resonant converter, which was covered in this paper. The gain computation also considered the leakage inductance on the transformer's secondary side.

REFERENCES

1. Akash Gangwar, N. J. Merlin Mary & Shelas Sathyan (2023) Design, modelling, and simulation of bridgeless SEPIC-fed three-level soft switched converter for EV battery charging, International Journal of Modelling and Simulation, DOI: 10.1080/02286203.2023.2169977.
2. Birand Erdogan, Adnan Tan, Murat Mustafa Savrun, Mehmet Ugras Cuma and Mehmet Tumay "Design and Analysis of a High-Efficiency Resonant Converter for EV Battery Charger" balkan journal of electrical & computer engineering, vol. 11, No. 2, April 2023.
3. Hang-Seok Choi "Design Consideration of Half-Bridge LLC Resonant Converter" Journal of Power Electronics, Vol. 7, No. 1, January 2007.
4. Hong Huang "Designing an LLC Resonant Half-Bridge Power Converter" SEM1900, Topic 3 TI Literature Number: SLUP263, 2010
5. Rahul Maurya & Radheshyam Saha "Design and Simulation of an Half-Bridge LLC Resonant Converter for Battery Charger in EV" 2022 IEEE Delhi Section Conference (DELCON).
6. Sumantra Bhattacharya, Caroline Willich and Josef Kallo "Design and Demonstration of a 540 V/28 V SiC-Based Resonant DC–DC Converter for Auxiliary Power Supply in More Electric Aircraft" Electronics 2022, 11, 1382.

Emerging Technologies and Applications in Electrical Engineering –
Prof. Dr. Anamika Yadav et al. (eds)
© 2024 Taylor & Francis Group, London, ISBN 978-1-032-82568-7

Three Level T-type Inverter based Grid Connected Active Power Filter

33

D. Suresh*

Assistant Professor, National Institute of Technology Raipur

ABSTRACT: In this article, the active power filter (APF) based on three level T-Type converter based is presented for grid connected operation. The control method of the T-type converter based APF for grid connected operation is developed through the application of load power based control scheme for reference generation. The reference generation method is proposed based on the separation of the fundamental and harmonics current. The fundamental current of the reference scheme is estimated using peak value of current and phased locked loop. The perturb and observe based maximum power point tracking (MPPT) is employed for tracking maximum power. The reference current is inetgrated with MPPT algorithm for the T-type MLI based APF for the harmonics elimination and injection of the active power from the PV source to the grid. The simulation study is conducted by employing MATLAB/Simunlink simpower system.

KEYWORDS: T-type MLI, Harmonics, Active power, Photovoltaic

1. Introduction

The power quality refers to reliability and stability of electrical supply in terms of volage, frequency, and waveform. It is influenced by various factors throughout the power system, including the power source, transmission and distribution systems, and the characteristics of the loads. Power quality issue can manifest as variations in voltage levels, frequency deviations, harmonics distortions, and other disturbances (Bhim Singh etal., 1999), (Suresh D etal., 2022). Mitigating power quality issues involves monitoring, analysis, and implementing solutions

---
*dsuresh.ee@nitrr.ac.in

DOI: 10.1201/9781003505181-33

such as voltage regulations, filters, and power factor correction devices to ensure a stable and high-quality electrical supply ().

In the literature (Bhim Singh etal., 1999) (Y. Komatsu etal., 2002) (N. Palla etal.,2020), various harmonics methods mitigating harmonics have been extensively reported. The passive filters filters uses the passive components like resistors, inductors, and capacitors to absorb shunt specific harmonics frequencies. The passive filter (PF) based approach is simple and cost-effective for specifics harmonics reduction. However, the PF based methods suffers from the resonance, fixed frequencies and aging problems. The active power filter (APF) based method utilize advance control algorithms and semi-conductor devices to dynamically compensate to harmonics currents. The APF based methods are suitable for wide range of harmonics frequencies and provide dynamic and precise compensation of harmonics current. The APF can be implemented using either two-level voltage source inverter (VSI) or a MLI based VSI. The choice between the two depends on several factors including the particular requirements of application, the complexity of the control system, and the desired performance criteria. In two level VSI, the inverter has two voltage levels such as postive and negative. The APF employing two level VSIs offers a solution characterized by simplicity, minimize control complexity and versatile applicability across a spectrum of power electronics applications. MLIs VSIs have more than two voltage levels, achieved by using multiple power semi-conductor devices and capacitors in series or parallel. The MLI VSIs are preferred in aplications demanding higher voltage quality, reduced harmonics content, and improved power factor correction. Both two level and MLI VSIs can be employed to realize APF, and the selection depends on the specific needs and performance criteria of the applications.

İn this article, three level T-type MLI based APF is implemenetd for dual purpose of harmonics eliminations and seamless integration with the grid power. Compared to the diode clamped multilevel inverter (DCMLI), the T-type MLI based VSI utilizes the bidirectional switch for the mid-point of the DC link and requires two diodes per bridge leg. İt offers effective alternate to three level active neutral point clamped inverter. The T-type MLI integrate aspects of two inverter and MLI VSIs such as low conduction losses, fewer component need and (D. Casadei etal., 2006), (M. Schweizer etal., 2013) (M. Liang, et al., 2010) . The computer simulation study utilizze Simpowersystem in MATLAB/Simulink, which offers collectios of blocks for modeling a T-type MLI based APF in grid connected scenario. The simulation results indicate that the T-type MLI based APF is capable of mitigating harmonic currents and integrating photo-voltaic power into the grid. This is achieved without the need for a separate VSI based APF

2. Topology of T-type MLI based APF

Figure 33.1 shows the circuit diagram of a T-type MLI-based APF developed for grid applications in a three-phase four-wire (3P4W) distribution system. The APF is connected to the grid using inductor at the point of common coupling. The coupling inductor size in a grid-connected APF linked at the PCC is determined by factors such as the T-type MLI's switching frequency, required isolation level, and overall design considerations. The APF's control

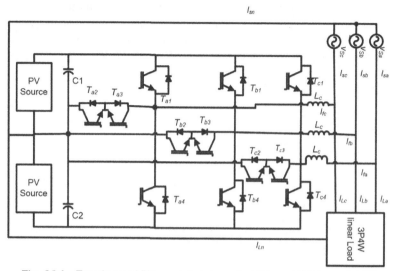

Fig. 33.1 Topology of PV module integrated T-type NPC based-APF

system can also impact inductor size. The switching frequency of T-type MLI influences rate at which the inverter components change switching states. The PV modules are connected on the DC side of the T-type MLI-based APF, a process known as DC coupling. This technology effectively transforms the solar panels' DC power output into alternating current (AC), which is acceptable for grid connection. Figure 33.1 depicts the T-type MLİ-based APF.

3. Control Scheme of APF

The load current of the source with compensation with T-type MLI based APF is expressed as follows

$$i_{TL}(t) = i_{T1}\cos(\phi_1)\sin(\omega t) + i_{T1}\sin(\phi_1)\cos(\omega t) + \sum_{n=2}^{\infty} i_{Tn}\sin(n\omega t + \varphi_n) \tag{1}$$

This load current can be further expressed as

$$i_{TL}(t) = i_{Tf}(t) + i_{Tq}(t) + i_{Th}(t) \tag{2}$$

The compensating current of T-type APF is computed as follows

$$i_{TL}(t) = i_{Tf}(t) + i_{TC}(t) \tag{3}$$

$$i_{TL}(t) - i_{Tf}(t) = i_{TC}(t) \tag{4}$$

Where i_{Tf} is the fundamental component of current and i_{TC} is the compensating current of the APF. The load current is consisting of fundamental components, reactive component, and harmonics. The T-type APF eliminates both reactive current and harmonics components of the current by injecting compensating currents to counteract the load demand. This results in compensation of the source harmonics component to improve the quality of the current waveform. The reference currents of the APF are computed by multiplying the sinusoidal voltage template from the phase lock loop. The reference currents of the grid connected APF are computed using (5) and phase lock loop. The maximum power tracking algorithm is implemented using the perturb and observe method. The reference voltage waveform is obtained with MPPT algorithm consequently added with reference current of the APF. The control scheme is shown in Fig. 33.2.

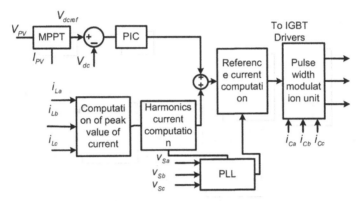

Fig. 33.2 Control scheme of T-type APF

4. Simulation Results and Discussion

The simulation study of the APF based on the T-type MLI is conducted using MATLAB software. The simulation results of the APF after harmonics suppression is shown in Fig. 33.3. The simulation study of APF connected with the grid has the capabilities of the harmonics eliminations. After the suppression of the harmonics using APF based on the T-type MLIs, the harmonics current present in the source reduces to great extent. The total harmonics distortion of the load is found to be 13.8 %. Even without PV modules on the DC side capacitors, the APF linked to the grid may successfully suppress current harmonics. The source current waveforms tend to be sinusoidal as a result of APF correction. The THD of the source current has been lowered to 3.34%. The source current waveform is in phase with the source voltage waveform. As seen in Fig. 33.3(a). Figure (b) shows that the neutral line current (Isn) has also been lowered significantly. Figure 33.3 (c) depicts a simulated waveform for the grid-connected T-type MLI. At t=1 second, the source current is in phase opposition to the source voltage waveform.

This type of APF is known as grid injection of PV electricity to the load. The perturb and observe approach is used to create maximum power point tracking. Figure 33.3(d) depicts the dc link voltage, PV current, and PV power variations. Figures 33.3(c) and 33.3(d) show that the control technique includes the MPPT of the grid-connected APF based on the T-type MLI. This integration allows the passage of active power generated from the PV source injected to the grid via the coupling inductor.

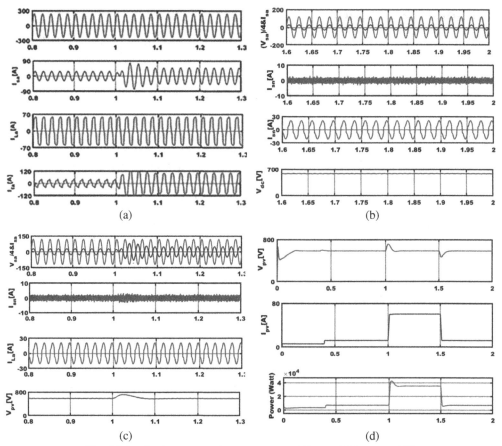

Fig. 33.3 Performance characteristics of T-type MLI based APF

The photo-voltaic panel current, voltage and power are shown in Fig. 33.3(d). The photo-voltaic current, voltage and power is gradually increase along with solar irradiation level. As the photo-voltaic power increase subsequently increases the compensation current, which in turn increases the grid current.

5. Conclusion

İn this article, the APF using T-type MLI is realized for the grid connected applications. The grid APF is used to eliminates the harmonics, reactive power of load and simultaneously inject the active power into the grid. The control method of T-type MLI based APF is implemented based on the separation of the fundamental components from the load and consequent estimation of harmonics current. The reference estimated is integrated with the perturb and observe method for MPPT algorithm. The simulation study is conducted using MATLAB/simulink software. The simulation results are presented.

Acknowledgements

Thank to National İnstitute of Technology Raipur for seed grant project (NITRR/Seed grant/2021-22/44)

REFERENCES

1. Bhim Singh, A. Chandra and K. Al-Haddad,(1999) "A review of active filters for power quality improvement", IEEE Transactions on Industrial Electronics, 46(5), pp. 112
2. M. Schweizer and J. W. Kolar,(2012) "Design and Implementation of a Highly Efficient Three-Level T-Type Converter for Low-Voltage Applications," in *IEEE Transactions on Power Electronics*, 28(2): 899-907, Feb. 2013, doi: 10.1109/TPEL.2012.2203151.
3. Y. Komatsu,(2002) "Application of the extension pq theory to a mains-coupled photovoltaic system", Proc. Power Convers. Conf. (PCC), vol. 2, pp. 816-821, 2002.
4. D. Casadei, G. Grandi and C. Rossi, (2006) "Single-phase single-stage photovoltaic generation system based on a ripple correlation control maximum power point tracking," in *IEEE Transactions on Energy Conversion*, . 21(2): 562-568:doi: 10.1109/TEC.2005.853784.
5. N. Palla and V. S. S. Kumar (2020), "Coordinated Control of PV-Ultracapacitor System for Enhanced Operation Under Variable Solar Irradiance and Short-Term Voltage Dips," in *IEEE Access*, 8:211809-211819,doi: 10.1109/ACCESS.2020.3040058.
6. Suresh. D., Kumar, V., Mahesh, M. (2022). Twelve Pulse-Based Battery Charger with PV Power Integration. In: Marati, N., Bhoi, A.K., De Albuquerque, V.H.C., Kalam, A. (eds) AI Enabled IoT for Electrification and Connected Transportation. Transactions on Computer Systems and Networks. Springer, Singapore. https://doi.org/10.1007/978-981-19-2184-1_9.
7. M. Liang, et al.,(2010) "Modeling and control of 100 kW three-phase gridconnected photovoltaic inverter," in 2010 5th IEEE Conference on industrial Electronics and Applications (ICIEA 2010), 825-830.

Emerging Technologies and Applications in Electrical Engineering –
Prof. Dr. Anamika Yadav et al. (eds)
© 2024 Taylor & Francis Group, London, ISBN 978-1-032-82568-7

Effect of HALBACH Array PM Arrangement on Surface Mounted PMSM

34

**Supriya Naik[1], Baidyanath Bag[2] and
Kandasamy Chandrasekaran[3]**
Electrical Engineering Department, National Institute of Technology, Raipur

ABSTRACT: This paper represents the Halbach magnetization effect on magnetic field of PMSM as compared to radial field. To achieve this, a 4-pole, 24 stator slots PMSM is considered. Rating of this motor is considered as 220 V with 1.6A. For comparison magnetic field, field intensity, current density output parameters are considered. All these observations have done in the platform of ANSYS Motor CAD. It has been achieved using inbuilt numerical method finite element analysis (FEA). The result described here is helpful to understand Halbach concept and appropriate utilization in many applications.

KEYWORDS: ANSYS motor CAD, Finite element analysis, Halbach, Magnetic flux density, PMSM

1. Introduction

Day by day necessary of high-speed motors, is inviting the application of PMSM. Its uses applicable to all type of field such as Air conditioners, Refrigerators, AC compressors, washing machines, which are direct-drive, Automotive electrical power steering, machine tools, large power systems to improve leading, and lagging power factor, control of traction etc. (L. Sepulchre, etc., 2016) (Mohd Zaihidee, etc. 2019). For solving this type of motor, finite element method (FEM) is so famous as compared to other numerical methods due to its advanced boundary conditions for different complex geometry (Wu, J.Y. and Lee, R. 1997). Nowadays ease availability of this method in many software attracted the thing. To reduces the size of motor, conventional motor designs such as axial flux motor (Duffy, K.P., 2016)(Jin, P

[1]snaik.phd2021.ee@nitrr.ac.in, [2]bbag.ee@nitrr.ac.in, [3]kchandrasekaran.ee@nitrr.ac.in

DOI: 10.1201/9781003505181-34

etc. 2015), brushless dc motor (M. J. Bala, 2016) has also involved in introduction of Halbach concept to it. Many 2-D, 3-D models of motors are observed to solve their magnetic field using FEM. As compared to other numerical method, for complex design FEA is preferable. Here, maxwell equations are involved to get the magnetic outputs. Most of the topology of motors are contributed radial and parallel type magnetization. To avoid saturation problem of core part of motor, and increase the field intensity on airgap section, Halbach concept of magnetization increases. So that, many fields such as electric vehicle, advance pump system, biological samples etc. discuss nowadays focused on this concept to improve the motor design and get more efficiency as much as possible (Lee, etc. 2004)(M. R etc. 2015)(Amri, L etc. 2023)(Kang, J.H, etc. 2016). Software is available to solve magnetic problem like COMSOL, FEMM, Magnet etc. (Sharma, S.V, etc. 2022)(Hampshire, D.P.,2018)(Ekreem, N.B etc. 2016) but due to easy way of learning Motor CAD is most suitable for initial study which is considered here also.

There is no literature discusses about Halbach magnetic array concept in surface mounted PMSM. Therefore, this study focuses on the introduction of Halbach magnetization in it. This paper also introduces a comparative study between radial and Halbach magnetization. Here, section 2 explains the motor design details taken of observation. Section 3 focuses on the Halbach concept in motor designs. Similarly, section 4 elaborates the result and discussion, and section 5 concludes about the comparative study.

2. Designing and Modelling of PMSM

Here a surface mounted PMSM having 220V DC with peak current 1.6 A is considered for observation. Due to easy construction unlike interior permanent magnet motor (IPM), this is focused here. Total 24 number of stator slots with 4 number of poles are taken in rotor part. Other design details such as slot design in stator, stator and rotor diameters, tooth details, magnet design details etc. are elaborated in Table 34.1 and 34.2.

Design of PMSM is done in ANSYS Motor CAD platform. Figure 34.1 represents the cross section of PMSM motor having different sections. Fig. 34.1(b) represents the magnetic field direction from the rotor to airgap zone.

Table 34.1 Stator parameters details

Stator parameters	Value (mm)
Slot number	24
Stator Lamination Diameter	120
Stator bore	75
Tooth width	4.7
Slot depth	18
Slot corner radius	0.5
Tooth tip depth	1
Slot opening	2.5
Tooth tip angle	30
Sleeve thickness	0

Table 34.2 Rotor parameters details

Rotor parameters	Value (mm)
Pole number	4
Magnet Thickness	4
Magnet Reduction	0
Magnet Arc	180
Magnet segment	3
Rotor diameters	74
Airgap	1
Banding thickness	0
Shaft diameter	26
Shaft hole diameter	0

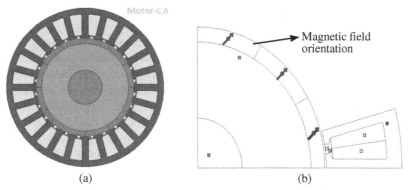

(a)   (b)

Fig. 34.1   (a) Modelling of PMSM and (b) magnetic field orientation

3. HALBACH Magnetization

As it is known that present of permanent magnet create magnetic field in PMSM which is reduce the applied excitation to motor. This magnetic field is equally distributed on both side of rotor section. It creates the magnetic saturation problem. To avoid unnecessary field distribution toward rotor inner section, Halbach array magnetization is introduced. After introducing Halbach magnetization, position of central magnet arc is $120°$. Basically, a Halbach array is a set of permanent magnets arranged so that one side has a stronger field while the other side has a weaker field. Typically, such an arrangement involves orienting the magnets in an out-of-phase configuration with their poles at 90 degrees apart. As a result of this orientation, the magnetic field below the structure (the "non-working" surface) is rerouted to the plane above it (the "working" surface), strengthening the working surface's magnetic field and reducing it to nearly zero on the non-working surface. A pictorial representation is given in Fig. 34.2. It describes the model of PMSM under Halbach magnetization concept. Arrow in Fig. 34.2(b) indicates the magnet direction under Halbach array concept.

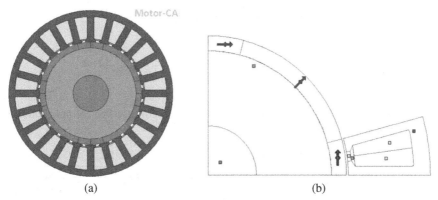

(a)   (b)

Fig. 34.2   Halbach effect on (a) model and (b) rotor magnet orientation

4. Result and Analysis

After FEM simulation, Fig. 34.3 represents output parameters related to radial magnetic field distribution. Figure 34.3(a) represents the magnetic flux density under radial condition. Like this Fig. 34.3(b) and (c) represents the field intensity and current density under same condition. It is also observed that under radial magnetic field, field distribution is non-uniform, which is the basic reason of saturation of iron core. Figure 34.4 represents the field properties under Halbach magnetization. Flux density distribution represents by Fig. 34.4(a) in this case. Whereas magnetic field intensity and current density represents in Fig. 34.4(b) and (c). Under Halbach magnetization, field distribution is more compared to radial. All field parameters values are clearly elaborated in Table 34.3. From this, it is observed that magnitude of all field parameters are same for both cases.

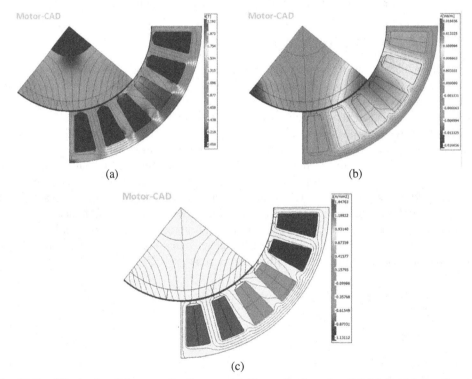

(a)　　　　　　　　　　(b)

(c)

Fig. 34.3 Effect of radial magnetization on (a) Magnetic flux density, (b) field intensity, and (c) current density

Table 34.3 Results of radial and Halbach effect

Parameters	Radial Magnetization	Halbach Magnetization
Magnetic flux density (T)	2.192	2.078
Magnetic field intensity (Wb/m)	0.016656	0.013627
Current density (A/mm2)	1.44703	1.44703

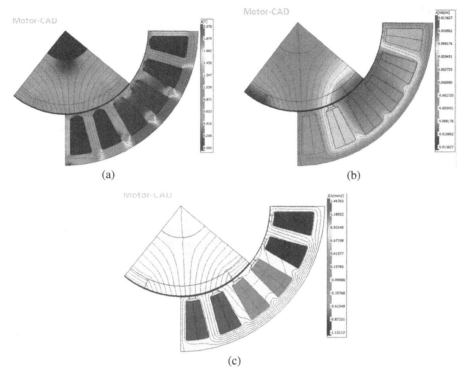

(a)

(b)

(c)

Fig. 34.4 Effect of Halbach magnetization on (a) Magnetic flux density, (b) Field intensity, and (c) Current density

5. Conclusion

In this paper, a low rating surface mounted PMSM is considered for comparative study between radial and Halbach magnetization. For this, three field parameters such as flux density, field intensity and current density are considered. From field analysis, it is observed that field lines are more uniform under Halbach condition compared to radial one. But magnitude of all parameters is same under both conditions of orientation. This study helpful for basic analysis of other types of motor to develop based on Halbach concept.

REFERENCES

1. L. Sepulchre, M. Fadel, M. Pietrzak-David and G. Porte (2016). Flux-weakening strategy for high speed PMSM for vehicle application. 2016 International Conference on Electrical Systems for Aircraft, Railway, Ship Propulsion and Road Vehicles & International Transportation Electrification Conference (ESARS-ITEC), Toulouse. France :1-7.
2. Mohd Zaihidee, F., Mekhilef, S. and Mubin, M.(2019). Robust speed control of PMSM using sliding mode control (SMC)—A review. Energies, 12(9):1669.

3. Wu, J.Y. and Lee, R., (1997). The advantages of triangular and tetrahedral edge elements for electromagnetic modeling with the finite-element method. IEEE Transactions on Antennas and Propagation, 45(9):1431-1437.

4. Duffy, K.P., (2016). Optimizing Power Density and Efficiency of a Double-Halbach Permanent-Magnet Ironless Axial Flux Motor. In 52nd AIAA/SAE/ASEE Joint Propulsion Conference :4712.

5. Jin, P., Yuan, Y., Xu, Q., Fang, S., Lin, H. and Ho, S.L., (2015). Analysis of axial-flux Halbach permanent-magnet machine. IEEE Transactions on Magnetics, 51(11), :1-4.

6. M. J. Bala, D. Roy and A. Sengupta (2020,)The Performance Enhancement of BLDC Motor Using Halbach Array Rotor. IEEE 1st International Conference for Convergence in Engineering (ICCE), Kolkata, India,: 405-409.

7. Lee, M.G., Lee, S.Q. and Gweon, D.G., (2004). Analysis of Halbach magnet array and its application to linear motor. Mechatronics, 14(1):115-128.

8. M. R. Dubois and J. P. Trovao (2015). Motor Drive with Halbach Permanent Magnet Array for Urban Electric Vehicle Concept," 2015 IEEE Vehicle Power and Propulsion Conference (VPPC), Montreal, QC, Canada: 1-6.

9. Amri, L., Zouggar, S., Charpentier, J.F., Kebdani, M., Senhaji, A., Attar, A. and Bakir, F., (2023). Design and Optimization of Synchronous Motor Using PM Halbach Arrays for Rim-Driven Counter-Rotating Pump. Energies, 16(7):3070.

10. Kang, J.H., Driscoll, H., Super, M. and Ingber, D.E., (2016). Application of a Halbach magnetic array for long-range cell and particle separations in biological samples. Applied Physics Letters, 108(21).

11. Sharma, S.V., Hemalatha, G. and Ramadevi, K., (2022). Analysis of magnetic field-strength of multiple coiled MR-damper using comsolmultiphysics. Materials Today: Proceedings, 66:1789-1795.

12. Hampshire, D.P., (2018). A derivation of Maxwell's equations using the Heaviside notation. Philosophical Transactions of the Royal Society A: Mathematical, Physical and Engineering Sciences, 376(2134):20170447.

13. Ekreem, N.B., Hassan, M.A., Shashoa, N.A.A. and Elmezughi, A.S.,(2016) Measurement and FEMM Modelling of Experimentally Generated Strong Magnetic Fields. International Journal of Research in Engineering Technology, 1(2).

Emerging Technologies and Applications in Electrical Engineering –
Prof. Dr. Anamika Yadav et al. (eds)
© 2024 Taylor & Francis Group, London, ISBN 978-1-032-82568-7

Hybrid Sliding Mode Observer-Based Sensorless Speed Control of IPMSM for Widest Speed Range

35

Uppalapati Sudheer Kumar[1]
PhD scholar, DoEE, SVNIT Surat

Anand Kumar[2]
M.Tech scholar, DoEE, SVNIT Surat

Sukanta Halder[3]
Assistant Professor, DoEE, SVNIT Surat

Nilanjan Das[4]
PhD scholar, DoEE, SVNIT Surat

Sk Bittu[5]
PhD scholar, DoEE, SVNIT Surat

Sujeet Kumar[6]
M.Tech scholar, DoEE, SVNIT Surat

ABSTRACT: Advancement in the transportation sector towards greener options result in huge growth of electric automobiles. Because of its superior power density, the interior permanent magnet synchronous motor (IPMSM) is the most commonly utilized motor in electric vehicles. However, these motors do have drawback towards the position sensor failures. To address these issues, the current work proposes a sensorless speed control system based on the hybrid sliding mode observer (SMO). An improved SMO is proposed for speed control of IPMSM at low and medium range with high precision rotor position estimation and low computational load. Secondly, the improved SMO along with high frequency signal injection is combined with the weighted switching algorithm for full speed operation and to increase the robustness. The whole system has been generated using the software developed by MATLAB, and the findings

[1]ds22el004@eed.svnit.ac.in, [2]anand.Kumar.388a@gmail.com, [3]sukanta.raj@gmail.com, [4]nilanjandas009@gmail.com, [5]skbittu693@gmail.com, [6]sujeet.ee14@gmail.com

DOI: 10.1201/9781003505181-35

show that the suggested model's robustness and accuracy are improved when compared to the standard SMO.

KEYWORDS: IPMSM, Sensorless control, SMO, Hybrid SMO, Electric vehicle

1. Introduction

World is shifting towards the greener transportation by adopting the Electric vehicles for both commercial and public transportation systems. The electric vehicles were mostly dependent on the IPMSM motors because of the higher density offered by these motors with educed size and low cost (Bhattacharjee et al., 2022). However, these motor has a drawback of position sensor failures with causes fault in the motor operation, to overcome these problem, sensorless control for IPMSM designed by (Wu and Selmon, 1991, 1006), has been emerged as a prominent area of research. Sensorless control strategy is of two types among which signal injection method is most commonly used method and some fundamental excitation methods elucidated by (Sepe and Lang, 1992, 1347). Further in the signal inject method two more classification were proposed by (Briz et al., 2004), That includes rotating signal with greater frequencies and pulsing signals that have high frequencies were discussed by (Ebrahimi et al., 2018). (Liu et al., 2022) offered a comparison for both of these concepts.

On the other hand, the fundamental excitation models were mostly based on the motor mathematical models, where the rotational positioning is calculated by using back electromagnetic fields from the motor's stator winding (Kumar et al., 2023). The Extended Kalman Filter was used in a model that was nonlinear to estimate the velocity as well as the position of the rotor by (Chi et al., 2009). (Sun et al., 2022), proposed novel strategy considering the flux state observer for the sensorless control is proposed while considering the stationery reference point. Further these fundamental models has been affected with the drawback like reduced accuracy associated at lower speeds where EMF is very weak to observe.

The SMO gain its strength in sensorless control because of robust parameter estimation and its stability towards the fluctuating values of control parameters. Traditional SMO were incorporated with low pass filter for compensating the chattering effect caused by the signum function, it has been overcomed by the model proposed by (Zhao et al., 2013) with the use of switching function replaced with signum function, Further advancing SMO control, (Zhang and Cheng, 2016) integrated the EMF sensor for positioning the rotor. In this paper, the hybrid SMO is proposed for sensor less speed control of IPMSM for EV application, in which the medium and high-speed ranges will be controlled through the proposed hybrid SMO and at lower speeds, the proposed SMO is integrated with high frequency injection method. Further, for improving the stability of the system during the transition from lower speeds to higher speed a composite weighted switching mechanism is proposed inline with the proposed hybrid SMO.

2. Design of Hybrid SMO for Rotor Position Detection

For the design of the hybrid SMO, initially the mathematical model of the SMO considering the evaluated values by reconstructing them with actual original system values as per (1).

$$\begin{bmatrix} k_\alpha \\ k_\beta \end{bmatrix} = R_s \begin{bmatrix} i_\alpha \\ i_\beta \end{bmatrix} + \begin{bmatrix} l_d & 0 \\ 0 & l_q \end{bmatrix} \frac{d}{dt} \begin{bmatrix} i_\alpha \\ i_\beta \end{bmatrix} + \begin{bmatrix} m_\alpha \\ m_\beta \end{bmatrix} \tag{1}$$

In the above equation m_α and m_β correlates to the motor's back EMF. The mathematical information to be analysed by the SMO is as follows;

$$\begin{bmatrix} m_\alpha \\ m_\beta \end{bmatrix} = \left[\left(l_d - l_q \right) \left(\omega_e i_d - \frac{di_q}{dt} \right) + \varphi_f \omega_e \right] \begin{bmatrix} -\sin \theta_e \\ \cos \theta_e \end{bmatrix} \tag{2}$$

From the Equation (2) it can be observed that all the relevant information related to the IPMSM magnetic properties, speed, current determination and polarity. Further rotor position can be evaluated such that rewriting the Equation (1) gives Equation (3).

$$\frac{d}{dt} \begin{bmatrix} i_\alpha \\ i_\beta \end{bmatrix} = \frac{1}{l_d} \begin{bmatrix} -R_s & -\omega_e \left(l_d - l_q \right) \\ -\omega_e \left(l_d - l_q \right) & -R_s \end{bmatrix} \begin{bmatrix} \hat{i}_\alpha \\ \hat{i}_\beta \end{bmatrix} + \frac{1}{l_d} \begin{bmatrix} k_\alpha \\ k_\beta \end{bmatrix} - \frac{1}{l_d} \begin{bmatrix} m_\alpha \\ m_\beta \end{bmatrix} \tag{3}$$

From the above state equation, the SMO is designed as:

$$\frac{d}{dt} \begin{bmatrix} \hat{i}_\alpha \\ \hat{i}_\beta \end{bmatrix} = \frac{1}{l_d} \begin{bmatrix} -R_s & -\omega_e \left(l_d - l_q \right) \\ -\omega_e \left(l_d - l_q \right) & -R_s \end{bmatrix} \begin{bmatrix} \hat{i}_\alpha \\ \hat{i}_\beta \end{bmatrix} + \frac{1}{l_d} \begin{bmatrix} k_\alpha \\ k_\beta \end{bmatrix} - \frac{1}{l_d} \begin{bmatrix} \phi_\alpha \\ \phi_\beta \end{bmatrix} \tag{4}$$

In the Equation (4) the $\hat{i}$ values are the evaluated values from the SMO and control function is defined as:

$$\begin{bmatrix} \phi_\alpha \\ \phi_\beta \end{bmatrix} = \varepsilon \begin{bmatrix} sat \left(\hat{i}_\alpha - i_\alpha \right) \\ sat \left(\hat{i}_\beta - i_\beta \right) \end{bmatrix} \tag{5}$$

Here ε is the SMO sliding gain. In the case of traditional SMO, the control function from the digital realm permits the integrated system to link smoothly. where the vibrations associated with transformative functions are eliminated but in the practical conditions saturation effects in the control function effects the system stability and leads to jitter problem when the gain value is set to higher value. Hence, in this proposed methodology the saturation function values were replaced with the segmented composite function as demonstrated by Equation (6).

$$y(z) = \begin{cases} 1 & z \ge a \\ \dfrac{1}{a^2} z^2 & 0 \le z < a \\ -\dfrac{1}{a^2} z^2 & -a < z < 0 \\ -1 & z \le -a \end{cases} \tag{6}$$

Where z is the state error associated with the system, which also defined as difference in the values of current and true values observed. Figure 35.3 illustrates the symbolic and power functions associated with the Equation (6) in establishing the smooth connection.

Variable thickness of the boundaries affected by the value of 'a' which has noticeable effect on jitter and control of the system, at zero point the proposed composite function exhibits the improved degree of jitter when compared with the conventional SMO, however the control effect is reduced. Such that the value of the 'a' is predominant factor in defining the segmented composite function. The enhancements for the hybrid SMO can be sated as follows:

$$\frac{d}{dt}\begin{bmatrix} \hat{i}_\alpha \\ \hat{i}_\beta \end{bmatrix} = \frac{1}{l_d}\begin{bmatrix} -R_s & -\omega_e\left(l_d - l_q\right) \\ -\omega_e\left(l_d - l_q\right) & -R_s \end{bmatrix}\begin{bmatrix} \hat{i}_\alpha \\ \hat{i}_\beta \end{bmatrix} + \frac{1}{l_d}\begin{bmatrix} k_\alpha \\ k_\beta \end{bmatrix} - \frac{\varepsilon}{l_d}\begin{bmatrix} y\left(\hat{i}_\alpha - i_\alpha\right) \\ y\left(\hat{i}_\beta - i_\beta\right) \end{bmatrix} \tag{7}$$

State equation for current error can be rewrites as

$$\frac{d}{dt}\begin{bmatrix} \hat{i}_\alpha \\ \hat{i}_\beta \end{bmatrix} = \frac{1}{l_d}\begin{bmatrix} -R_s & -\omega_e\left(l_d - l_q\right) \\ -\omega_e\left(l_d - l_q\right) & -R_s \end{bmatrix}\begin{bmatrix} \hat{i}_\alpha \\ \hat{i}_\beta \end{bmatrix} + \frac{1}{l_d}\begin{bmatrix} m_\alpha - hy\left(\hat{i}_\alpha - i_\alpha\right) \\ m_\beta - hy\left(\hat{i}_\beta - i_\beta\right) \end{bmatrix} \tag{8}$$

The enhanced SMO is obtained by replacing the saturation function of the traditional SMO with Equation (6) such that back EMF can be easily estimated and thus rotor position also estimated with an ease. The proposed hybrid SMO structure is shown in the Figure 35.1.

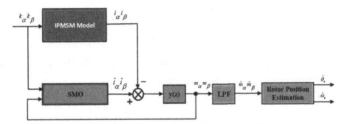

Fig. 35.1 Schematic of hybrid SMO

Source: Authors Compilation

The Lyapunov function is integrated for sliding surface such that system stability is maintained at equilibrium point:

$$V = \frac{1}{2}\left(s_\alpha^2 + s_\beta^2\right) \tag{9}$$

The stability conditions are as follows:

$$\begin{cases} \lim_{|S| \to \infty} s_\alpha^2 V = \infty \\ V < 0 \quad s \neq 0 \end{cases} \tag{10}$$

From the Equation (9) & (10) the stabilization is expressed as follows:

$$V = -\frac{R_s}{l_d}\left(\hat{i}_\alpha^2 + \hat{i}_\beta^2\right) + \frac{1}{l_d}\left[\hat{i}_\alpha\left(m_\alpha - \varepsilon y\left(\hat{i}_\alpha\right)\right) + \hat{i}_\beta\left(m_\beta - \varepsilon y\left(\hat{i}_\beta\right)\right)\right] \tag{11}$$

Further, the estimated back EMF estimates the rotor position, The low pass filtering system, on the other hand, is used to remove the higher frequency components from the signal, which results in the phase delay of rotor estimation hence rotor position compensation is needed as follows:

$$\Delta\theta_e = \tan\left(\frac{\hat{\omega}_e}{\omega_c}\right) \qquad (12)$$

By applying inverse to tangent, rotor position is estimated as:

$$\begin{cases} \hat{\omega}_e = \rho_n\omega_m \\ N_r = \dfrac{30}{\pi}\omega_m \end{cases} \qquad (13)$$

Since the above designed model is robust in estimating the rotor position at medium and top speeds but fails at lower speeds due to the degraded back EMF strength in lower speeds. As a result, for functioning at lesser speeds, a high frequency signal injection approach is used. In order to club these models a softer switching mechanism is designed for the control system. The linear weighting approach will be utilized to switch between different speed levels, with the High frequency injection approach employed at slower speeds and the proposed hybrid SMO method employed at medium to high speeds.

3. Simulation and Results

Figure 35.2 depicts the proposed hybrid SMO control block schematic, and the associated simulation for the suggested framework is embedded in the MATLAB software. Furthermore, the proposed model's effectiveness is compared to the standard model in the following section.

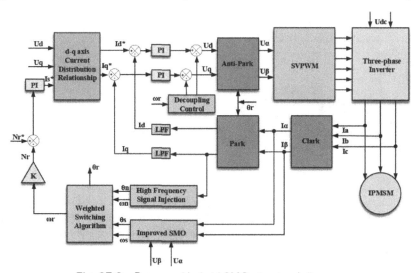

Fig. 35.2 Proposed hybrid SMO structural diagram

Source: Authors Compilation

The motor parameters consider for the simulation were shown in the Table 35.1.

Speed Performance comparison

The proposed hybrid SMO model's speed performance is compared to the standard SMO, and the relevant results are shown in Fig. 35.4. From the results depicted in the Fig. 35.3 it is observed that Fig. 35.3(a) when the traditional SMO speed performance the jitter error is existed and instability in the

Table 35.1 Simulation parameter

Variable	Value
No. of poles	4
Stator resistance	3.85 ohms
Flux Linkage	0.009 Wb
Inductance D-axis & Q axis	4.985 mH & 11.75 mH
Inertia	0.029 Kg.m^2

Source: Authors Compilation

speed can be observed where as in the case of proposed method jitter error is nullified. Further when the error percentage values as shown in Fig. 35.3(a) & (b) were analysed it can be observed that traditional SMO has an error ranging from -5 to 12 rpm but the proposed system has an error range of 2 to 1 rpm.

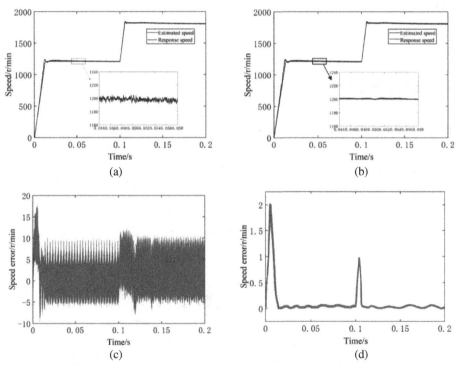

Fig. 35.3 Speed performance comparison (a) Speed of traditional SMO, (b) Speed of Hybrid SMO, (c) Speed error of traditional SMO, (d) Speed error of Hybrid SMO

Source: Authors Compilation

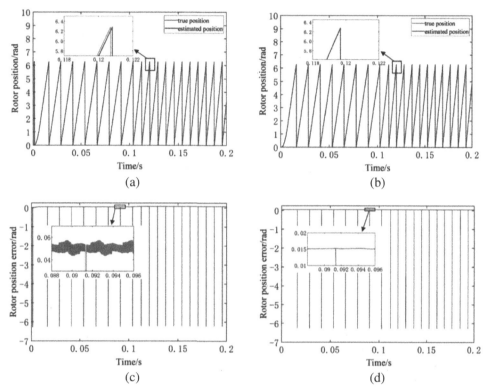

Fig. 35.4 Effectiveness of the rotor's position estimation (a) Rotor position for traditional SMO, (b) Rotor position for Hybrid SMO, (c) Rotor position error for traditional SMO, (d) Rotor position error for Hybrid SMO,

Source: Authors Compilation

Rotor position estimation performance

Figure 35.4 depicts the rotor position estimate performance of the standard and proposed hybrid SMOs. When the results shown in Fig. 35.4(a) and (b) are seen, the rotor position calculated by the suggested hybrid SMO improves significantly. Furthermore, the inaccuracies in standard SMO's assessment of rotor position are about 0.05 rad. The proposed hybrid SMO model error is improved and recorded as 0.0149 rad, such that observation accuracy is greatly improved.

Weighted Switching Algorithm Performance at wider speed ranges

Figure 35.5 depicts the computer-simulated outcomes pertaining to the Weighted Switching Algorithm. Initially the speed set at 1199 r/min and observed the effectiveness of the proposed algorithm in transformation of speeds from zero to rated speed at a load of 18 N-m. From the results, it is observed that smooth transition occurred with proposed model.

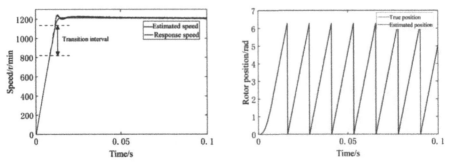

Fig. 35.5 Weighted switching algorithm performance graphs

Source: Authors Compilation

4. Conclusion

In this paper, the combined hybrid SMO model for the speed control of the IPMSM motor employed for EV application is implemented and simulated in the MATLAB software. At the lower speed high frequency injection model designed and at medium & high speeds the hybrid SMO model is implemented. Further, for smoother transition between lower and higher speed weighted algorithm is utilised for integrated the high frequency injection model and hybrid SMO. From the results it is elucidated that the proposed model improved speed tracking error from 5 to 12 rpm error to 2 to 1 rpm. Further, the rotor position estimation error from the proposed is improved from 0.05 rad to 0.0149 rad. Such that the proposed hybrid SMO model is both robust and most stable in speed transitions when compared with traditional SMO.

REFERENCES

1. Briz, F. and Degner, M. W. and Garcia, P. and Lorenz, R. D., Comparison of Saliency-Based Sensorless Control Techniques for AC Machines, IEEE Trans. on Industry Applications, Vol.40, No.4, 1107-1115, 2004.
2. Chi, Song and Zhang, Zheng and Xu, Longya, SlidingMode Sensorless Control of Direct-Drive PM Synchronous Motors for Washing Machine Applications, IEEE Trans. on Industry Applications, Vol.45, No.2, 582-590, 2009.
3. Ebrahimi, N.; Ozgoli, S.; Ramezani, A. Model-free sliding mode control, theory and application. *Proc. Inst. Mech. Eng. Part. I J. Syst. Control. Eng.* 2018, *232*, 1292–1301.
4. Liu, S.; Wang, Q.; Zhang, G.; Wang, G.; Xu, D. Online Temperature Identification Strategy for Position Sensorless PMSM Drives with Position Error Adaptive Compensation. *IEEE Trans. Power Electron.* 2022, *37*, 8502–8512.
5. S. Bhattacharjee, S. Halder, Y. Yan, A. Balamurali, L. V. Iyer and N. C. Kar, "Real-Time SIL Validation of a Novel PMSM Control Based on Deep Deterministic Policy Gradient Scheme for Electrified Vehicles," in IEEE Transactions on Power Electronics, vol. 37, no. 8, pp. 9000-9011, Aug. 2022, doi: 10.1109/TPEL.2022.3153845.
6. S. Halder, P. Agarwal and S. P. Srivastava, "Comparative analysis of MTPA and ZDAC control in PMSM drive," 2015 Annual IEEE India Conference (INDICON), New Delhi, India, 2015, pp. 1-5, doi: 10.1109/INDICON.2015.7443809.

7. Sepe, R. B. and Lang, Real-time observer-based (adaptive) control of a permanent-magnet synchronous motor without mechanical sensors, IEEE Trans. on Industry Applications, Vol.28, No.6, 1345 - 1352, 1992.

8. Sun, X.; Cai, F.; Yang, Z.; Tian, X. Finite Position Control of Interior Permanent Magnet Synchronous Motors at Low Speed. *IEEE Trans. Power Electron.* 2022, *37*, 7729–7738.

9. U. S. Kumar, K. Bhuvir, S. Halder, S. Tolani, S. Bhattacharjee and A. Panda, "Performance Analysis of GaN Inverter fed Electric Traction Drive System for EV Application," 2023 International Conference on Computer, Electronics & Electrical Engineering & their Applications (IC2E3), Srinagar Garhwal, India, 2023, pp. 1-6, doi: 10.1109/IC2E357697.2023.10262722.

10. Wu, R. and G, Slemon, A permanent magnet motor drive without a shaft sensor, IEEE Trans. on Industry Applications, Vol.27, No.5, 1005-1011, 1991.

11. Zhao, Yue and Qiao, Wei and Wu, Long, An Adaptive Quasi-Sliding-Mode Rotor Position Observer-Based Sensorless Control for Interior Permanent Magnet Synchronous Machines, IEEE Trans. on Power Electronics, Vol.28, No.12, 5618-5629, 2013.

12. Zhang, Y.; Cheng, X.-F. Sensorless Control of Permanent Magnet Synchronous Motors and EKF Parameter Tuning Research. *Math. Probl. Eng.* 2016, *2016*, 3916231.

Emerging Technologies and Applications in Electrical Engineering –
Prof. Dr. Anamika Yadav et al. (eds)
© 2024 Taylor & Francis Group, London, ISBN 978-1-032-82568-7

Torque Distortion Minimization of BLDC Motor by Modified DTC Techniques

36

Avismit Dutta*

Research Scholar, NIT Sikkim

Aurobinda Panda

Assistant Professor, NIT Sikkim

ABSTRACT: In the current era of climate change, there is a pressing need for an alternative solution in the field of transportation. One promising option is the electric vehicle (EV). The EV utilizes a BLDC motor as its primary source of rotation, offering numerous advantages. However, a major drawback of the BLDC motor is torque distortion. To address this issue, a modified 4-level direct torque controller (DTC) is proposed in this research paper. The effectiveness of this system is verified through the development of a Simulink model using MATLAB and a physical prototype for experimental verification. The modelling and investigation results reveal that the proposed system effectively reduces torque ripple, thereby improving overall performance.

KEYWORDS: BLDC motor, Electric vehicle, DTC

1. Introduction

During the initial half of the current decade, there has been a remarkable surge in electric vehicle adoption across various countries. The sales of electric vehicles witnessed a significant increase, resulting in a total of over 10 million electric vehicles now being actively used worldwide. Notably, China accounts for approximately 47% of this total. Furthermore, in several countries, electric vehicles constitute more than 1% of the overall market share. When considering lightweight electric vehicles, Brushless DC (BLDC) motors are a popular choice,

*phee210003@nitsikkim.ac.in

DOI: 10.1201/9781003505181-36

especially BLDC hub motors, which are extensively used in electric scooters because of their retrofitting convenience. These hub motors can be controlled using either sensor-based or sensor-less motor controllers.

A BLDC motor is a motor that utilizes permanent magnets and trapezoidal excitation instead of sinusoidal excitation. BLDC motors are widely utilized in applications involving two- and three-wheeled vehicles due to their impressive torque-to-weight ratio(Kakodia and Dyanamina 2023). Nevertheless, in spite of their many benefits, BLDC motors suffer from a notable disadvantage referred to as torque ripple. The problem occurs because of variations in the non-commutation current flowing through the motor (Huang et al. 2022; Yao et al. 2019). This torque distortion leads to mechanical vibrations and noise during the operation of electric vehicles. Consequently, these fluctuations in torque cause a reduction in driving smoothness and overall efficiency of the electric vehicle (Lu, Zhang, and Qu 2008).

Various control strategies have been employed to mitigate torque ripple in these types of motors. These control techniques include the Modified Pulse Width Modulation (PWM) control technique (Lu et al. 2008), DC bus voltage control technique (de Castro et al. 2018; Huang et al. 2022), a current control-based technique (Liu, Zhou, and Hua 2019; Yao et al. 2019), phase conduction method-based control technique (Park et al. 2019), Model Predictive Control (MPC) technique (Xia et al. 2020), Direct Torque Control (DTC) technique (Chen et al. 2017; Khazaee et al. 2021; Li et al. 2016; Masmoudi, El Badsi, and Masmoudi 2014a, 2014b; Mozaffari Niapour et al. 2011), Field-Oriented Control (FOC) technique (Basu, Prasad, and Narayanan 2009; Dae-Woong Chung and Sul 1998; Lara, Xu, and Chandra 2016; Sumega, Rafajdus, and Stulrajter 2020), Model Adaptive Control technique (Fang, Li, and Han 2012; Wang et al. 2022; Xia et al. 2014; Yin et al. 2020), and Soft Computing techniques (Dutta et al. 2022; Heidari and Ahn 2024), among others. Each of these techniques has its own set of advantages and disadvantages.

DTC is an excellent and straightforward approach for reducing torque ripple because of its reduced computing complexity, and lack of dependence on pulse width modulation (PWM) methods (Mozaffari Niapour et al. 2011).

The presented paper introduces a modified 4-level Direct Torque Control (DTC) technique aimed at alleviating torque distortion in BLDC motors. To validate the efficacy of the system, a Simulink model is developed in the MATLAB platform. Also, a hardware prototype model is developed for experimental authentication. The modelling and investigational findings demonstrate that the proposed system has significantly reduced the torque ripple while obtaining the necessary input dc bus voltage.

2. Brushless DC Motor

The BLDC motor, also known as a permanent magnet motor, operates using permanent magnets. It is characterized by a non-sinusoidal excitation waveform, specifically a trapezoidal shape, which is generated through the excitation of two windings. The motor is powered by a Voltage Source Inverter (VSI) (Shen and Peng 2008). The voltage supplied to the motor can be represented mathematically as follows:

$$V_{abc} = R.i_{abc} + L . \frac{di_{abc}}{dt} + E_{abc} \tag{1}$$

The electromagnetic torque of the BLDC motor can be denoted by the following expression:

$$T_e = \frac{\sum E_x i_x}{\omega}, \tag{2}$$

In the given expression for the electromagnetic torque, the symbols x = a, b, c represents the motor phases, while "ω" represents the rotor speed.

To minimize torque fluctuations, a 4 level modified Direct Torque Control (DTC) is utilized. DTC combines the principles of direct self-control and field focus to enable a practical variable frequency motor drive. By selecting the optimal switching modes for the inverter, DTC achieves rapid torque response and maximum efficiency (Khazaee et al. 2021). Moreover, this control method directly regulates the stator flux, the deviation between the actual and desired torque, and constrains the flux within predefined hysteresis bands.

The electromagnetic torque of the BLDC motor in the stationary α-β frame can be mathematically expressed as.

$$T_{em} = \frac{3}{2}\frac{p}{2}\frac{1}{\omega_e}\left[e_\alpha i_{s\alpha} + e_\beta i_{s\beta}\right] = \frac{3}{2}\frac{p}{2}\left[k_{\alpha(\theta_e)} i_{s\alpha} + k_{\beta(\theta_e)} i_{s\beta}\right] \tag{3}$$

In the equation you provided, the variables represent the following:
- "p" represents the number of poles in the motor.
- "θ_e" represents the position of the rotor.
- "ω_e" represents the rotor speed.
- "$k_\alpha(\theta e)$" and "$k_\beta(\theta e)$" are the back electromotive force constants in the stationary reference frame (ab-axes).
- "e_α" and "e_β" represent the motor back electromotive forces.
- "$i_{s\alpha}$" and "$i_{s\beta}$" denote the stator currents.

By utilizing this equation, the torque can be determined at any given speed, taking into account the number of poles, rotor position, rotor speed, back EMF constants, back electromotive forces, and stator currents.

3. Modified 4 Level Direct Torque Control

Direct torque technique is a very popular technique to maintain a fixed torque for several motor. In conventional work a 2-level torque comparator has been used. But in this proposed work 4 level torque comparator has been used which gives better torque response as well as better speed tracking capability.

3.1 Conventional DTC

According to Fig. 36.1, the deviation between the reference speed and the actual speed is input to a Proportional-Integral (PI) controller, which produces a reference torque. This reference

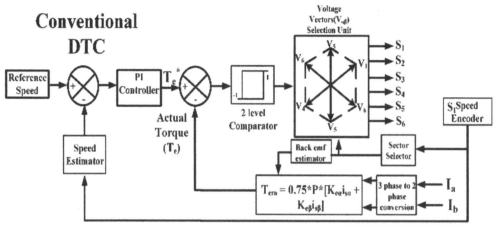

Fig. 36.1 Block diagram of conventional DTC

torque is subsequently compared to the actual torque, computed using the equation provided in Equation 3.

To calculate the actual torque, several parameters are required. These include the transformed values of "$i_{s\alpha}$," "$i_{s\beta}$," "$k_{\alpha}(\theta_e)$," and "$k_{\beta}(\theta_e)$." The transformed values of "$i_{s\alpha}$" and "$i_{s\alpha}$" are obtained through the 3-phase to α-β transformation of the 3-phase currents. On the other hand, the values of "$k_{\alpha}(\theta_e)$" and "$k_{\beta}(\theta_e)$" are evaluated using the sector and back EMF selector, as shown in Fig. 36.2. These calculations and transformations allow for the determination of the actual torque based on the given equation and system configuration.

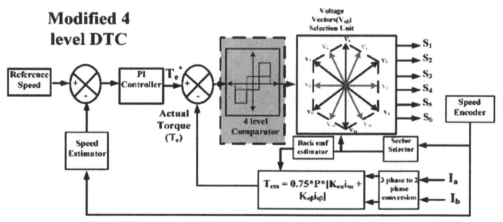

Fig. 36.2 Block diagram of 4 level DTC

To process the torque error and minimize torque ripple, a hysteresis controller, specifically a two-level torque comparator, is employed. This controller selects the appropriate switching sequence from Table 36.1 to generate a voltage vector with minimum torque ripple. It is crucial to emphasize that the choice of the switching sequence relies on the rotor's position, which is determined using Table 36.2.

Table 36.1 Switching sector of DTC

T_{st}	F_{st}	Sector					
		θ_1	θ_2	θ_3	θ_4	θ_5	θ_6
T_I	F_I	V_1	V_2	V_3	V_4	V_5	V_6
T_d		V_6	V_1	V_2	V_3	V_4	V_5
T_I	F	V_2	V_3	V_4	V_5	V_6	V_1
T_d		V_5	V_6	V_1	V_2	V_3	V_4
T_I	F_d	V_3	V_4	V_5	V_6	V_1	V_2
T_d		V_4	V_5	V_6	V_1	V_2	V_3

Table 36.2 Sector and Back emf co-efficient selection

Hall sensor			Selection of K_e	Sector
H_a	H_b	H_c		
1	0	0	K_a	VI
1	0	1	K_a	V
1	1	0	K_b	I
0	1	1	K_c	III
0	0	1	K_c	IV
0	1	0	K_b	II

Reduction in torque (TD), increment in torque (TI), reduction in flux (FD), increment (FI), and constant flux (F).

V1 = (110000), V2 = (011000), V3 = (001100), V4 = (000110), V5 = (000011), V6 = (100001)

The sector of the rotor and the back emf is calculated with the help of the hall effect sensors and is shown in Table 36.2.

In the case of a BLDC motor with a permanent magnet rotor, the change in flux is not taken into consideration when selecting the switching sequence. Instead, the highlighted rows in Table 36.3 are used to determine the voltage vector for achieving the desired control objectives. By utilizing this approach, the hysteresis controller ensures effective torque control and minimizes torque fluctuations in the BLDC motor.

3.2 Modified 4 Level DTC

This modified DTC techniques have 4 level torque comparators, where 4 kinds of errors are injecting to the comparator i.e., ±1, ±2. The switching sequences are described in Table 36.3.

Table 36.3   Sector and back emf co-efficient selection

Sector	T_{error}	Switching States					
		S_1	S_4	S_3	S_6	S_5	S_2
Sector 1	+2	1	0	1	0	0	1
	+1	0	0	1	0	0	1
	-1	0	0	0	1	1	0
	-2	1	0	0	1	1	0
Sector 2	+2	0	1	1	0	0	1
	+1	0	1	1	0	0	0
	-1	1	0	0	1	0	0
	-2	1	0	0	1	0	1
Sector 3	+2	0	1	1	0	1	0
	+1	0	1	0	0	1	0
	-1	1	0	0	0	0	1
	-2	1	0	1	0	0	1
Sector 4	+2	0	1	0	1	1	0
	+1	0	0	0	1	1	0
	-1	0	0	1	0	0	1
	-2	0	1	1	0	0	1
Sector 5	+2	1	0	0	1	1	0
	+1	1	0	0	1	0	0
	-1	0	1	1	0	0	0
	-2	0	1	1	0	1	0
Sector 6	+2	1	0	0	1	0	1
	+1	1	0	0	0	0	1
	-1	0	1	1	0	0	0
	-2	0	1	0	1	1	0

In Fig. 36.2 the overall block diagram of 4-level DTC is depicted and in Fig. 36.3 the 4-level torque comparator is also shown. Here the sector is chosen as the conventional control techniques.

4. Results and Discussion

To assess the effectiveness of the suggested modified direct torque control method for a BLDC motor, MATLAB Simulink is employed. Figure 36.3 illustrates the speed tracking capability and torque response of a conventional DTC-based BLDC motor and modified DTC approach. To validate the speed tracking capability within the Simulink environment, a speed change

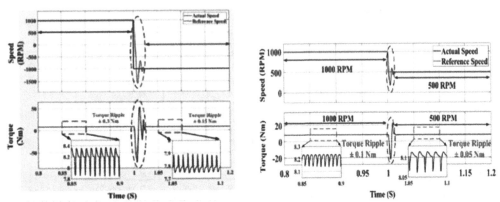

Fig. 36.3 Speed tracking and torque response of BLDC motor for conventional DTC and modified DTC

from 1000 RPM to 500 RPM is selected, with a reference torque of 8 Nm. The simulation parameters are detailed in Table 36.4. The simulation results reveal that the torque distortion in the modified DTC is lower by 2.5% when compared to the conventional DTC.

Table 36.4 System parameters for simulation

System parameters	Brushless DC motor parameters
• V_{dc} = 300 V • Fs =5 kHz • Voltage ripple = 4% • Current ripple = 10% • Sampling Time 20 μS	• Power rating = 1000 W • Voltage rating = 310V • Current rating = 4.33 A • Pole number = 8 • Speed rating = 3000 rpm • Torque at rated speed = 9 N-m • Winding resistance = 0.34Ω • Winding inductance = 3mH • Back EMF co-efficient = 0.69 N-m/A

To verify the dynamic response, a 10-second drive cycle is selected, illustrating three modes of operation: forward motoring, forward braking, and steady states. The speed tracking and torque responses for these modes are presented in Fig. 36.4. The drive cycle comprises three distinct states: forward motoring states (0-5 seconds), forward braking states (6-9 seconds), and steady states (5-6 seconds and 9-10 seconds).

To validate the simulation results, a prototype model for controlling a BLDC motor with a brake load is constructed in the laboratory. The hardware setup is depicted in Fig. 36.5.

To implement control algorithms in real-time, a dSPACE DS1104 Digital Signal Processor (DSP) was employed. The controllers for the proposed system were developed in Simulink using MATLAB's Real-Time Workshop (RTW) and interfaced with the dSPACE-Real-Time DS1104 functionality. The control strategy was initially constructed and optimized within the MATLAB/Simulink program, and then C-code generated by MATLAB's RTW facilitated real-time implementation.

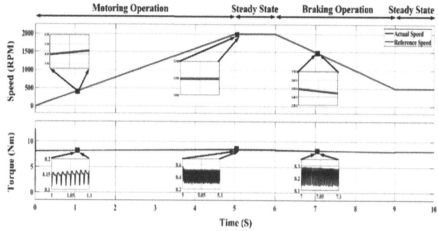

Fig. 36.4 Dynamic speed tracking of modified DTC based BLDC motor

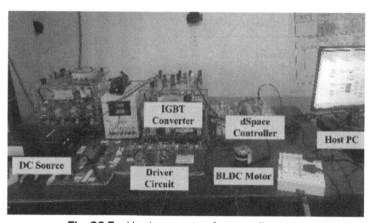

Fig. 36.5 Hardware setup for overall system

Gate pulses were generated using master bit I/Os, and the measured currents were interfaced through Analog to Digital Converters (ADCs). The driver board included an isolation and amplification circuit along with an inverter, while a regulated DC power supply served as the power source. For isolation, six TLP-250 units were employed, and an IGBT Semikron inverter module was utilized for the converter..

To enable current and voltage sensing, LA-55P and AD202JY isolation amplifiers were used, respectively. The hardware specifications are detailed in Table 36.5.

The speed tracking capability of the proposed controller is also validated experimentally and the result for the same is given in Fig. 36.6. It can be observed that the actual speed is tracking the reference speed accurately. With this proposed control technique, the BLDC motor is able to reach a speed of 2000 rpm within 5 Sec. in forward motoring mode. Similarly,

Table 36.5 Hardware specification

Components	Parameter	Values
Brushless DC motor	Power rating(W)	350
	Voltage rating (V)	60
	Current rating (A)	7
	Pole number	24
	Rated Speed(rpm)	7000
	Maximum Torque	3
Converters Module (IGBT)	Maximum input DC(V)	600
	Output AC voltage(V)	415
	Output AC current(A)	30
	Maximum switching frequency (KHz)	2
Driver and amplification Circuit (TLP 250)	Supply current (mA)	11
	Supply voltage(V)	10-35
	Output current(V)	±1.5

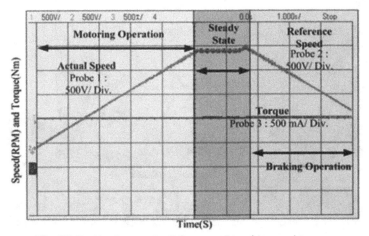

Fig. 36.6 Hardware result for speed tracking and torque

during forward braking operation, the speed is reduced to 500 rpm from 2000 rpm in 3.2 Sec. Here, the prototype system reached to its maximum speed of 2000 rpm within 5 Sec. and in forward braking operation the system reached to 500 rpm from 2000 rpm within 3.2 sec. The electromagnetic torque corresponds to aforementioned operations is also given in Fig. 36.6 which has very low ripple.

5. Conclusion

The present study introduces a modified Direct Torque Controller (DTC) that is applied to a BLDC motor. The key benefit of this controller is its ability to achieve greater torque distortion reduction compared to the conventional DTC, leading to improved overall system performance.

To validate the efficacy of the proposed controller, a simulation model is developed using MATLAB/Simulink and is further verified through a hardware prototype model.

REFERENCES

1. Basu, K., J. Prasad, and G. Narayanan. 2009. "Minimization of Torque Ripple in PWM AC Drives." IEEE Transactions on Industrial Electronics 56(2):553–58. doi: 10.1109/TIE.2008.2004391.
2. Chen, Wei, Yapeng Liu, Xinmin Li, Tingna Shi, and Changliang Xia. 2017. "A Novel Method of Reducing Commutation Torque Ripple for Brushless DC Motor Based on Cuk Converter." IEEE Transactions on Power Electronics 32(7):5497–5508. doi: 10.1109/TPEL.2016.2613126.
3. de Castro, Allan Gregori, William Cesar Andrade Pereira, Thales Eugenio Portes de Almeida, Carlos Matheus Rodrigues de Oliveira, Jose Roberto Boffino de Almeida Monteiro, and Azauri Albano de Oliveira. 2018. "Improved Finite Control-Set Model-Based Direct Power Control of BLDC Motor With Reduced Torque Ripple." IEEE Transactions on Industry Applications 54(5):4476–84. doi: 10.1109/TIA.2018.2835394.
4. Dae-Woong Chung, and Seung-Ki Sul. 1998. "Analysis and Compensation of Current Measurement Error in Vector-Controlled AC Motor Drives." IEEE Transactions on Industry Applications 34(2):340–45. doi: 10.1109/28.663477.
5. Dutta, Avismit, Nilanjan Das, Aurobinda Panda, and Sukanta Halder. 2022. "Mitigation of Torque Ripple of ZSI-Based BLDC Motor with Direct Torque Control Techniques." Pp. 1–6 in 2022 IEEE 19th India Council International Conference (INDICON). IEEE.
6. Fang, Jiancheng, Haitao Li, and Bangcheng Han. 2012. "Torque Ripple Reduction in BLDC Torque Motor With Nonideal Back EMF." IEEE Transactions on Power Electronics 27(11):4630–37. doi: 10.1109/TPEL.2011.2176143.
7. Heidari, Reza, and Jin-Woo Ahn. 2024. "Torque Ripple Reduction of BLDC Motor with a Low-Cost Fast-Response Direct DC-Link Current Control." IEEE Transactions on Industrial Electronics 71(1):150–59. doi: 10.1109/TIE.2023.3247732.
8. Huang, Ching-Lon, Feng-Chi Lee, Chia-Jung Liu, Jyun-You Chen, Yi-Jen Lin, and Shih-Chin Yang. 2022. "Torque Ripple Reduction for BLDC Permanent Magnet Motor Drive Using DC-Link Voltage and Current Modulation." IEEE Access 10:51272–84. doi: 10.1109/ACCESS.2022.3173325.
9. Kakodia, Sanjay Kumar, and Giribabu Dyanamina. 2023. "Improved Federal Test Procedure (FTP75) Driving Cycle Performance for PMSM-fed Hybrid Electric Vehicles Using Artificial Neural Network." International Journal of Circuit Theory and Applications. doi: 10.1002/cta.3786.
10. Khazaee, Amir, Hossein Abootorabi Zarchi, Gholamreza Arab Markadeh, and Hamidreza Mosaddegh Hesar. 2021. "MTPA Strategy for Direct Torque Control of Brushless DC Motor Drive." IEEE Transactions on Industrial Electronics 68(8):6692–6700. doi: 10.1109/TIE.2020.3009576.
11. Lara, Jorge, Jianhong Xu, and Ambrish Chandra. 2016. "Effects of Rotor Position Error in the Performance of Field Oriented Controlled PMSM Drives for Electric Vehicle Traction Applications." IEEE Transactions on Industrial Electronics 1–1. doi: 10.1109/TIE.2016.2549983.
12. Li, Xinmin, Changliang Xia, Yanfei Cao, Wei Chen, and Tingna Shi. 2016. "Commutation Torque Ripple Reduction Strategy of Z-Source Inverter Fed Brushless DC Motor." IEEE Transactions on Power Electronics 31(11):7677–90. doi: 10.1109/TPEL.2016.2550489.
13. Liu, Kai, Zhiqiang Zhou, and Wei Hua. 2019. "A Novel Region-Refinement Pulse Width Modulation Method for Torque Ripple Reduction of Brushless DC Motors." IEEE Access 7:5333–42. doi: 10.1109/ACCESS.2018.2888630.

14. Lu, Haifeng, Lei Zhang, and Wenlong Qu. 2008. "A New Torque Control Method for Torque Ripple Minimization of BLDC Motors With Un-Ideal Back EMF." IEEE Transactions on Power Electronics 23(2):950–58. doi: 10.1109/TPEL.2007.915667.

15. Masmoudi, Mourad, Bassem El Badsi, and Ahmed Masmoudi. 2014a. "Direct Torque Control of Brushless DC Motor Drives With Improved Reliability." IEEE Transactions on Industry Applications 50(6):3744–53. doi: 10.1109/TIA.2014.2313700.

16. Masmoudi, Mourad, Bassem El Badsi, and Ahmed Masmoudi. 2014b. "DTC of B4-Inverter-Fed BLDC Motor Drives With Reduced Torque Ripple During Sector-to-Sector Commutations." IEEE Transactions on Power Electronics 29(9):4855–65. doi: 10.1109/TPEL.2013.2284111.

17. Mozaffari Niapour, S. A. KH., S. Danyali, M. B. B. Sharifian, and M. R. Feyzi. 2011. "Brushless DC Motor Drives Supplied by PV Power System Based on Z-Source Inverter and FL-IC MPPT Controller." Energy Conversion and Management 52(8–9):3043–59. doi: 10.1016/j.enconman.2011.04.016.

18. Park, Joon Sung, Ki-Doek Lee, Sung Gu Lee, and Won-Ho Kim. 2019. "Unbalanced ZCP Compensation Method for Position Sensorless BLDC Motor." IEEE Transactions on Power Electronics 34(4):3020–24. doi: 10.1109/TPEL.2018.2868828.

19. Shen, Miaosen, and Fang Zheng Peng. 2008. "Operation Modes and Characteristics of the Z-Source Inverter With Small Inductance or Low Power Factor." IEEE Transactions on Industrial Electronics 55(1):89–96. doi: 10.1109/TIE.2007.909063.

20. Sumega, Martin, Pavol Rafajdus, and Marek Stulrajter. 2020. "Current Harmonics Controller for Reduction of Acoustic Noise, Vibrations and Torque Ripple Caused by Cogging Torque in PM Motors under FOC Operation." Energies 13(10):2534. doi: 10.3390/en13102534.

21. Wang, Tingting, Hongzhi Wang, Huangshui Hu, Xiaofan Lu, and Siyuan Zhao. 2022. "An Adaptive Fuzzy PID Controller for Speed Control of Brushless Direct Current Motor." SN Applied Sciences 4(3):71. doi: 10.1007/s42452-022-04957-6.

22. Xia, Changliang, Youwen Xiao, Wei Chen, and Tingna Shi. 2014. "Torque Ripple Reduction in Brushless DC Drives Based on Reference Current Optimization Using Integral Variable Structure Control." IEEE Transactions on Industrial Electronics 61(2):738–52. doi: 10.1109/TIE.2013.2254093.

23. Xia, Kun, Yanhong Ye, Jiawen Ni, Yiming Wang, and Po Xu. 2020. "Model Predictive Control Method of Torque Ripple Reduction for BLDC Motor." IEEE Transactions on Magnetics 56(1):1–6. doi: 10.1109/TMAG.2019.2950953.

24. Yao, Xuliang, Jicheng Zhao, Jingfang Wang, Shengqi Huang, and Yishu Jiang. 2019. "Commutation Torque Ripple Reduction for Brushless DC Motor Based on an Auxiliary Step-Up Circuit." IEEE Access 7:138721–31. doi: 10.1109/ACCESS.2019.2943411.

Emerging Technologies and Applications in Electrical Engineering –
Prof. Dr. Anamika Yadav et al. (eds)
© 2024 Taylor & Francis Group, London, ISBN 978-1-032-82568-7

Genetic Algorithm based Extended Kalman Filter for Precise State of Charge Estimation of Lithium-Ion Batteries

37

Nirmala[1], Subhojit Ghosh[2]
Department of Electrical Engineering,
National Institute of Technology, Raipur
Raipur (C.G.), India

ABSTRACT: In the context of advancing EV development, lithium-ion battery serve as a predominant technology in the field of energy storage. Precise SOC estimation is imperative to guarantee the secure and dependable execution of battery charging and discharging processes. SOC, being inherently unmeasurable, exhibits variability attributed to parameters such as temperature and noise characteristics. This paper employs a combination of equivalent circuit modeling, Extended Kalman Filter and Coulomb-counting techniques for estimating the SOC of lithium-ion battery. We leverage the dataset obtained from the ALM12V7 lithium-ion battery to execute genetic algorithm for fine-tuning the process noise(Q) and measurement noise(R) parameters. The obtained results indicate that the propose GA-EKF outperforms the classical EKF in estimating the SOC.

KEYWORDS: State of charge, Coulomb counting method, Genetic algorithm, Li-ion battery, Extended kalman filter

1. Introduction

The extent fossil fuels utilised for transportation has increased significantly during the last few decades. The use of electric vehicles negates the environmental effects of fossil fuels. Battery packs are necessary for electric vehicles to store energy. Lithium-ion batteries offer longer life cycles, a lower self-discharge rate, and a comparatively high specific energy density

[1]nirmala.mtech2023.ee@nitrr.ac.in, [2]sghosh.ele@nitrr.ac.in

DOI: 10.1201/9781003505181-37

as compared to conventional Lead-acid batteries. Li-ion battery is therefore best suited for a variety of applications. For Li-ion battery, SOC is considered the most crucial indicator for designing charging and discharging strategies and ensuring the battery's longevity, dependability, efficiency, and safety. The SOC is expressed as a ratio of the available capacity of a battery to its maximum possible charge, which is commonly referred to as the nominal capacity. The value indicates the remaining charge capacity of the battery. Understanding the state of charge is crucial in safeguarding batteries against over-discharging and over-charging. Calculation of SOC necessitates the utilization of battery data such as current, voltage, and temperature values.

The SOC can be estimated using a variety of methods. There are two types of SOC estimating approaches: model-based and non-model based. The Coulomb counting (Mohammadi 2022), open circuit voltage (Lee, et al. 2008), voltage-based (Mussi, et al. 2021), machine learning-based (Gupta and Mishra 2023), are examples of common non-model-based techniques. Observer-based approaches (Xu, et al. 2013), particle filters (Jiani, et al. 2013), Kalman filter (Spagnol, et al. 2011), Extended kalman Filter (Disci, et al. 2017), Unscented Kalman Filter (Wang, et al. 2018), as well as other model-based techniques, are examples of model-based techniques. The non-model derived methodology is unable to account for external disturbances, measurement noise, and erroneous starting SOC due to the absence of feedback. The estimation accuracy of machine learning-based methodologies is critically influenced by the magnitude of training data. Due to the lengthy training data gathering process, online SOC estimate systems can only apply a limited number of non-model derived and open loop techniques. The model-derived approaches are more appealing for precise estimation and have been extensively studied in last few years due to their close-loop characteristic and capacity to offset errors brought on by erroneous starting SOC and measurement noise.

The Kalman filter is a computational method employed to estimate unknown variables by examining observed measurements over a period of time. It is primarily designed for linear systems. However, for nonlinear systems, EKF and UKF are employed. The EKF follows a three-step process: Initialization, Correction, and Prediction. EKF locally approximates nonlinear functions using the linear equation obtained from the Taylor expansion, utilizing only the first-order term. On the other hand, the UKF employs nonlinear transformations on a set of sigma points determined deterministically. This process is known as unscented transformation. The mean and covariance matrix of the transformed points are accurate up to the second order of the Taylor series expansion.

In this paper, a first order RC equivalent network of ALM12V7 li-ion battery data is used. Coulomb counting and Extended kalman filter methods are used to estimate the SOC of the lithium-ion battery. The EKF and Coulomb counting method uses battery's current and voltage profile. A common limitation with kalman filter SOC estimation techniques refers to the selection of the noise covariance matrices. It is generally carried out using a trial-and-error approach because of the unavailability of precise information of the noise statistics. In this regard, in the present work, the noise parameters (Q, R) of EKF are updated in real time by using Genetic Algorithm (GA). Test results illustrate that the inclusion of GA for parameter tuning improves the accuracy of SOC estimation. Unlike (Ting, et al. 2014), in the present work the impact of temperature on the model parameter has been incorporated.

2. Battery Model

The effectiveness of battery model-based estimating approaches is heavily reliant on the battery type. The same emphasizes the need of carefully choosing the model used to improve estimation efficiency. For the purpose of our analysis, we employed the ALM12V7 lithium-ion battery dataset to conduct simulations and selecting the RC model configuration. Figure 37.1 depicts an equivalent circuit model of representing a battery, whereby the battery's dynamics are characterized by a one RC model.

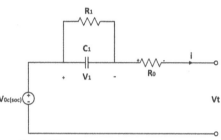

Fig. 37.1 Equivalent battery model

The first order RC circuit model is given below:

$$\frac{dSOC}{dt} = -\frac{i}{3600AH} \tag{1}$$

$$\frac{dV_1}{dt} = \frac{i}{C_1(SOC,T)} - \frac{V_1}{R_1(SOC,T)C_1(SOC,T)} \tag{2}$$

$$V_t = V_0(SOC,T) - iR_0 - V_1 \tag{3}$$

Where, SOC represents the state of charge, i denotes the current, V_0 is the no-load voltage. V_t is the terminal voltage, AH is the ampere-hour rating, R_1 and C_1 are dynamic resistance and capacitance of parallel RC network, T represents the temperature, V_1 is the voltage across parallel RC network. R_0 is the terminal resistance. State Space model of the equivalent circuit is given below:

$$x = \begin{bmatrix} SOC & V_1 \end{bmatrix}^T \tag{4}$$

$$f(x,i) = \begin{bmatrix} -\dfrac{i}{3600AH} \\ \dfrac{i}{C_1(SOC,T)} - \dfrac{V_1}{R_1(SOC,T)C_1(SOC,T)} \end{bmatrix} \tag{5}$$

$$h(x,i) = V_0(SOC,T) - iR_0 - V_1 \tag{6}$$

Figure 37.2 illustrates the battery current and terminal voltage characteristics.

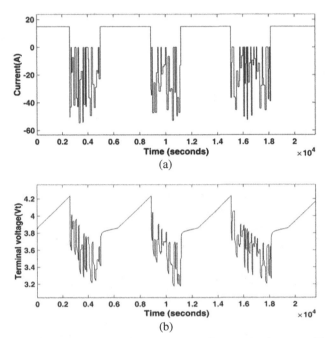

Fig. 37.2   (a) Battery current profile, (a) Terminal voltage profile

3. Genetic Algorithm Tuned EKF Based SOC Estimation

3.1 Extended Kalman Filter (EKF)

The EKF utilises the algorithm of first-order discrete-time kalman filter to estimate the states of a discrete-time nonlinear system. It offers an effective approach for handling nonlinear systems. The application of the Extended Kalman Filter technique is predicated in the execution of linearization at every iteration in order to approximate the dynamics of the non-linear system. In order to achieve the process of linearization at each time step, the method performs real-time calculations of the Jacobian matrices. The Simulink model for EKF based SOC estimation is shown in Fig. 37.3.

The Extended Kalman Filter is commonly described as a discrete-time method. After the discretization process, the computation of the Jacobian matrices pertaining to state of charge estimation for the cell is carried out.

$$F_d = \begin{bmatrix} 1 & 0 \\ 0 & e^{\frac{-T_s}{R_1 C_1}} \end{bmatrix} \tag{7}$$

$$H_d = \begin{bmatrix} \dfrac{\partial V_{oc}}{\partial SOC} & -1 \end{bmatrix} \tag{8}$$

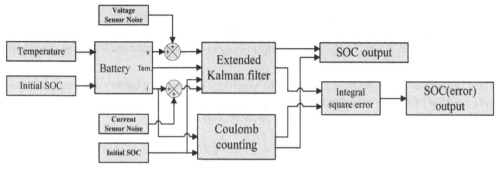

Fig. 37.3   Simulink model of EKF based SOC estimation

The error covariance equation is represented as:

$$\hat{P}(k+1|k) = F_d(k)\hat{P}(k|k)F_d^T(k) + Q \tag{9}$$

The EKF approach comprises of two fundamental steps: prediction and correction. The revised equations Kalman gain and error covariance are presented below:

$$K(k+1) = \hat{P}(k+1|k)H_d^T(k)\left(H_d(k)\hat{P}(k+1|k)H_d^T(k) + R\right)^{-1} \tag{10}$$

$$\hat{P}(k+1|k+1) = (I - K(k+1)H_d)\hat{P}(k+1|k) \tag{11}$$

Where, Q and R indicate the covariance matrix associated with process noise and measurement noise.

3.2 Genetic Algorithm

The Genetic Algorithm is a widely employed search methodology that is extensively utilized to generate efficient solutions in optimization and search challenges. Evolutionary algorithms, such as the genetic algorithm, are optimization methodologies that draw inspiration from the principles of biological evolution. The techniques employed in generating solutions to optimisation issues are derived from natural evolution, drawing inspiration from processes including as inheritance, mutation, selection, and crossover. The algorithms in problem function by operating on a population of people and utilising methods of natural selection to enhance the overall fitness of the population. Within the framework of function minimization, a collection of possible solutions is randomly generated and subsequently subjected to evaluation using a fitness function that evaluates the quality of each solution. After that, candidates of higher quality are selected for the crossover and mutation processes. Crossover comprises of combining of genetic information from two individuals to generate two offspring, whereas mutation encompasses the introduction of alterations to an individual's genetic profile, resulting in the creation of a novel offspring. The recently formed candidates displace the previous ones, so triggering the subsequent generation. The previously stated iterative procedure persists until a minimum objective function of satisfactory quality is ascertained, or until a predetermined number of iterations has been attained.

3.3 SOC Estimation using EKF

This study uses Coulomb counting and the Extended Kalman Filter to estimate the SOC of a lithium battery. Additionally, the Genetic Algorithm is utilised to optimise the noise parameter and minimise the estimation error.

The Coulomb counting method measures the discharge current of a battery and integrates this over a period of time to estimate the state of charge. To calculate the current SOC, this method evaluates the discharging current and previous SOC estimates.

$$SOC = SOC(t_0) + \frac{1}{C_{rated}} \int_{t_0}^{t_0+\tau} I_{batt} \cdot dt \tag{12}$$

Where C_{rated} represents the rated capacity, and I_{batt} denotes the current across the load.

In the extended kalman filter algorithm, it is essential to provide initial value for the process noise Q and the measurement noise R. Following the procedure described above, the SOC of the extended Kalman filter contrasts with the coulomb counting method, hence yielding an appreciable estimation error. The high error in State of charge estimation of the EKF can be attributed to the presence of noise. The genetic algorithm was implemented with the Extended Kalman Filter structure to optimise the values for Q and R, while simultaneously minimizing the SOC error (Fig. 37.4). Specifically, the different between the SOC obtained from EKF denoted as $\hat{y}$ and the SOC derived from Coulomb counting denoted as y is identified. The integral square error (ISE) of both EKF and

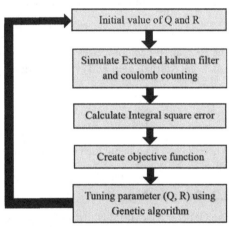

Fig. 37.4 Flow chart of GA tuned EKF SOC estimation

Coulomb counting is utilized as the fitness function (f(x)) for the genetic algorithm. The ISE is defined as follows:

$$ISE = \int (y - \hat{y})^2 \, dt \tag{13}$$

The execution of the GA aims at finding the optimal value of the variable $x = [Q \ R]$ for which the fitness function is minimized i.e.

$$\min_x f(x) = ISE \tag{14}$$

4. Results and Discussion

The parameters used for executing the GA is depicted in Table 37.1.

Inclusion of GA within the EKF framwork leads to improved estimation of SOC. Table 37.2 reports error in the calculated SOC using classical EKF and GA-EKF. For the classical EKF the noise covarience matrices were selected on a trial-and-error basis while for GA-EKF the solution derived post-convergance of GA was used. It can be observed that the inclusion of the GA solution significantly reduces the ISE.The SOC and deviation(error) from the coulomb counting method for both EKF and GA-EKF is shown in Fig. 37.5(A) and 5(B) respectively.The results over multiple cycles of charging and discharging cleary illustrates the superiority of the proposed scheme.

Table 37.1 Parameter value of GA

Maximum Generations	200
Population Size	50
Crossover probability	0.8
Mutation probability	0.001

Table 37.2 Noise parameters (Q, R) and integral square error value recorded after simulation

S. No.	Output	Process Noise (Q)	Measurement Noise(R)	Ise Value
1	EKF	[0.001,0;0,0.001]	0.7	4.608
2	GA-EKF	[0.0005972,0;0,0.0005972]	0.6415	3.917

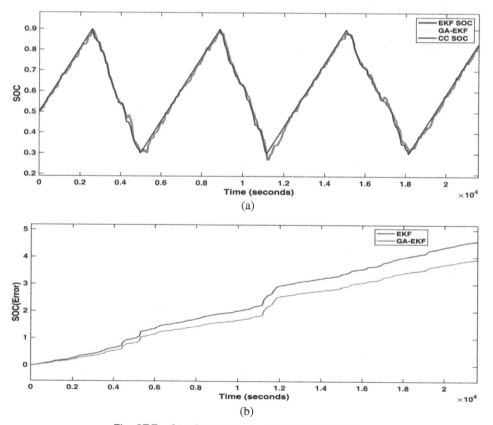

Fig. 37.5 Simulation results (a) SOC, (b) SOC(Error)

5. Conclusion

In this work a scheme for an approach based on the combined framework of equivalent circuit modeling, EKF and GA has been proposed. The limitation of the classical EKF of high dependents of the final estimate value on the selection of noise covariance matrices has been addressed by formulating the matrix. Selection task as an optimization problem and further it using GA the same avoid execution of multiple execution of EKF to determine the covariance matrices. Test result over multiple cycles of charging and discharging reflect the efficacy of the proposed scheme in estimating SOC over differencing operating condition of the lithium-ion battery. Further work in this direction could be concentrated on the implementation of the proposed scheme on a digital platform.

REFERENCES

1. Dişçi, F. N., El-Kahalout, Y., & Balıkçı, A. (2017). Li-ion battery modeling and SOC estimation using extended Kalman filter. 10th International Conference on Electrical and Electronics Engineering (ELECO). Bursa, Turkey:166-169.

2. Gupta, S., & Mishra, P. K. (2023). Machine Learning based SoC Estimation for Li-Ion Battery. 5th International Conference on Energy, Power and Environment: Towards Flexible Green Energy Technologies (ICEPE), Shillong:1-6.

3. Jiani, D., Youyi, W., & Changyun, W. (2013). Li-ion battery SOC estimation using particle filter based on an equivalent circuit model. 10th IEEE International Conference on Control and Automation (ICCA). Hangzhou, China:580-585.

4. Lee, S., Kim, J., Lee, J., & Cho, B. H. (2008). State-of-charge and capacity estimation of lithium-ion battery using a new open-circuit voltage versus state-of-charge. Journal of Power Souces, 185(2): 1367-1373.

5. Mohammadi, F. (2022). Lithium-ion battery State-of-Charge estimation based on an improved Coulomb-Counting algorithm and uncertainty evaluation. Journal of energy storage, 48:104061.

6. Mussi, M., Pellegrino, L., Restelli, M., & Trovò, F. (2021). A voltage dynamic-based state of charge estimation method for batteries storage systems. Journal of Energy Storage,44:103309.

7. Spagnol, P., Rossi, S., & Savaresi, S. M. (2011). Kalman Filter SoC estimation for Li-Ion batteries. IEEE International Conference on Control Applications (CCA) Part of 2011 IEEE Multi-Conference on Systems and Control. Denver, CO, USA:587-592.

8. Ting, T. O., Man, K. L., Lim, E. G., & Leach, M. (2014). Tuning of Kalman Filter Parameters via Genetic Algorithm for State-of-Charge Estimation in Battery Management System. The Scientific World Journal: 1–11.

9. Wang, W., Wang, X., Xiang, C., Wei, C., & Zhao, (2018). Unscented Kalman Filter-Based Battery SOC Estimation and Peak Power Prediction Method for Power Distribution of Hybrid Electric Vehicles. IEEE Access ,6: 35957-35965.

10. Xu, J., Mi, C. C., Cao, B., Deng, J., Chen, Z., & Li, S. (2013). The State of Charge Estimation of Lithium-Ion Batteries Based on a Proportional-Integral Observer. In IEEE Transactions on Vehicular Technology, 63(4): 1614-1621.

Emerging Technologies and Applications in Electrical Engineering –
Prof. Dr. Anamika Yadav et al. (eds)
© 2024 Taylor & Francis Group, London, ISBN 978-1-032-82568-7

Communication Protocols for Sensor Data Transmission in Smart Grid

38

Brundavanam Seshasai[1]
Research Scholar, Department of EE, NIT Raipur
Ebha Koley[2]
Associate Professor, Department of EE, NIT Raipur
Subhojit Ghosh[3]
Professor, Department of EE, NIT Raipur

ABSTRACT: The effective transfer of sensor data is a crucial component in the quickly changing smart grid environment, influencing the dependability and functionality of these intelligent energy distribution networks. This research examines many communication protocols designed especially for smart grid sensor data transmission. Reviewing the interface between technology and energy infrastructure, it covers protocols like DNP3, IEC61850, and Modbus and assesses their suitability, benefits, and drawbacks when it comes to smart grid sensor networks. Modbus, which is easy to use and reasonably priced, is examined in conjunction with DNP3, which is excellent at data prioritization and strong error checking. While IEC61850 is investigated for its cutting-edge features and smooth interaction with power systems, the worldwide standard IEC60870 is examined for its standardization and flexibility. This research provides insights into the fundamental characteristics of various protocols, which helps to optimize the communication systems in smart grids. Ensuring smooth sensor data flow is the goal in order to improve operational efficiency and enable well-informed decision-making in smart grid networks.

KEYWORDS: Smart grids, Sensor data transmission, Communication protocols, Integration with power systems, Optimization

---
[1]brundavanam.phd2022.ee@nitrr.ac.in., [2]ekoley.ele@nitrr.ac.in, [3]sghosh.ele@nitrr.ac.in

DOI: 10.1201/9781003505181-38

1. Introduction

The introduction of DER (Distributed Energy Resources) into electrical networks has caused a paradigm change in the field, posing new difficulties for the management of contemporary power distribution networks. Due to the significant degree of fluctuation in DERs' fundamental resources, information and communication technologies (ICTs) can be used to boost system efficiency, flexibility, and dependability in order to add more capabilities. Communications are a part of conventional power systems at several levels, from control centres to Supervisory Control and Data Acquisition (SCADA) (McLaughlin et al., 2016). Transmission line protection systems also require communications. These communications are currently trending toward standardization, particularly in intelligent substations. Power distribution networks have fewer communication systems. Typically, these systems have unidirectional flow and are passive.

DERs do, however, introduce bidirectional flows and necessitate communication for effective primary resource coordination. Active distribution networks are systems that possess these features. In microgrids, communications are also necessary. There exists a wide range of communication methods utilized in industrial applications. However, wired technologies are widely utilized in sustaining power systems (Choobkar and Rahmani, 2019). This work provides a comprehensive analysis of the aforementioned technology, with a particular focus on its application within power distribution networks and smart grids.

2. Advanced Metering Infrastructure

Advanced Metering Infrastructure (AMI) uses cutting-edge communication technology to transform utility administration. AMI facilitates real-time data interchange by allowing utility providers and smart meters to connect seamlessly, with an emphasis on two-way communication. To guarantee reliable connectivity, the system combines wired solutions like power line communication with wireless ones like radio frequency and cellular networks. With the help of this combination, utilities may read meters remotely and gather precise and efficient consumption data without the need for human participation (Cleveland, 2008). The core competency of AMI is real-time data analytics, which gives utilities a constant flow of insightful information on peak demand times and consumption trends. Utilities may improve overall operational efficiency, allocate resources optimally, and make well-informed decisions using this data-driven strategy.

3. Standard Protocols

From a data networking perspective, the most significant protocols in this field are Modbus, IEC61850, DNP3, Modbus Plus and IEC 60870 and. At the electronic utility level, these protocols are still in use. Systems can't replace conventional, fast communication technologies like Ethernet since they usually offer specialized services. One of the most significant networking technologies is Ethernet, its use and integration with industry communication standards is a hot topic that brings with it a wealth of assets. An important feature of Ethernet is that it is a part of standard networking models like ISO/OSI and TCP/IP protocol stack

3.1 MODBUS

Over time, the MODBUS protocol has evolved into the industry standard for building automation systems across a broad range of industrial applications. These days, Modbus supports a wide range of networking technologies, such as TCP/IP improvements, serial communication, optical or radio networks, RS-232, RS-422, and RS-485 serial communication. Within the protocol stack paradigm, MODBUS functions at several layers based on the transport technology. Modbus operates under a Master-Slave model at the operational level. The sequence code mentioned in the protocol description presents the required function. The use of this protocol is expanded with the adoption of the TCP/IP protocol stack. Reserved system port 502 was gained by Modbus for the internet connection. Modbus/TCP is just a straightforward way to wrap a Modbus frame into a TCP frame. Transmission Control Protocol, as opposed to other industrial or network technologies, represents the connection-oriented and dependable method. Consequently, Modbus is able to leverage the benefits of internetworking technologies, which aligns with its master and slave architecture (Guo et.al,2012).

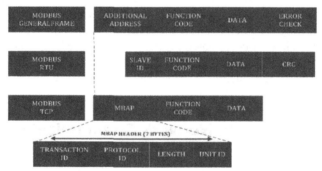

Fig. 38.1 MODBUS TCP/IP Protocol

3.2 DNP 3

The distributed network protocol, or DNP3, is a series of communication protocols or a protocol stack that connects automation systems. usually found in the power industry's SCADA systems and Intelligent Electronic Devices, or IEDs as the IEEE TC97 group refers to them. Other industries do not employ DNP3 very often. The defined frame (FT3) of IEC 60870-5 is used by DNP3. Although not exactly the same, the FT3 frame is fairly comparable. The two primary distinctions are CRC checking and optimum upgrades. DNP3 is primarily the layer 2 protocol in networking parlance, as defined by the ISO/OSI reference model. The DNP3 packet loses its own logical context, the relationship with data units, and substation transport events when seen from the perspective of the transport and application layers. Certain improvements were created, such as the UCA 2.0 (Ortega et.al,2013).

3.3 IEC 60870

The interface definitions for RTU (Remote Terminal Units) and IED (Intelligent Electronic Device) are provided by the structured substandard IEC 60870-5-10. It guarantees system interoperability and includes all the parts and profile specifications required for vendor development. The relationship model of ISO/OSI states that the communication profiles and procedures are independent of technology. The choice of compatible RS-232 and RS-485 standards is supported at the physical layer, which also supports fibre optic interfaces. The frame standard offers the highest level of efficiency and necessary data integrity for workable implementations (Musil and Mlynek, 2020).

3.4 IEC 61850

IEC 61850 represents a response to the limitations imposed by prior standards. It offers several advantages for the advancement and execution of technology. The standardization process encompasses various aspects such as object modeling and programming conventions, the incorporation of contemporary networking technologies, the establishment of command schemas, data representation and transfer protocols, as well as the implementation of encapsulation and other related elements. The IEC 61850 standard is extensive and encompasses numerous substandard. The communication and data transfers in the context of IEC 61850 can be achieved through the utilization of both serial and contemporary computer network technologies, employing the TCP/IP model and Ethernet encapsulation techniques. There are two distinct classifications of communication that we acknowledge: vertical and horizontal. The IEC 61850 standards encompass a range of guidelines and protocols that pertain to the integration of devices, encapsulation of data, and provision of network services (Janssen et.al, 2011).

4. Substation and Field Network Communication

Regarding field devices and substation communications, Fig. 38.2. shows several distinct functional groups. Field device communication is included in the first category. Often, there is relatively little available network bandwidth for this population. There must be as few delays as possible at the same time. Typical uses include concentrators collecting measuring data or commanding a circuit breaker to trip (Kalalas et.al,2016). Data availability and integrity are therefore the most crucial prerequisites. Time synchronization may be necessary for field equipment like phasor measurement units (PMUs) to function. Figure 38.1 illustrates various functional groups with respect to field devices and substation communications. Field device communication is included in the first category. Often, there is relatively little available network bandwidth for this population. Simultaneously, delays must be kept to a minimum. Typical uses include concentrators collecting measuring data or commanding a circuit breaker to trip. As a result, data integrity and availability are the most important requirements. Time synchronization may be necessary for field equipment like phasor measurement units (PMUs) to function. Low-level protocols like C37.118 and Modbus are most frequently used in this group. Through the use of a field area network, substations manage field equipment. Time synchronization may be necessary for field equipment like phasor measurement units (PMUs) to function. Low-level protocols like C37.118 and Modbus are most frequently used in this group. Through the use of a field area network, substations manage field equipment. Substations can also connect with one another over the same network. Time-critical communications

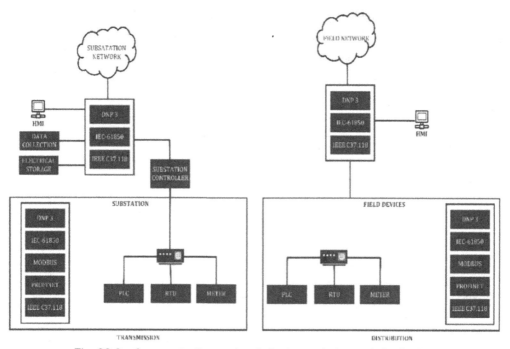

Fig. 38.2 Communication protocols for transmission and distribution

in substations are used to coordinate protective systems; non-time-critical communications are used to update configurations or handle faults Multiple protocols may be utilized for the various communication lines that are present within a substation, as demonstrated in Fig. 38.2. This is due to the fact that the needs are different for each line.

5. Comparison of Standard Protocols

Key characteristics of the industrial communication protocols MODBUS, DNP3, IEC 60870-5-101 (IEC 60780), and IEC 61850 are highlighted in the comparison Table 38.1. This brief summary makes it easier to choose the best protocol for a given set of industrial or power system requirements.

6. Conclusions

In summary, unique characteristics that address various automation and power system requirements are revealed by comparing the industrial communication protocols MODBUS, DNP3, IEC 60870-5-101 (IEC 60780), and IEC 61850. Although it lacks sophisticated security and real-time features, MODBUS, an early industrial automation standard, offers simplicity. The particular needs of industrial or power system applications determine which protocol is best. Although MODBUS and DNP3 are adequate in some situations, IEC 61850 is preferred

Table 38.1 Comparison of standard protocols

Feature	MODBUS	DNP3	IEC 60870-5-101	IEC 61850
Purpose	Industrial Automation	Electric Power Systems	Substation Automation	Substation Automation
Open Standard	Yes	Yes	Yes	Yes
Year of Standardization	1979	1993	1994	2003
Protocol Type	Master-Slave	Master-Slave	Master-Slave	Client-Server
Topology	Serial and Ethernet	Serial and Ethernet	Serial and Ethernet	Ethernet-based
Addressing	Binary, ASCII, RTU	Point-to-Point	Point-to-Point	Logical Nodes
Data Representation	Binary, ASCII	Binary	Binary	XML-based (MMS)
Transport Layer	TCP, UDP, Serial	TCP, UDP	TCP/IP, ISO Transport Class 1	TCP/IP, ISO Transport Class 4
Error Detection	CRC	CRC	CRC	CRC
Security Features	Limited	Limited	Limited	Enhanced security features
Device Addressing	Slave IDs	Outstation IDs	Common Address Space	Logical Device IDs
Real-time Data Exchange	Limited	Yes	Yes	Yes
Scalability	Limited	Yes	Yes	Yes
Typical Applications	PLCs, RTUs, HMI	SCADA Systems	RTUs, Gateways	Intelligent Electronic Devices

by the changing environment due to its sophisticated characteristics, especially in critical infrastructure where instantaneous and secure communication is crucial.

REFERENCES

1. S. McLaughlin et al., "The Cybersecurity Landscape in Industrial Control Systems," in Proceedings of the IEEE, vol. 104, no. 5, pp. 1039-1057, May 2016, doi: 10.1109/JPROC.2015.2512235.
2. S. Choobkar and M. Rahmani, "Communication Routes for DER Interconnection with Power Grid," 2019 International Power System Conference (PSC), Tehran, Iran, 2019, pp. 696-701, doi: 10.1109/PSC49016.2019.9081490.
3. F. M. Cleveland, "Cyber security issues for Advanced Metering Infrasttructure (AMI)," 2008 IEEE Power and Energy Society General Meeting - Conversion and Delivery of Electrical Energy in the 21st Century, Pittsburgh, PA, USA, 2008, pp. 1-5, doi: 10.1109/PES.2008.4596535.
4. Yaguang Guo, B. X. Du, Y. Gao, Xiaolong Li and H. B. Li, "On-line monitoring system based on MODBUS for temperature measurement in smart grid," IEEE PES Innovative Smart Grid Technologies, Tianjin, China, 2012, pp. 1-5, doi: 10.1109/ISGT-Asia.2012.6303180.

5. A. Ortega, A. A. Shinoda and C. M. Schweitzer, "Performance analysis of smart grid communication protocol DNP3 over TCP/IP in a heterogeneous traffic environment," 2013 IEEE Colombian Conference on Communications and Computing (COLCOM), Medellin, Colombia, 2013, pp. 1-6, doi: 10.1109/ColComCon.2013.6564828.

6. P. Musil and P. Mlynek, "Overview of Communication Scenarios for IEC 60870-5-104 Substation Model," 2020 21st International Scientific Conference on Electric Power Engineering (EPE), Prague, Czech Republic, 2020, pp. 1-4, doi: 10.1109/EPE51172.2020.9269173.

7. M. C. Janssen, P. A. Crossley and L. Yang, "Bringing IEC 61850 and Smart Grid together," 2011 2nd IEEE PES International Conference and Exhibition on Innovative Smart Grid Technologies, Manchester, UK, 2011, pp. 1-5, doi: 10.1109/ISGTEurope.2011.6162749.

8. C. Kalalas, L. Thrybom and J. Alonso-Zarate, "Cellular Communications for Smart Grid Neighborhood Area Networks: A Survey," in IEEE Access, vol. 4, pp. 1469-1493, 2016, doi: 10.1109/ACCESS.2016.2551978.

Emerging Technologies and Applications in Electrical Engineering –
Prof. Dr. Anamika Yadav et al. (eds)
© 2024 Taylor & Francis Group, London, ISBN 978-1-032-82568-7

Arduino-Based Online Estimation of Waveform Attributes

39

Alim Khan[1], Akash Mallela[2], Jai Kumar Verma[3], Subhojit Ghosh[4]

Electrical Engineering, National Institute of Technology Raipur

ABSTRACT: In the realm of electronic instrumentation, having a knowledge of the attributes of voltage waveform is of significant importance. Traditionally, physical oscilloscopes have been the go-to solution for visualizing electrical signals, offering invaluable insights into circuit behavior. The issues related to bulkiness, cost and ease of operation of standard digital Oscilloscope (DSO) has been addresses in the present work using a microcontroller-based prototype for online based estimation of signal attributes. The components used are LCD display, a bread board, resistors, jumper wires and in this prototype a ATMEGA328 micro controller board(Arduino uno). The effectiveness of the developed prototype has been evaluated in accurately estimating the average values, RMS values and Fourier transform of different periodic waveforms.

KEYWORDS: Digital storage oscilloscope, Arduino board, Periodic waveform, Signal attributes

1. Introduction

An oscilloscope is an essential tool for waveform analysis, providing valuable insights into the behaviour of electrical signals. Over the years, advancements in technology have led to the development of increasingly sophisticated oscilloscopes, incorporating enhanced features and functionalities. In this research paper, a novel concept of Arduino board-based Oscilloscope. This innovative approach leverages the capabilities of existing oscilloscopes to create a powerful and versatile instrument for signal analysis and measurement. By exploring the

[1]samsoon7789@gmail.com, [2]akashmallela2504@gmail.com, [3]vermajai2001@gmail.com, [4]sghosh.ele@nitrr.ac.in

DOI: 10.1201/9781003505181-39

potential synergies between multiple oscilloscopes, the work aims to unlock new opportunities in the field of waveform analysis and provide researchers and engineers with a unique tool for their applications. Through a comprehensive analysis of the design principles, implementation challenges, and potential benefits, this research paper sheds light on the prospects of this intriguing concept and its implications in the field of oscilloscope technology.

Periodic electric signals are typically measured and quantified in terms of its peak voltage, frequency, phase difference, pulse width, delay time, etc. Oscilloscopes are distinct from other measurement tools because they can visualize electrical signals, allowing the user to conduct in-depth analysis and get a deeper understanding.

In addition to Rms values and average value, the Fourier transform provides information about the intricacies of a periodic signal. The Fast Fourier Transform (FFT) is a powerful algorithmic technique that efficiently computes the Discrete Fourier Transform (DFT) of a sequence, reducing the computational complexity. In the realm of signal processing, the FFT plays a pivotal role in analyzing the frequency content of signals. By identifying specific frequencies in time-domain signals, it facilitates crucial applications such as signal filtering, modulation, and demodulation. Its efficiency is particularly beneficial in real-time applications where rapid analysis of large datasets is essential. Moreover, the FFT is extensively used in communication systems for spectrum analysis, enabling the determination of frequency components in signals. Its versatility makes it indispensable in diverse applications, ensuring its prominence in the field of signal processing where understanding and manipulating frequency information are paramount. In this regard, the present research aims at the development of a micro controller-based oscilloscope for estimation of average, RMS and FFT.

2. Microcontroller Based Oscilloscope

This paper addresses the creation of a low-cost handheld digital oscilloscope utilizing an embedded microcontroller system. The goal of this study is to provide a low cost and modular alternative to standard DSO.

1. The present work aims at enabling a personal computer (PC) to function as a flexible and adaptable oscilloscope for signal analysis and measurement.
2. The project will focus on utilizing affordable components, ATMEGA and open-source software, so as to keep the overall cost of the oscilloscope as low as possible without compromising functionality and performance.
3. The Software interface provides a user-friendly environment on the PC, enabling control, visualization, and analysis of acquired signals.
5. By utilizing Arduino and PC, the constructed wireless oscilloscope offers a high degree of flexibility and customizability. Users can extend the functionality of the oscilloscope by modifying the software and adding additional modules to the Arduino platform, enabling tailored solutions for specific measurement requirements or application areas.

Oscilloscopes have evolved over time due to continuous technological advancements. Development efforts focus on improving key features such as bandwidth, sampling rate, and resolution to enable more accurate and detailed waveform analysis. Manufacturers invest in research and development to integrate advanced signal processing algorithms, user-friendly interfaces, and wireless connectivity options, enhancing the usability and versatility of oscilloscopes.

The use of Arduino board in the present work is motivated by the fact that is extensibly used in various projects, providing a programmable hardware platform for controlling and interacting with electronic components. It offers a user-friendly development environment, a wide range of built-in input/output pins, and a vast library of pre-existing code, enabling users to easily prototype and implement projects involving sensors, actuators, and communication interfaces. Arduino's versatility and simplicity make it a popular choice for beginners and experienced enthusiasts alike in the field of electronics and embedded systems.

Implementing the Fast Fourier Transform (FFT) in a digital oscilloscope using Arduino enhances the device's capabilities for real-time frequency analysis. This feature allows users to visualize and identify the frequency components of input signals, offering valuable insights into the nature of waveforms.

Moreover, the implementation broadens the oscilloscope's utility as a DIY spectrum analyzer, making it suitable for radio frequency experimentation and other projects where a dedicated spectrum analyzer might be impractical.

Figure 39.1 illustrates the block diagram representation of the proposed Arduino based Oscilloscope.

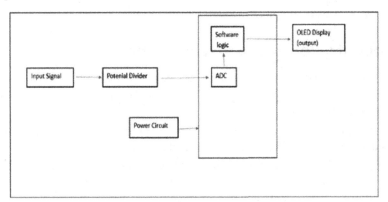

Fig. 39.1 Block diagram of oscilloscope using arduino

Any circuits output is used as the input signal, which in this case is a periodic signal. The analog read command is used to read the input when analogue pins are provided. The processor is then provided with the input read after that. The subsequent procedure, which is the conversion to digital output, is carried out in the processor. Here, digital write is utilized (this is like print command). To perform the sampling, quantization, and coding steps, an ADC is connected to the processor. In an OLED display, the output is displayed based on the values at the location of the coordinates. If, for example, the value in the first row and first column at position 1,1 is high, it indicates that we have read some input there. The result will be presented in accordance with sampling, which creates columns and quantization, which creates rows. When you take 0 for no design and 1 for a design value for code.

The major components used in the prototype oscilloscope are discussed henceforth.

1. The Atmega328P: This is the core microcontroller used in several Arduino development boards, including the popular Arduino Uno. Its compatibility with the Arduino ecosystem makes it highly accessible and user-friendly, allowing developers to leverage a vast library of pre-existing code, shields, and resources to quickly prototype and develop projects. Its availability, affordability, and community support make the Atmega328P a preferred choice for Arduino enthusiasts and beginners in the field of embedded systems.

2. LCD display(16x2): The 16x2 LCD display is a compact, two-line alphanumeric display commonly used in electronics, featuring 16 characters per line. Widely employed in embedded systems and DIY projects, it serves as a versatile interface for displaying information such as sensor readings, system status, and alphanumeric characters, providing a user-friendly visual output.

The code used for estimating the RMS, average and FFT values of the input signal is given below

```
#include <LiquidCrystal.h>
#include "arduinoFFT.h"
#define numberOfSamples 128
#define samplingFrequency 1000
arduinoFFT FFT = arduinoFFT();
unsigned int samplingPeriod;
unsigned long microseconds;
double Real[numberOfSamples];
double Imag[numberOfSamples];
LiquidCrystal lcd(12, 11,5, 4, 3, 2);  // RS, E, D4, D5, D6, D7
int contrast = 75;
int ip = A2; // Define the analog pin you are using
int mult_factor = 5.8;
double offset;
void setup() {
  // Your setup code here
  analogWrite(6, contrast);
  lcd.begin(16, 2);  // Set the LCD dimensions (16x2)
  lcd.clear();
  lcd.setCursor(0, 0);
  lcd.print("RMS: ");
  lcd.setCursor(9, 0);
  lcd.print(" V ");
  lcd.setCursor(0, 1);
```

```
lcd.print("Avg: ");
lcd.setCursor(10,1);
lcd.print(" V ");  // Set the LCD dimensions (16x2)
Serial.begin(9600);
unsigned long startTime = millis();
double samples = 0;
double volt = 0;
while (millis() - startTime < 1000) {
  double currVolt = analogRead(ip) * 5.0 / 1023.0;
  currVolt *= mult_factor;
volt += currVolt;
  samples++;}
offset = volt / samples;
samplingPeriod = round(1000000*(1.0/samplingFrequency));}
void loop() {
unsigned long startTime = millis(); // Get the current time in milliseconds
double avg = 0;
double rms = 0;
double samples = 0;
while (millis() - startTime < 1000) { // Run for 1 second (1000 milliseconds)
  double currVolt = analogRead(ip) * 5.0 / 1023.0;
  currVolt *= mult_factor;
  currVolt -= offset;
  avg += currVolt;
  rms += currVolt * currVolt;
  samples++;
}
avg /= samples;
rms /= samples;
rms = sqrt(rms);
// Display the RMS voltage on the first line
lcd.setCursor(5, 0);
lcd.print(rms);  // Display with 2 decimal places
// Display the average voltage on the second line
lcd.setCursor(5, 1);
lcd.print(avg);  // Display with 2 decimal places
//SAMPLING
  for(int i=0; i<numberOfSamples; i++)
  {
    microseconds = micros();
    Real[i] = analogRead(A2)*5.0/1023.0;
    Imag[i] = 0;
    while(micros() < (microseconds + samplingPeriod)){}}
```

```
//FFT
    FFT.Windowing(Real, numberOfSamples, FFT_WIN_TYP_HAMMING, FFT_
FORWARD);
    FFT.Compute(Real, Imag, numberOfSamples, FFT_FORWARD);
    FFT.ComplexToMagnitude(Real, Imag, numberOfSamples);
        double peakAmplitude = FFT.MajorpeakAmplitude(Real, numberOfSamples,
samplingFrequency);
    double maxAmplitude = 0;
    double maxIndex = 0;
    for(int i=0; i<(numberOfSamples/2); i++)
    {double amplitude = Real[i];
      if(amplitude > maxAmplitude){
        maxAmplitude = amplitude;
        maxIndex = i;}}
    Serial.print("Freq");   Serial.print(" ");
    Serial.println("Amp");
    Serial.print(peakAmplitude);
    /PRINT RESULTS/
    Serial.print(" ");
    Serial.println(maxAmplitude);
    for(int i=1; i<(numberOfSamples/2); i++)
    {
      Serial.print((i * 1.0 * samplingFrequency) / numberOfSamples, 1);
      Serial.print(" ");
      Serial.println(Real[i], 1);
    }
    while(1);   }
```

3. Results

Some of the test results obtained using the proposed Arduino-based oscilloscope are dealt with in this section. The breadboard-based Hardware setup with the necessary connections and LCD display is shown in Fig. 39.2.

It can be observed that for all the periodic signals the attributes are accurately calculated. The spectrum derived for a sinusoidal signal A = sin 100πt + 2 (f=50hz) is shown in Fig. 39.3. (Frequency in X-axis and Amplitude in Y-axis)

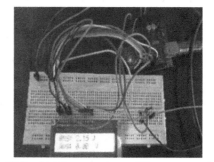

Fig. 39.2 Hardware set-up of the proposed prototype for estimating periodic signal attributes

Table 39.1 Displays the signal attributes estimated foe different periodic signals.

S. No	Signal type	Offset (v)	Magnitude (v)	Frequency (hz)	RMS Estimated (v)	Average Value Estimated (v)
1	Sinusoidal	2	2	20	1.49	2.00
2	Sinusoidal	2	3	20	2.15	2.00
3	Triangular	4	3	30	1.75	4.01
4	Triangular	3	2	30	1.16	3.01

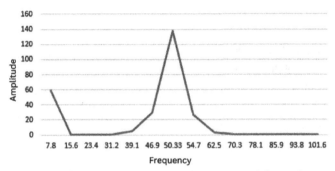

Fig. 39.3 Spectrum derived for a sinusoidal signal

4. Conclusion

A low cost and portable alternative the standard DSO has been proposed in the present work for online estimation of signal attributes. Using a Arduino uno board the input periodic signal is acquired and further processed to derive the Average value, RMS value and FFT and further display the estimated values on the LCD screen. The use of generalised formulas avoids restricting the utility of the proposed prototype to specific signal types. Future work in this direction would be concentrated on improving the bandwidth and increasing the precision of the spectrum derived using FFT.

REFERENCES

1. I Fushshilatand D.Barmana," Low Cost Handheld Digital Oscilloscope", ISMEF 2017.
2. Wildan maulana,Nyoman Bogi Adithya Karna, Ridha Muldina Negara, "Product design of portable oscilloscope using Arduino to analyse signal", December 2020.
3. Theophilus Ezra Nugroho Pandin, Application of oscilloscope technology in the early 21st century,12 December 2021.
4. Rohit Tare, Rushikesh Bhonde, Prof. Suyog Guptha," PC based oscilloscope using Arduino" ICEMESM-2018.
5. Sidhant s.Kulkarni and S.K.Guptha "Portable Oscilloscope", Volume 2, Issue 4 april 2017.
6. C. Bhunia, S. Giri, S. Kar, S. Haldar and P. Purkait," A low-cost PC based virtual oscilloscope", in IEEE transactions on education, May 2004.
7. Ishtiak Ahmed Karim," A low-cost portable oscilloscope based on Arduino and GLCD", IEEE 2014.